The History of Astronomy and Astrophysics

A Biographical Approach

H. Thomas Milhorn, MD, PhD

"The History of Astronomy and Astrophysics: A Biographical Approach," by H. Thomas Milhorn, MD, PhD. ISBN 978-1-60264-258-4.

Published 2008 by Virtualbookworm.com Publishing Inc., P.O. Box 9949, College Station, TX 77842, US.

Manufactured in the United States of America.

Preface

Astronomy is the oldest of the natural sciences, with its origins in the religious, mythological, and astrological practices of pre-history. Originally, peopled gazed at the sky and speculated about the nature of the Moon, the Sun, and the stars. The behavior and nature of the world and celestial phenomena came to be explained by invoking the actions of the various gods. Then, one god became the explanation for both. And then explanations for earthly events began to be accepted based on speculation as to their cause and nature, while celestial events were still felt to be of a divine nature, requiring no explanation. Eventually, human beings began to look for scientific explanations for both. Early on, the Earth was considered to be at the center of a celestial sphere, with the Sun, Moon, stars, and planets located on the surface of the sphere.

The invention of the telescope about 400 years ago opened new avenues of research and speculation. The Sun and the Moon could be scrutinized in greater detail, and new planets came to be identified and studied. Slowly, the concept that Earth was not the center of the universe took hold, and eventually the notion that the Milky Way is but one of millions of galaxies became accepted.

Speculation about life on the Moon and other planets became a hot topic. Some believed it a possibility; others felt it to be a fact, identifying what they felt to be canals on Mars and cities and even buildings on the Moon.

New investigative tools, such as spectrometry and photometry, were developed for determining spectral characteristics and brightness, respectively, of cosmological objects. Then scientists began to look at the skies, detecting electromagnetic radiation other than visible light, such as infrared and X-rays. Others studied cosmic radiation, including background microwave radiation, neutrinos produced within the Sun, and ions of various elements.

Theories about the evolution of the universe were developed, the two most prominent being the steady-state theory and the Big Bang theory. According to the steady-state theory, the universe has no ending and no beginning. As the universe expands, new matter is created so that the average density stays the same. The Big Bang theory, on the other hand, views the universe as being created approximately 13.7 billion years ago by a violent explosion. At the point of this event, all of the matter and energy of space was contained at one point, a singularity. The universe is seen to still be expanding from the explosion.

Although the great body of knowledge we now call astronomy and astrophysics has come into being because of the work of astronomers and astrophysicists, many individuals from other disciplines have contributed, including those from mathematics, physics, engineering, and chemistry. Many of the earlier astronomers made significant contributions without having the benefit of a formal education.

To those who made significant contributions to astronomy and astrophysics, and I inadvertently left them out of this book, I humbly apologize.

Those individuals whose contributions have been mainly physics are notably absent. An additional volume, *The History of Physics*, was published earlier this year.

I have chosen to approach the history of astronomy and astrophysics from a biographical point of view, feeling that people are more interesting than things, and the combination of the two are more interesting than the sum of the individual parts. After a brief overview of astronomy and astrophysics, 297 one-page biographies of individuals who have made significant contributions to the field of astronomy and astrophysics are presented.

H. Thomas Milhorn, MD, PhD
Meridian, Mississippi

Contents

Introduction

Astronomy is the study of the universe and the celestial bodies, gas, and dust within it. It includes observations and theories about the Solar System, the stars, the galaxies, and the general structure of the universe. Astronomy also includes cosmology, which is the study of the universe as a whole. It includes the universe's past, present, and future. Because astronomy now uses many ideas from physics, in recent years the terms astronomy and astrophysics have become synonymous. I will use them as such in this book.

Astronomy is the oldest science, dating back thousands of years. In ancient Egypt, the first appearance of certain stars each year marked the onset of the seasonal flood, an important event for agriculture. In 17th-century England, astronomy provided methods of keeping track of time that were especially useful for accurate navigation at sea.

Much of today's research in astronomy involves basic work to satisfy curiosity about the universe and the objects in it. Astronomy has resulted in our understanding of the stars, the phases of the Moon, day and night, the seasons, and much more

Astronomers first observe astronomical objects by guiding telescopes and other instruments, such as spectrometers and photometers, to collect the appropriate information. They then analyze the images and data. After the analysis, they compare their results with existing theories to determine whether their observations match with what theories predict or whether the theories must be improved or abandoned entirely. Some astronomers work solely on observation and analysis, and some work solely on developing new theories.

Astronomy is such a broad topic that astronomers specialize in one or more parts of the field. For example, the study of the Solar System is a different area of specialization than the study of stars or the galaxies. Astronomers who study the Milky Way often use techniques different from those used by astronomers who look for planets in distant galaxies. Astronomers who study the Sun use different instruments than nighttime astronomers who study the Moon and planets. Theoretical astronomers may never use telescopes or other instruments at all. Instead, they use existing data or previous theoretical results to develop and test theories. In computational astronomy, astronomers use computers to simulate astronomical events.

Astronomers usually learn about astronomical objects by observing the energy they emit in the form of electromagnetic radiation. This radiation travels throughout the universe in the form of waves, which can range from radio waves with very long wavelengths to gamma rays with extremely short wavelengths. The entire range of these different wavelengths makes up the electromagnetic spectrum.

Classification of Astronomy

Astronomy can be classified as observational astronomy or theoretical astronomy. The definitions of many of the terms used in this discussion can be found in the section, *Selected Astronomy Terms*, which follows immediately after this section.

Observational Astronomy

Observational astronomy can be divided into those branches that are dependent on particular wavelengths in the electromagnetic spectrum and those that are not. The former include radio and radar astronomy, microwave astronomy, infrared astronomy, visual light astronomy, ultraviolet astronomy, X-ray astronomy, and gamma ray astronomy. The latter include cosmic-ray astronomy, neutrino astronomy, and gravitational wave astronomy.

Dependent on the Electromagnetic Spectrum

The electromagnetic spectrum includes the entire range of radiation, extending in frequency from about three Hz to approximately 10^{23} Hz. In corresponding wavelengths, it extends from approaching infinity to 10^{-13} cm. In order of increasing frequency, and decreasing wavelength, the electromagnetic spectrum includes radio waves, microwaves, infrared radiation, visible light, ultraviolet radiation, X-rays, and gamma rays.

The Earth's atmosphere complicates studies by absorbing many wavelengths of the electromagnetic spectrum. Ideally, many of these studies need to be done above the Earth's atmosphere.

Radio Wave Astronomy. Both radio and radar astronomy involve radio waves, which range in frequency from about three Hz to 3×10^9 Hz.

Radio Astronomy. Radio waves emanating from celestial bodies are received by specially constructed antennas, called radio telescopes. In the most common design, a parabolic dish replaces the mirror of the reflecting optical telescope. This dish serves to focus the radio waves into a concentrated signal that is then filtered, amplified, and finally analyzed using a computer.

Radio wave astronomy is different from most other forms of observational astronomy in that the observed radio waves, because of their long wavelengths, can be treated as waves rather than as discrete photons. Hence, it is relatively easier to measure both the amplitude and phase of radio waves, whereas this is not as easily done at shorter wavelengths.

A wide variety of objects are observable at radio wavelengths, including supernovae, interstellar gas, pulsars, and active galactic nuclei.

Although some radio waves are produced by astronomical objects in the form of thermal emission, most of the radio emission that is observed from Earth is seen in the form of synchrotron radiation, which is produced when electrons oscillate around magnetic fields. Additionally, a number of spectral lines produced by interstellar gas, notably the hydrogen spectral line at 21 cm, are observable at radio wavelengths.

Radar Astronomy. Radar is a method of detecting distant objects, such as planets and the Moon, and determining their position, velocity, or other characteristics by analysis of radio waves reflected from their surfaces.

Radar works by transmitting a short burst of radio waves in the direction of the object under study. The object reflects the radio waves back to Earth where they are detected by the same antenna that sent the signal. Since radio waves travel with the speed of light, the roundtrip distance from the Earth to the object and back is then easily computed.

The first yield of radar astronomy was a much-improved value for the distance from Earth to the Moon. Using more powerful transmitters, the distances to Venus and Mercury were also determined, as well as their rotational periods and gross surface properties. In addition, much of the surface of Venus has been mapped by unmanned probes using radar altimeters to penetrate the cloud cover.

Even greater precision is obtained by replacing the radio transmitter with a laser. During Project Apollo, special reflectors were installed on the Moon. Subsequently, by bouncing laser light off the Moon, the distance from the Earth to the Moon was determined within centimeters. Radar observations are also useful for studying asteroids and comets whose orbits take them relatively near the Earth.

Microwave Astronomy. Microwave astronomy involves examining the cosmic microwave background, which is the radiation left over from when the universe was first created by the Big Bang. This allows a look at what the universe was like near its birth.

The microwave region of the electromagnetic spectrum lies between that of radio waves and the infrared region, or 3×10^{9} and 3×10^{11} Hz.

Microwaves are considered to be high-energy radio waves.

Because microwaves are hard to observe from the ground, space-based telescopes are used. There have been a number of probes launched to map the entire sky in the microwave band, with each probe having progressively better resolution. The result from the most recent space-born microwave telescope, the Wilkinson Microwave Anisotropy Probe, helped determine the age of the universe (13.7 billion years) with improved accuracy, and also gave an idea of what the universe looked like less than 400,000 years after the Big Bang.

Infrared Astronomy. Infrared astronomy is the study of celestial objects by means of the infrared radiation (IR) they emit. The infrared region of the electromagnetic spectrum lies between that of microwaves and visible light, or from about 3.9×10^{11} to 3.9×10^{14} Hz.

Except at wavelengths close to visible light, IR is heavily absorbed by the atmosphere, and the atmosphere produces significant infrared emission. Consequently, infrared observatories have to be located in high, dry places or in space. Infrared astronomy is particularly useful for observation of galactic regions cloaked by dust and for studies of molecular gases.

All objects emit IR. The study of such radiation from celestial objects is of importance for several reasons: (1) Cosmic dust particles obscure parts of the visible universe, such as the center of the Milky Way, but this dust is transparent in the IR wavelengths. (2) Most of the energy radiated by objects, ranging from interstellar matter to planets, lies in the IR wavelengths. IR observations are therefore of value for studying asteroids, comets, planetary satellites, and interstellar dust clouds where stars are forming. (3) The expansion of the universe shifts energy to longer wavelengths. The amount of the shift is directly proportional to how far away an object is. As a result, the distant objects we observe today are as they appeared when the Universe was younger. Since the Universe is expanding, most of the visible and ultraviolet radiation from astronomical objects has shifted into the IR range.

Visible Light Astronomy. Visible light astronomy, also called optical astronomy, is the oldest form of astronomy. It is the use of visible light to study celestial objects. The visible light region of the electromagnetic spectrum lies between that of infrared radiation and the ultraviolet region, or 3.9×10^{14} to 7.9×10^{14} Hz.

Early astronomers simply looked at the sky with the naked eye and noted what they saw. The invention of the refractor, and later the reflector, telescope opened up broad, new pathways of study.

Optical images were originally drawn by hand. In the late 19th century and most of the twentieth century, images were made using photographic equipment. Modern images are made using digital detectors, particularly detectors using charge-coupled devices (CCD). A CCD is a light-sensitive integrated circuit that stores and displays the data for an image in such a way that each pixel in the image is converted into an electrical charge, the intensity of which is related to a color in the color spectrum.

Ultraviolet Astronomy. Ultraviolet astronomy is the study of celestial objects by means of the ultraviolet radiation they emit. The ultraviolet region of the electromagnetic spectrum lies between that of visible light and the X-ray region, or 7.9×10^{14} to 3.4×10^{16} Hz. Radiation at these frequencies is absorbed by the Earth's atmosphere, so observations must be performed from the upper atmosphere or from space.

Ultraviolet astronomy is best suited to the study of thermal radiation and spectral emission lines from hot blue stars that are very bright in this wave band. Other objects commonly observed in ultraviolet light include planetary nebulae, supernova remnants, and active galactic nuclei.

Ultraviolet light is easily absorbed by interstellar dust, a phenomenon known as extinction. Measurement of the ultraviolet radiation from objects has to be corrected for this.

X-ray Astronomy. X-ray astronomy is the study of celestial objects by means of the X-rays they emit. The X-ray region of the electromagnetic spectrum lies between that of ultraviolet radiation and the gamma ray region, or 3.4×10^{16} to 5×10^{19} Hz.

Typically, objects emit X-ray radiation as synchrotron emission, which is produced by electrons oscillating around magnetic field lines, and thermal emission above 10^7 Kelvin from both thin and thick gases.

Since X-rays are absorbed by the Earth's atmosphere, all X-ray observations must be done from high-altitude balloons, rockets, or spacecraft. Much of the data in X-ray astronomy is now gathered by orbiting satellites.

X-ray sources include X-ray binaries, pulsars, supernova remnants, elliptical galaxies, clusters of galaxies, and active galactic nuclei.

Gamma Ray Astronomy. Gamma ray astronomy is the study of astronomical objects by analysis of the most energetic electromagnetic radiation they emit. The gamma ray region of the electromagnetic spectrum lies above 5×10^{19} Hz.

Ninety percent of gamma-ray emitting sources produce gamma-ray bursts lasting only a few milliseconds to thousands of seconds before fading away. The other 10 percent of gamma rays come from non-transient sources. Gamma ray emitters include pulsars, neutron stars, and black holes.

Gamma rays may be observed directly by satellites, such as the Compton Gamma Ray Observatory, or by specialized telescopes called atmospheric Cherenkov telescopes. The Cherenkov telescopes do not actually detect gamma rays directly, but instead detect the flashes of visible light produced when gamma rays are absorbed by the Earth's atmosphere.

Not dependent on the electromagnetic spectrum

Other than electromagnetic radiation, a few things may be observed from Earth that originate from great distances.

Neutrino Astronomy. The neutrino is an elusive particle that plays a major role in the physics of stellar interiors. They are so nonreactive with matter that that they can pass through the Earth as if it were not there. In neutrino astronomy, astronomers use special underground facilities to avoid the false detection of cosmic background radiation. Neutrino detectors typically detect only one out of every 10^{22} neutrinos passing through them.

Cosmic Ray Astronomy. Cosmic rays are energetic particles originating from space that impinge on Earth's atmosphere. Almost 90 percent of all the incoming cosmic ray particles are protons, about nine percent are helium nuclei (alpha particles), and about one percent are electrons. The term "ray" is a misnomer, since cosmic particles arrive individually, not in the form of a ray or beam of particles.

The supermassive black holes at the heart of the Milky Way and other galaxies are the likely sources of cosmic rays. Additionally, some future neutrino detectors will be sensitive to the neutrinos produced when cosmic rays hit the atmosphere.

Gravitational Wave Astronomy. Gravity waves are disturbances in a gravitational field similar to the waves that spread out from the disturbance created when a pebble is tossed on the quiet surface of a pond. Although we have yet to detect gravity waves, there is strong, although indirect, evidence of their existence.

A few gravitational wave observatories have been constructed, such as the Laser Interferometer Gravitational Observatory (LIGO). LIGO operates two gravitational wave observatories in unison, one in Livingston, Louisiana and the other near Richland, Washington. These sites are separated by 1,876 miles. Since gravitational waves are expected to travel at the speed of light, this distance corresponds to a difference in gravitational wave arrival times of up to 10 milliseconds. Through the use of triangulation, the difference in arrival times can be used to determine the source of the wave in the sky.

Other

Celestial Mechanics. One of the oldest fields in astronomy, and in all of science, celestial mechanics is the measurement of the positions of celestial objects. Historically, accurate knowledge of the positions of the Sun, Moon, planets, and stars has been essential in celestial navigation at sea.

Careful measurement of the positions of the planets has led to an understanding of gravitational perturbations and an ability to determine past and future positions of the planets with great accuracy. In addition, the tracking of near-Earth objects will allow predictions of close encounters and potential collisions with the Earth

The measurement of stellar parallax of nearby stars provides a baseline in the cosmic distance ladder that is used to measure the scale of the universe. Parallax measurements of nearby stars provide a baseline for the properties of more distant stars because their properties can be compared. Measurements of radial velocity and proper motion show the kinematics of these systems. Measurements of positions and movements of celestial bodies are also used to determine the distribution of dark matter in the galaxy.

During the 1990s, the astrometric technique of measuring the stellar wobble was used to detect large extrasolar planets orbiting nearby stars.

Theoretical Astronomy

Theoretical astronomers endeavor to create theoretical models and study the observational consequences of those models using digital computers. They also modify models to take into account new data. In some cases, a large amount of inconsistent data over time may lead to total abandonment of a model and development of a new one.

Topics studied by theoretical astronomers include stellar dynamics and evolution; galaxy formation; large-scale structure of matter in the universe; origin of cosmic rays; and general relativity and physical cosmology, including string cosmology and astroparticle physics. String cosmology is a relatively new field that tries to apply equations of string theory from theoretical physics to solve the questions of early cosmology. Astroparticle physics is a term used to indicate that branch of particle physics that studies elementary particles of astronomical origin and their relation to astrophysics and cosmology.

Sub-fields of Astronomy

Sub-fields of astronomy include solar astronomy, planetary astronomy, stellar astronomy, galactic astronomy, extragalactic astronomy, cosmology, and interdisciplinary studies.

Solar astronomy

The most frequently studied star is the Sun, a typical main-sequence star about 4.6 billion years in age. The study of the Sun includes its (1) location within the galaxy; (2) life cycle; (3) structure, including its core, radiative zone, convection zone, photosphere, and atmosphere; (4) chemical composition; (5) sunspot cycles, and (6) magnetic field and solar wind.

The Sun is studied using telescopes, photometry, spectroscopy, radio telescopes, the Doppler effect, and techniques for measuring its electromagnetic spectrum.

Planetary Astronomy

Planetary astronomy examines objects such as planets, moons, dwarf planets, comets, asteroids, and other bodies orbiting the Sun, as well as extrasolar planets.

Planets orbiting distant stars are located by six methods: (1) Spectrographic radial velocity, (2) transit, (3) gravitational lensing, (4) pulsar timing, (5) circumstellar disks, and (6) direct imaging.

Spectrographic Radial Velocity. This method has been the most successful in locating planets orbiting around distant stars. As a planet rotates around a star, the planet's gravity causes the position of the star to wobble back and forth ever so slightly. By observing the Doppler shift of the star, astronomers can sometimes detect this wobble, and infer the properties of the planet—mass and distance from the star.

Transit. If a planet should pass in front of a distant star, astronomers would see the light from the star diminish by a tiny amount. To date, astronomers have been able to identify only larger planets by this method.

In 2008, NASA planned to launch an orbiting spacecraft, called Kepler, with photometric instruments able to detect transits of smaller planets, such as Earth-size ones.

Gravitational Lensing. The gravitational field of an unseen planet bends light, much as a lens does, in such a way that it causes the star to brighten temporarily while the planet passes in front of it.

Pulsar Timing. Pulsars emit radio waves extremely regularly as they rotate. The timing of its observed radio pulses can be used to track changes in the pulsar's motion caused by the presence of planets.

Circumstellar Disks. This method deals with the disks of space dust that surround many stars. These disks can be used to detect extra solar planets because they absorb ordinary starlight and re-emit it as infrared radiation. Features in dust disks sometimes suggest the presence of full-sized planets.

Direct Imaging. Since planets are extremely faint light sources, current telescopes can only get a direct image of extra solar planets under certain circumstances. These circumstances exist only if the planet is considerably larger than Jupiter, is a good distance from its parent star, and is young so that it emits intense infrared radiation.

Planetary astronomy has benefited from direct observation in the form of spacecraft and sample return missions. These include fly-by missions with remote sensors; landing vehicles that can perform experiments on the surface materials; impactors that allow remote sensing of buried material; and sample return missions that allow direct, laboratory examination.

Stellar Astronomy

Stellar astronomy is the study of stars and the phenomena exhibited by the various forms and developmental stages of stars. Stellar astronomy has two main approaches, observational and theoretical.

Observational methods for studying distant suns include telescopes, photometry, spectroscopy, radio telescopes, the Doppler effect, and the various methods for studying their electromagnetic spectra.

Theoretical stellar astronomy consists of mathematical modeling of stellar phenomena, with solutions obtained with the use of digital computers.

Galactic Astronomy

Galactic astronomy is the study of the Milky Way, which is a prominent member of the Local Group of galaxies. The Local Group of galaxies comprises over 35 galaxies. It's gravitational center is located somewhere between the Milky Way and the Andromeda Galaxy. The galaxies of the Local Group cover a 10 million light-year diameter.

The Milky Way is a rotating mass of gas, dust, stars and other objects held together by mutual gravitational attraction. Because the Earth is located within the dusty outer arms, large portions of the Milky Way are obscured from view. In the center of the Milky Way is the core, believed to be a supermassive black hole.

The Milky Way is studied by a combination of techniques already described under solar astronomy, planetary astronomy, and stellar astronomy.

Extragalactic Astronomy

The study of objects outside of the Milky Way is a branch of astronomy concerned with the formation and evolution of galaxies and their morphology and classification. It examines active galaxies and groups and clusters of galaxies.

Most galaxies are organized into distinct shapes that allow a classification scheme. They are commonly divided into spiral, elliptical, and irregular galaxies, depending on their appearance.

The large-scale structure of the universe is represented by groups and clusters of galaxies. This structure is organized in a hierarchy of groupings, with the largest being the superclusters. The collective matter is formed into filaments and walls, leaving large voids in between.

Objects outside the Milky Way are studied by methods already discussed under solar astronomy, planetary astronomy, and solar astronomy.

Cosmology

Cosmology is the study of the universe as a whole. It attempts to use the scientific method to understand the origin, evolution, and ultimate fate of the universe. Observations of the large-scale structure of the universe, a branch known as physical cosmology, have provided an understanding of the formation and evolution of the universe.

Fundamental to cosmology is the Big Bang theory, which states that the universe began with an enormous explosion, and thereafter expanded over the course of 13.7 billion years to its present condition. The universe continues to expand.

In the course of this expansion, the universe underwent several evolutionary stages. In the very early moments, it is theorized that the universe experienced a very rapid cosmic inflation. Thereafter, nucleosynthesis produced the elemental abundance of the early universe.

When the first atoms formed, space became transparent to radiation, releasing the energy viewed today as the microwave background radiation. The expanding universe then underwent a dark age due to the lack of stellar energy sources.

A hierarchical structure of matter began to form from minute variations in the mass density. Matter accumulated in the densest regions, forming clouds of gas and the earliest stars. These massive stars triggered the re-ionization process, and are believed to have created many of the heavy elements in the early universe.

Gradually, organizations of gas and dust merged to form the first primitive galaxies. Over time, these pulled in more matter, and became organized into groups and clusters of galaxies, and then into larger-scale superclusters.

Fundamental to the structure of the universe is the existence of dark matter and dark energy. These are now thought to be the dominant components, forming 96 percent of the density of the universe.

Interdisciplinary Studies

Astronomy and astrophysics have developed significant interdisciplinary links with other major scientific fields. Interdisciplinary studies include archaeoastronomy, astrobiology, archeochemistry, and cosmochemistry.

Archaeoastronomy. Archaeoastronomy is the study of the astronomical knowledge, practices, celestial lore, mythologies, religions, and world-views of all ancient cultures regarding celestial objects or phenomena. It is considered to be the anthropology of astronomy to distinguish it from the history of astronomy. It has its roots in the study of Stonehenge. Archaeostronomy is sometimes called anthropological astronomy.

Astrobiology. Astrobiology is the interdisciplinary study of life in the Universe, combining aspects of astronomy, biology, and geology. It seeks to understand the origin and evolution of life on earth, to determine if life exists elsewhere in the universe, and to predict the future of life on earth and in the rest of the universe.

Astrochemistry. Astrochemistry is the study of the chemical elements found in outer space, particularly in molecular gas clouds. But chemical elements may also appear in low temperature stars, brown dwarfs, and planets. The field includes the study of the formation, interaction, and destruction of these chemicals. As such, it represents an overlap of the disciplines of astronomy and chemistry. On the Solar System scale, the study of chemical elements is usually called cosmochemistry.

Cosmochemistry. Cosmochemistry is the study of the chemicals found within the Solar System, including the origins of the elements and variations in the isotope ratios. It frequently involves direct measurement of physical samples in laboratories on the Earth. Samples are most often come from meteorites and micrometeorites, which include material that originated on the Moon, Mars, many different asteroids, and quite possibly comets, as well as samples returned from the Moon by manned and robotic missions.

Amateur Astronomy

Amateur astronomy is a hobby whose participants oserve and study celestial objects for the enjoyment of it. Amateur astronomers do not depend on the field of astronomy as a primary source of income or support, and do not have a professional degree or advanced academic training in the field.

Common tools for amateur astronomy include portable telescopes and binoculars. Common targets include the Moon, planets, stars, comets, meteor showers, and a variety of deep sky objects, such as star clusters, galaxies, and nebulae.

Amateur astrophotography involves the taking of photos of the night sky. It has become more popular for amateurs in recent years as relatively sophisticated equipment has become more affordable.

Most amateur astronomers work at visible wavelengths, but a small minority experiment with wavelengths outside the visible spectrum. This includes the use of infrared filters on conventional telescopes, and also the use of radio telescopes.

Amateur astronomers also use star charts that may range from simple charts to detailed maps of very specific areas of the night sky.

A range of astronomy software is available to amateur astronomers, and the Internet has become an essential tool. Most astronomy clubs have a website.

Selected Astronomy Terms

Asteroid. A small, rocky object that orbits the Sun. Tens of thousands of these objects make up the Asteroid Belt, located between the orbits of Mars and Jupiter.

Asteroid Belt. The region of the Solar System between the orbits of Mars and Jupiter where most of the asteroids are found.

Astrology. The study of the positions and aspects of celestial bodies in the false belief that they have an influence on the course of natural earthly occurrences and human affairs.

Astronomical Unit. Unit of length based on the distance from the Earth to the Sun, nearly 150 million kilometers.

Big Bang Theory. A theory that describes the origin of the universe as an enormous explosion at the beginning of time. The explosion was followed by a cooling and thinning out of the universe. The universe is still expanding.

Binary Star. Two stars that are paired by virtue of being gravitationally attracted to each other. They orbit about a common center of mass.

Black Hole. An object predicted to exist by the theory of general relativity. It is maximally gravitationally collapsed, and not even light can escape.

Black Hole Singularity. The object of zero radius into which the matter of a black hole is comprised.

Blazar. A very compact and highly variable energy source associated with a supermassive black hole at the center of a host galaxy. Blazars are among the most violent phenomena in the universe.

Celestial Mechanics. The branch of astrophysics that deals with the motions of celestial objects.

Celestial Sphere. An imaginary sphere around the Earth on which the Sun, Moon, stars, and planets appear to be placed.

Cepheid. A member of a particular class of variable stars, notable for a fairly tight correlation between their period of variability and absolute luminosity.

Chromosphere. A layer of the Sun's atmosphere lying above the photosphere. It has a width of about 2000 km. It has a lower gas density than the photosphere, but a higher temperature. The temperature is sufficiently high to ionize hydrogen gas and produce emission lines, notably the red Balmer line that gives the chromosphere a pinkish color.

Comet. An object composed of rock, ice, dust, and gases moving about the Sun in an elliptical orbit. A comet has three distinct components—the nucleus, made up of rock and ice; the coma, consisting of gases and dust; and the tail, formed when gases and dust spread out from the nucleus or coma. Short period comets complete their orbits in less than 200 years; long-period comets may take thousands of years to revolve around the Sun. Comets are thought to originate in the Oort cloud, a region or space that lies outside our Solar System.

Constellation. An arbitrary formation of stars perceived as a figure or design. One of 88 recognized groups named after characters from classical mythology and various common animals and objects.

Core Collapse. Catastrophic gravitational infall of the center of a star when it no longer can generate sufficient pressure to maintain hydrostatic equilibrium.

Corona. The atmosphere of the Sun, composed of hot, very thin gas and extending out from the Sun for a substantial distance. This gas emits light, but normally that light can't be seen against the direct glare of the Sun. During a total eclipse this direct light is blocked by the Moon and the white glow of the corona becomes visible.

Cosmic Background Radiation. The blackbody radiation, mostly in the microwave band, which consists of photons left over from the very hot, early phase of the Big Bang.

Cosmic Rays. Very high energy atomic particles (mostly protons) traveling through space at high speeds close to that of light. When they hit atoms in the upper atmosphere of the Earth they generate short-lived exotic particles in much the same way that experimental particle accelerators do.

Cosmological Principle. The principle that there is no center to the universe, and that the universe is the same in all directions. This principle means that what we observe of the universe from our specific location is representative of the true nature of the universe.

Cosmology. The study of the universe, its origin and evolution, the distribution and behavior of the matter and energy in it, and the laws that govern these factors.

Dark Matter. Matter that is thought to exist in the universe but has not yet been observed. Its existence is based on measurements of unexplained gravitational effects on visible matter.

Declination. The angular distance to a point on a celestial object, measured north or south from the celestial equator.

Distance Ladder. The chain of overlapping methods by which astronomers establish a distance scale for objects in the universe. At every step of the distance ladder, errors and uncertainties creep in. Each step inherits all the problems of the ones below.

Doppler Effect. The change in frequency of wave motion resulting from motion of the sender or the receiver. Things moving toward you have their emitted wavelengths shortened. Things moving away have their emitted wavelengths lengthened.

Dwarf Planet. A celestial body that orbits the sun and is large enough to assume a nearly round shape, but that does not clear the neighborhood around its orbit and is not a satellite of a planet.

Dwarf Star. White dwarfs are collapsed stars that are still hot and shining; black dwarfs are cold, dead stars; and brown dwarfs are not massive enough to be able to fuse hydrogen in their cores.

Ecliptic. The annual path of the Sun in a circular path in the sky.

Event Horizon. In the case of a black hole, an event horizon is that surface surrounding the region out of which light itself cannot escape.

Exosphere. The fifth atmospheric layer above the Earth's surface, extending from the thermosphere upward and out into interplanetary space.

Full Moon. The phase of the Moon in which its sunlit side is the side facing Earth.

Fusion, Nuclear. The release of nuclear energy by the fusing of light elements to form a heavier element. The Sun obtains its central power from the fusing of four hydrogen atoms into one helium atom.

Galaxy. A large system of stars, usually containing between a million and a trillion stars, along with clouds of gas and dust. Galaxies are sometimes classified according to their shapes as elliptical, irregular, or spiral.

Galaxy Cluster. Pertains to a group of more than one galaxy.

Galaxy Super Cluster. A group of an enormous number of galaxies.

Galaxy, Elliptical. Elliptical galaxies range in size from the relatively rare giant ellipticals, which can have a trillion stars, to the very common dwarf ellipticals, which can be as small as a million stars.

Galaxy, Irregular. A galaxy with a chaotic appearance and with large clouds of gas and dust, but without spiral arms.

Galaxy, Spiral. A galaxy consisting of a flattened, rotating disk of stars; a central bulge; and a surrounding halo. The disk is prominent due to the presence of young, hot stars which are often arrayed in spiral patterns. The Milky Way is a spiral galaxy.

Giant Molecular Cloud. A region of dense interstellar medium that is sufficiently cold that molecules can form. The molecules in these molecular clouds emit radio radiation which can be detected on Earth. These regions are believed to be where new stars can form.

Globular Cluster. A nearly spherical, dense cluster of hundreds of thousands to millions of stars.

Gravitational Collapse. The contraction of a star when the pressure of thermonuclear reactions can no longer sustain the force of self-gravitation. Collapse occurs at the end of a star's life when its fuel of hydrogen and other elements is depleted. Depending on its original mass, the star may evolve into a white dwarf, a neutron star, or a black hole, or it may explode as a supernova.

Gravitational Radiation. The theory of general relativity predicts that if one changes the distributions of masses (which generate gravitational fields) in certain ways one can get propagating waves of gravity in a manner analogous to the propagating waves of electric and magnetic fields. Gravitational radiation carries energy and travels at the speed of light.

H II region. A cloud of glowing gas and plasma, sometimes several hundred light-years across, in which star formation is taking place.

H-R (Hertzsprung-Russell) Diagram. A graph that plots star luminosity versus star surface temperature. When so plotted, most stars fall about a main sequence line, with exotic stars above or below these.

Hubbles's Law. The farther a galaxy, the faster it is receding from Earth. $V = H_0D$, where V is velocity, D is distance, and H_0 is Hubble's constant.

Interstellar Extinction. As light from a star travels through interstellar space it encounters some amount of dust. This dust scatters some of the light, causing the total intensity of the light to diminish. The more dust, the dimmer the star will appear.

Interstellar Medium. The name given to the material that floats in space between the stars. It consists of gas (mostly hydrogen) and dust. Even at its densest, the interstellar medium is emptier than the best vacuum we can create in the laboratory. Because space is so vast, the interstellar medium adds up to a huge amount of mass.

Interstellar Reddening. As light from a star travels through interstellar space it encounters some amount of dust. This dust scatters some of the light, mainly the short wavelength (blue) components. The spectrum of the light that remains is increasingly dominated by the long wavelength (red) end of the spectrum. Hence the light is "reddened" as it travels through space.

Kuiper Belt. Disk-shaped region of the sky beyond Neptune, populated by many icy bodies. A source of short-period comets.

Lunar Eclipse. The phenomenon whereby the shadow of the Earth falls upon the Moon, producing relative darkness of the full Moon.

Mesophere. The third atmospheric layer above the Earth's surface, extending from the top of the Stratosphere to a distance of 80 km.

Meteor. A meteoroid that produces a streak of light as it enters the Earth's atmosphere and is vaporized by the resulting friction. Meteors are commonly called shooting stars.

Meteorite. A meteor that passes through the outer layers of the Earth's atmosphere and strikes the planet's surface. Meteorites are classified as siderites (containing only metals, chiefly nickel and iron), aerolites (stony objects consisting of a variety of mineral elements), and siderolites (meteorites composed of both metal and stone).

Meteoroid. A small rock in interplanetary space.

Milky Way. The spiral galaxy in which our Solar System is located. It contains about 150 billion stars, has a diameter of 500,000 light-years, and is about 13.6 billion years old.

Moon Phases. The cycle of change of the face of the Moon, changing from new to waxing, to full, to waning, and back to new.

Near Earth Object (NEO). Asteroids, comets, and large meteoroids whose orbit intersects Earth's orbit, and which may therefore pose a collision danger.

Nebula. A concentration of gas and dust in the galaxy.

Nebula, Absorption. A nebula seen in silhouette because it is absorbing or blocking light from behind it (also called a dark nebula).

Nebula, Emission. A glowing cloud of hot interstellar gas. The gas is energized by nearby or embedded hot young stars. The gas is mainly hydrogen, and the light mainly hydrogen emission, but other elements are also present and give off their own emission lines.

Nebula, Planetary. A gaseous shell thrown off by a dying star just before the star settles down to become a degenerate white dwarf.

Nebula, Reflection. A nebula composed of dust particles that scatter and reflect incident light (rather than glowing from their own intrinsic emission). Dust preferentially scatters short wavelengths, so reflection nebulae have a characteristic blue appearance.

Neutrino. Any of three species of very weakly-interacting leptons with an extremely small, possibly zero, mass. Electron neutrinos are generated in the interior of the Sun and other stars. Generally, such neutrinos do not interact with matter and stream out through the Sun. In 1987, neutrinos from a Supernova in the Large Magellanic Cloud were detected in terrestrial neutrino experiments.

Neutron Star. A dead star supported by neutron degeneracy pressure. A neutron star is the core remnant left over after a supernova explosion.

Nova. A star which suddenly increases in brightness thousands of times, then fades back to near its original intensity.

Nuclear Fusion. The combining of nuclei of light atoms to form heavier nuclei, with the release of much energy.

Nucleosynthesis. The process by which heavier chemical elements are synthesized from hydrogen nuclei in the interiors of stars or in the Big Bang.

Nucleus, Galactic. The central region of a galaxy, characterized by high densities of stars. The nucleus may also contain a supermassive black hole and may be the source of considerable high-energy, nonstellar luminosity.

Nutation. A slight irregular motion in the axis of rotation of a largely axially symmetric object, such as a planet.

Oort Cloud. The region beyond the Kuiper Belt populated by trillions of icy bodies, and a source of long-period comets.

Parallax. An apparent shift in the position of an object, such as a star, caused by a change in the observer's position that provides a new line of sight. In astronomy, the term is used for several techniques for determining distance.

Perturbation. A local gravitational disturbance in the uniform motion of a body because of the gravitational influence of another object.

Photosphere. The surface layer of the Sun where the continuous blackbody-type spectrum is produced that we directly observe when we look at the Sun. It is as far into the Sun as we can directly see.

Planetoid. A mostly obsolete term, usually used to describe the larger remnants of rocks left over from the formation of the Solar System.

Planets. The major bodies orbiting the Sun, massive enough for their gravity to make them spherical, but small enough to avoid nuclear fusion in their cores.

Plasma. Matter in the form of electrically charged particles; the state in which most of the universe exists.

Proper Motion. The measurement of a star's change in position in the sky over time after improper motions are accounted for. Improper motion is the change of a star's coordinates on the sky not originating from the motion of the star itself.

Protostar. A forming star prior to settling down to the main sequence and burning hydrogen in its core.

Pulsar. A rotating, magnetized neutron star that produces regular pulses of radiation when observed from a distance. A pulse is produced every time the rotation brings the magnetic pole region of the neutron star into view. In this way the pulsar acts much as a light house, sweeping a short precisely-timed beam of radiation through space.

Quasar. A contraction of the word quasi-stellar used to describe celestial objects with a star-like appearance. Quasars are the most distant objects known. They have large red shifts, indicating great recessional velocities, and emit energy that is more than a thousand times that of an average galaxy.

Radial Velocity. The speed with which a star moves toward or away from the Earth. It is determined from the red or blue shift in the star's spectrum.

Radio Galaxy. A galaxy that is a powerful source of radio waves.

Radio Jets. Narrow, collimated beams of plasma that are producing radio (synchrotron) emission. These jets emerge from the cores of radio galaxies and can extend outward across regions of space larger than the size of the galaxy itself. These jets are believed to be powered and launched from an accretion disk orbiting a supermassive black hole at the galaxy's core.

Red Giant. A star with low surface temperature (thus red) and large size (giant). These stars are found on the upper-right hand side of the H-R diagram. The red giant phase in a star's life occurs after it has left the main sequence. The Sun will become a red giant in about five billion years.

Redshift. Shift of a spectrum of light toward long, red wavelengths. Due to the Doppler effect resulting from the recession of a star. The faster an object recedes from Earth, the greater the shift of its light toward the red end of the spectrum.

Satellite. A projectile or small celestial body that orbits a larger celestial body.

Solar Wind. A stream of particles, primarily protons and electrons, that constantly flow outward from the Sun.

Space-time. A four-dimensional way of describing events and locations with three units of distance and one of time. Under the influence of gravity, space-time can actually warp and bend.

Spectroscope. An instrument used to determine the spectrum or wavelength of a ray of light emanating from an object.

Spectrum. Radiation (usually visible light) broken into its component wavelengths.

Star. A spherical celestial body consisting of a large mass of hot gas held together by its own gravity. It is self-luminating because of extensive internal nuclear fusion. Our Sun is a typical star.

Star Cluster. Groups of stars held together by mutual gravitational attraction.

Steady-state Theory. Proposes that matter is being continuously created to allow the density of the universe to remain constant as it expands.

Stratosphere. The region of the atmosphere above the troposphere and below the mesosphere.

Sunspots. Temporary, relatively cool and dark regions of the Sun's surface. Due to intense magnetic fields.

Supercluster. A large conglomeration of galaxy clusters and galaxy groups, typically more than 100 million light-years in size and containing tens of thousands to hundreds of thousands of galaxies.

Supermassive Black Hole. An extremely large black hole that is believed to be at the center of many large galaxies. Has a mass ranging from a few million to more than several billion solar masses.

Supernova. Supernova come in two types: Type I is caused by sudden nuclear burning in a white dwarf star. Type II is caused by the collapse of the core of a super-massive star at the end of its nuclear-burning life. In either case, the star is destroyed and the light given off in its explosion briefly rivals the total light given off by a whole galaxy.

Supernova Remnant. The material blown off during a supernova, seen as a great glowing cloud expanding into space.

Telescope. (a) Reflector uses a single or combination of curved mirrors that reflect light and form an image. (b) Refractor uses a lens as its objective to form an image.

Troposphere. The atmospheric layer closest to the Earth's surface, containing 90 percent of the atmosphere's mass and essentially all of its water vapor and clouds.

T Tauri Star. A member of a class of very young, optically visible, solar-mass stars with peculiarities such as variability and evidence for mass loss.

Universe. The entirety of all that is known to exist. The size of the observable universe is limited to the distance light has traveled since the Big Bang.

Variable Star. A star that undergoes significant periodic variation in its luminosity.

References

1. Amateur Astronomy, Wikipedia, http://en.wikipedia.org/wiki/Amateur_astronomy.
2. Calvin, Hamilton J., Terms and Definitions, www.solarviews.com/eng/terms.
3. Carroll, Bradley W. and Dale A. Ostlie, Addison Wesley, Second Edition, San Francisco, An Introduction to Modern Astrophysics, 2007.
4. Cosmic ray mystery solved, Telegraph, www.telegraph.co.uk/earth/main.jhtml?xml=/earth/2007/11/08/scicosmic108.xml.
5. Exrasolar Planets, What is an extrasolar planet?, www.users.muohio.edu/weaksjt.

6. Field, George Brooks, Astrophysics, Microsoft Encarta Encyclopedia Online, http://encarta.msn.com/ encyclopedia_761561292/Astrophysics, 2008.
7. Gamma-ray Astronomy, The Columbia Encyclopedia, Sixth Edition, www.bartleby.com/65/ ga/gammaray, 2007.
8. Gravitational wave astronomy, Einstein, www.einstein-online.info/en/elementary/gravWav/ gw_astronomy/index.
9. Hawley, John F., Astronomy 124, Glossary of Terms, www.astro.virginia.edu/class/hawley/ astr124/ glossary, 1999.
10. Hester, Jeff, George Blumenthal, Bradford Smith, David Burstein, Ronald Greeley, and Howard G. Voss, 21st Century Astronomy, W.W. Norton and Company, New York, 2007.
11. Hewitt, Paul G., John Suchoki, and Leslie A. Hewitt, Conceptual Physical Science, Addison Wesley, San Francisco, 2003.
12. Infrared Astronomy, The Columbia Encyclopedia, Sixth Edition, www.bartleby.com/65/ir/IR-astro, 2007.
13. Pasachoff, Jay M., Astronomy, Microsoft Encarta Encyclopedia Online, http://encarta.msn.com/ encyclopedia_1741502444/ Astronomy, 2008.
14. Radar Astronomy, The Columbia Encyclopedia, Sixth Edition, www.bartleby.com/65/ra/radarast, 2007.
15. Radio Astronomy, The Columbia Encyclopedia, Sixth Edition, www.bartleby.com/65/ra/radioast, 2007.
16. Sheriff, Lucy, Black holes blamed for super-charged cosmic rays, The Register, November 9, 2007.
17. Sommers, Paul and Stefan Westerhoff, Cosmic-ray Astronomy, Astrophysics, http://arxiv.org/ abs/ 0802.1267v1.
18. Ultraviolet Astronomy, The Columbia Encyclopedia, Sixth Edition, www.bartleby.com/65/ uv/UV-astro, 2007.
19. X-ray Astronomy, The Columbia Encyclopedia, Sixth Edition, www.bartleby.com/65/xr/ Xrayastr, 2007.

Early Astronomers
624 BC to 1599

Thales (c.624-c.546 BC)

His questioning approach to the understanding of heavenly phenomena was the beginning of Greek astronomy

Thales was born about 624 BC in the city of Miletus, an ancient Ionian seaport on the western coast of Asia Minor. He visited Egypt, and probably Babylon, bringing back knowledge of astronomy and geometry.

He investigated almost all areas of knowledge—philosophy, history, science, mathematics, engineering, geography, and politics. He proposed theories to explain many of the events of nature.

Thales attempted to find naturalistic explanations of the world without reference to the supernatural or mythological. He explained earthquakes by hypothesizing that the Earth floats on water, and that earthquakes occur when the Earth is rocked by waves.

His questioning approach to the understanding of heavenly phenomena was the beginning of Greek astronomy. He set the seasons of the year and divided the year into 365 days. He estimated the size of the Sun at 1/720th of its path and that of the Moon at the same ratio of its smaller path. He incorrectly believed that the yearly flooding of the Nile was caused by seasonal winds blowing upstream.

Thales was known for his innovative use of geometry, both theoretical and practical. He understood similar triangles and right triangles, and he was familiar with ratio of the run to the rise of a slope. It was said that he measured the height of the pyramids by their shadows at the moment his own shadow was equal to his height. A right triangle with two equal legs is a 45-degree right triangle, all of which are similar. The length of the pyramid's shadow, measured from the center of the pyramid at that moment, must have been equal to its height.

Attributed to Thales is the discovery that a circle is bisected by its diameter, that the base angles of an isosceles triangle are equal, and that when two pairs of angles are formed by two intersecting lines the vertical angles are equal to each other.

Many philosophers followed Thales' lead in searching for explanations in nature (logos) rather than in the supernatural (mythos). He is considered by many to be the father of science.

Thales is said to have died of dehydration about 546 BC while watching a gymnastic contest.

1. Thales of Miletus (62?-546), The Internet Encyclopedia of Philosophy, www.iep.utm.edu/t/thales.
2. Thales of Miletus (634-546 BC), Eric Weisstein's World of Biography, Wolfram Research, http://scienceworld.wolfram.com/ biography/Thales.
3. Thales of Miletus, Wikipedia, http://en.wikipedia.org/wiki/Thales.

Anaximander (c.610-c.540 BC)

Often called the founder of astronomy

Anaximander, a Greek philosopher of the Milesian School, was born in the city of Miletus, an ancient Ionian seaport on the western coast of Asia Minor, in about 610 BC. He was a student of Thales (c.624–c.546 BC), also of Miletus. He succeeded Thales, becoming the second master of that school.

Anaximander was an early proponent of science and tried to observe and explain different aspects of the universe, with a particular interest in its origins. He claimed that nature is ruled by laws just like human societies.

He tried to describe the mechanics of celestial bodies in relation to the Earth. He was possibly the first person to speculate on the existence of other worlds. He felt that there might be an indefinite number of worlds existing throughout time, worlds that are born and perish within an eternal or ageless infinity. He taught, therefore, that there was not a multitude of planets in space, but of an endless temporal succession of planets.

Anaximander conceived of the universe as a number of concentric cylinders, of which the outermost is the Sun, the middle is the Moon, and the innermost is the stars. Thus, he pioneered the notion that the Earth is not flat, suggesting it was cylindrical and that it floated free, unsupported, at the exact center of the universe. He postulated that people lived on one of the flat ends.

Like many thinkers of his time, Anaximander contributed to many disciplines in addition to astronomy. In physics, he postulated that the indefinite (or apeiron) was the source of all things. His knowledge of geometry allowed him to introduce the concept of the gnomon, which is an ancient Greek word meaning indicator—one who discerns or that which reveals. For instance, the gnomon is the part of a sundial that casts the shadow to reveal the time.

Anaximander created a map of the world that contributed greatly to the advancement of geography. He was also involved in the politics of Miletus, as he was sent as a leader to one of its colonies.

He asserted that physical forces, rather than supernatural means, created order in the universe. He can be considered the first true scientist, having conducted the earliest recorded scientific experiment.

Anaximander is often called the founder of astronomy. He died about 540 BC.

1. Anaximander of Miletus (c.610-c.540 BC), The Encyclopedia of Science, www.daviddarling.info/encyclopedia/A/Anaximander.
2. Anaximander, Wikipedia, http://en.wikipedia.org/wiki/Anaximander.

Anaximenes (c.585-c.525 BC)

Believed that the Earth, Moon, and Sun came from air

Anaximenes, a Greek philosopher, was born in Miletus about 585 BC. He was said to be a student of Anaximander (c.610-c.540 BC), and like him he sought to give a quasi-scientific explanation of the world. He is best known for his doctrine that air is the source of all things. This claim contrasts with the view of Thales (c.624-c.546 BC) that water is the source, and with the view of Anaximander that all things came from an unspecified boundless material.

For Anaximenes, all things, including the gods, came from air, and ultimately were air. Because air is infinite and perpetually in motion it could produce all things without being produced by anything. He believed air came in threads. Very close threads was a solid. These threads came together by a process called felting, which is condensation and rarefaction. Felting was the mechanism whereby air was transformed into all things. Compressed air becomes first water and then Earth; when rarefied, it becomes fire. All other things are composites of varying degrees of these four elements; as such, they are the effect of the varying densities of air.

Like Anaximander, Anaximenes accounted for various natural phenomena—lightning and thunder result from wind breaking out of clouds, rainbows are the result of the rays of the Sun falling on clouds, and earthquakes are caused by the cracking of the Earth when it dries out after being moistened by rains. He gave an essentially correct account of hail as frozen rainwater.

According to Anaximenes, Earth, a flat disk, was formed from air. The Earth floated on a cushion of air. From evaporations from the Earth, fiery bodies arose, which came to be the heavenly bodies. The Sun and the Moon, also flat bodies, floated on streams of air. In one account, the heaven was said to be like a felt cap that turns around the head, the stars fixed to this surface like nails. In another account, the stars were like fiery leaves floating on air. The Sun was believed to circle around the Earth.

Anaximenes was the first to have a theory of the soul. He felt that the soul, being air, holds us together.

He wrote his philosophical views in a book, which survived well into the Hellenistic period.

Anaximenes died about 525 BC. The **Anaximenes crater** on the Moon was named in his honor.

1. Anaximenes, Wikipedia, http://en.wikipedia.org/wiki/Anaximenes_of_Miletus.
2. Anaximenes, www.abu.nb.ca/Courses/GrPhil/Anaximenes, September 14, 2006.
3. Graham, Daniel G., The Internet Encyclopedia of Philosophy, www.iep.utm.edu/a/anaximen, 2006.

Pythagoras (c.570- 500 BC)

One of the first to think that the Earth was round

Pythagoras was born about 570 BC. He spent his early years on the island of Samos off the coast of modern Turkey. At the age of 40, he immigrated to the city of Croton in southern Italy. Most of his philosophical activity occurred there. Pythagoras wrote nothing, and it is hard to say how much of the doctrine we know as Pythagorean is due to him and how much is later development.

During his life, and for years afterward, Pythagoras was said to be (1) an expert on the fate of the soul after death who thought that the soul was immortal and went through a series of reincarnations; (2) an expert on religious ritual; (3) a wonder-worker who had a thigh of gold and who could be in two places at the same time; and (4) the founder of a strict way of life that emphasized dietary restrictions, religious ritual, and rigorous self discipline.

Pythagoras was one of the first to think that the Earth was round, that all planets have an axis, and that all the planets travel around one central point. He originally identified that point as Earth, but later renounced it for the idea that the planets revolve around a central "fire" that he never identified as the Sun. He also believed that the Moon was another planet that he called a counter-Earth.

Pythagoras presented a cosmos that was structured according to moral principles and significant numerical relationships. It may have been akin to conceptions of the cosmos found in Platonic myths. In such a cosmos, the planets were seen as instruments of divine vengeance and the Sun and Moon as the isles of the blessed, where we may go if we live a good life. Thunder functioned to frighten the souls being punished in Tartarus. He felt that the heavenly bodies moved in accordance with the mathematical ratios that govern the concordant musical intervals to produce a "music of the heavens," which in the later tradition developed into "the harmony of the spheres." It is doubtful that Pythagoras himself thought in terms of spheres, and the mathematics of the movements of the heavens was not worked out in detail.

He is best known for the **Pythagorian theorem** named after him—the sum of the squares of the two lesser sides of a right triangle is equal to the square of the hypotenuse. Pythagoras died in 500 BC.

1. Pythagoras (fl. 530 BCE.), Internet Encyclopedia of Philosophy, www.iep.utm.edu/p/pythagor.
2. Pythagoras of Samos, MacTutor, www-groups.dcs.st-and.ac.uk/~history/Biographies/Pythagoras, January 1999.
3. Pythagoras, Stanford Encyclopedia of Philosophy, http://plato.stanford.edu/entries/pythagoras, October 18, 2006.

Nabu-rimanni (c.560- c.480 BC)

Devised the so-called System A, a group of tables giving the positions of the Moon, Sun, and planets at any given moment

Nabu-rimanni was born about 560 BC in Babylonia. He was the earliest Chaldean astronomer known by name. In Babylon, a settlement was set apart for the local philosophers, the Chaldeans as they were called, who were concerned mostly with astronomy.

From the times of king Nabonassar, who ruled from 747 to 732 BC, many Sumerian astronomical written records were preserved. From them the Sumerians, after long-standing observations of solar and lunar eclipses, identified a cycle of 18 years and 11 days, which was named the Saros cycle. Afterward, the Earth, Sun, and Moon return to nearly the same relative positions. At this time, the cycle of lunar and solar eclipses begins to repeat itself.

The Sumerians were not just able to determine astronomical conjunctions of the Sun and the Moon, but were aware of the changes of the lunar movement and the changes of the apparent angular velocity of the Sun and even the planets. They were able to define periods of these irregularities, which we now know as consequences of elliptical orbits of celestial bodies. By the early sixth century BC they had determined relative movements of the Sun and Moon and lunar perigee and apogee and their nodes.

Nabu-rimanni devised the so-called System A, a group of tables giving the positions of the Moon, Sun, and planets at any given moment. Although based on the centuries of observation by astronomers, these tables were nonetheless somewhat crude and were superseded about a century later by System B of Kidinnu (c.400-c.310 BC).

Nabu-rimanni determined the solar year to be 365 days, 6 hours, 15 minutes, and 41 seconds. He used a water clock to measure the days, months and the length of the solar year.

He also calculated the length of the synodic month to be 29 days, 12 hours, 44 minutes, and 5.05 seconds (29.530614 days) as compared with the modern value of 29 days, 12 hours, 44 minutes, and 3.49 seconds (29.530596 days), with an error of +1.56 seconds. He also showed how an apparent magnitude of lunar eclipse can be calculated from the deviation of the lunar nodes.

Nabu-rimanni died about 480 BC.

1. Nabu-rimanni Biography, biographybase, www.biographybase.com/biography/Nabu-rimanni.
2. Nabu-rimanni, EconomicExper.com, www.economicexpert.com/a/Nabu:rimanni.
3. Nabu-ri-man-nu, BookRags, www.bookrags.com/wiki/Naburimannu

Anaxagoras (c.500- c.428 BC)

Correctly explained the phases of the Moon and eclipses of the Moon and the Sun in terms of their movements

Anaxagoras was born about 500 BC in Clazomenae in Asia Minor. In early manhood he went to Athens, which was rapidly becoming the center of Greek culture. There he is said to have remained for 30 years.

Anaxagoras correctly explained the phases of the Moon and eclipses of the Moon and the Sun in terms of their movements. He believed that heaven and Earth were brought into existence by the same processes and were composed of the same materials. The heavenly bodies, he asserted, were masses of stone torn from the Earth and ignited by rapid rotation.

Because his theories brought him into conflict with the popular faith of the day, Anaxagoras's views on heavenly bodies were considered dangerous by the Church. He was arrested and charged with contravening the established religion. On release from custody, he was forced to move from Athens to Lampsacus in Ionia.

Anaxagoras believed that all things have existed from the beginning, but originally they existed in infinitesimally small fragments of themselves, endless in number and inextricably combined throughout the universe. All things existed in this mass, but in a confused and indistinguishable form. There were the seeds (miniatures of corn, flesh, gold, and so forth) in the primitive mixture; but the parts of like nature with their wholes had to be eliminated from the complex mass before they could receive a definite name and character. The existing species of things having thus been transferred were multiplied endlessly in number by reducing their size through continued subdivision. At the same time, each thing, as Anaxagoras saw it, was so indissolubly connected with every other that the keenest analysis could never completely sever them.

In trying to explain the processes of nutrition and growth, Anaxagoras theorized that for the food an animal eats to turn into bone, hair, flesh, and so forth, it must already contain all of those constituents within it.

Anaxagoras marked a turning-point in the history of philosophy. With him, scientific speculation passed from the colonies of Greece to settle at Athens. By the theory of minute constituents of things and his emphasis on mechanical processes in the formation of order, he helped pave the way for atomic theory. Anaxagoras died in Lampsacus in about 428 BC.

1. Anaxagoras of Clazomenae (ca. 500-ca. 428 BC, Eric Weisstein's World of Biography, Wolfram Research, chttp://scienceworld.wolfram.com/biography/Anaxagoras.
2. Anaxagoras, NNDB, www.nndb.com/people/884/000087623.
3. Anaxagoras, Stanford Internet Encyclopedia of Philosophy, http://plato.stanford.edu/entries/ Anaxagoras.

Democritus (c.460-370 BC)

Among the first to propose that the universe contains many worlds, some of them inhabited

Democritus was born at Abdera in Thrace, a northern territory of Greece, about 460 BC. He was a student of Leucippus (c.500-450 BC) and co-originator with him of the belief that all matter is made up of various imperishable, indivisible elements, which he called atoma, or indivisible units, from which we get the English word atom.

Democritus and Leucippus contended that the human soul is composed of exceedingly fine and spherical atoma. They held that these atoma move because it is their nature never to be still, and that as they move they draw the whole body along with them and set it in motion. Democritus viewed soul atoma as being similar to fire atoma—capable of penetrating solid bodies.

Democritus explained the three senses in terms of atoma as well. He hypothesized that different tastes are a result of differently shaped atoma in contact with the tongue. Smells and sounds were explained similarly. Vision, according to Democritus, was due to the eye receiving effluences of bodies, which emanated from the bodies.

Democritus's theory argued that atoma had several properties—size, shape, and perhaps weight. He considered all other properties, such as color and taste, to be the result of complex interactions between the atoma in our bodies and the atoma of the matter that we are examining. Furthermore, he believed that the properties of atoma determine the perceived properties of matter—something that is solid is made of small, pointy atoma, while something that has water-like properties is made of large, round atoma. He suggested that some types of matter are particularly solid because their atoma have hooks to attach to each other, and some are oily because they are made of very fine, small atoma which can easily slip past each other.

Democritus was also a pioneer of mathematics. He was among the first to observe that a cone or pyramid has one-third the volume of a cylinder or prism, respectively, with the same base and height. He realized that the celestial body we perceive as the Milky Way is formed from the light of distant stars. Other philosophers, including Aristotle, later argued against this. Democritus was among the first to propose that the universe contains many worlds, some of them inhabited. He died in 370 BC.

1. Berryman, Sylvia, Stanford Encyclopedia of Philosophy, http://plato.stanford.edu/entries/democritus, August 15, 2004.
2. O'Connor, J.J. and E F Robertson, Democritus, MacTutor, www-groups.dcs.st-and.ac.uk/~history/ Mathematicians/Democritus.

Eudoxus (c.400-c.350 BC)

Postulated that each crystalline sphere in Plato's model had its poles set to the next sphere

Eudoxus was born in Cnidos on the Black Sea in about 400 BC. He studied mathematics in Tarentum, and then studied medicine on Sicily. At 23 years of age he went to Plato's academy in Athens to study philosophy and rhetoric. Some time later he went to Egypt to learn astronomy at Helopolis. Afterward, he established a school at Cyzicus on the sea of Marmora and had many pupils. In 365 BC, he returned to Athens with his pupils, where he became a colleague of Plato (427-347 BC) .

In astronomy, Eudoxus accepted Plato's notion of the rotation of the planets around the Earth on crystalline spheres, but noticed discrepancies with observations. He tried to adjust Plato's model by postulating that each crystalline sphere had its poles set to the next sphere. In Eudoxus's model, the spherical Earth is at rest at the center; around this center 27 concentric spheres rotate. The exterior one caries the fixed stars; the others account for the Sun, Moon, and five planets; and each planet requires four spheres, the Sun and Moon three each.

There were problems, however, with his model. First of all, each hippopede produced by the superposition of the motions of two spheres produced the same curve, yet the retrogressions of planets were observed to exhibit differing shapes. The hippopede is a spiric section in which the intersecting plane is tangent to the interior of the torus. Secondly, his models predicted tolerable retrogressions for Jupiter and Saturn, but not for Mars or Venus. Thirdly, his model did not account for the observed differences in the lengths of the seasons. Finally, the model failed to account for variations in the observed diameter of the Moon or changes in the brightness of planets, which were correctly interpreted to indicate that their distances were changing. Eudoxus was the first Greek to make a map of the stars.

Eudoxus also excelled as a mathematician. His work on ratios formed the basis for *Book V* of Euclid's *Elements*, and anticipated in a number of ways the notion of algebra, which is otherwise absent from ancient Greek mathematics. He also constructed many geometric proofs and developed the method of exhaustion, later extended by Archimedes (287-212 BC).

In about 350 BC, at the age of 53, Eudoxus died in Cnidos. He was the leading mathematician and astronomer of his day.

1. Allen, Don, Eudoxus of Cnidus, www.math.tamu.edu/~dallen/history/eudoxus/eudoxus.
2. Eudoxus of Cnidus (ca. 400-ca. 347 BC), Eric Weisstein's World of Scientific Biography, http://scienceworld.wolfram.com/biography/Eudoxus.

Kidinnu (c.400-c.310 BC)

Made an accurate estimate of the synotic month

Kidinnu, a Chaldean astronomer and mathematician, was born about 400 BC in Babylon. He became principal of the astronomical school in the Babylonian city of Sippar in Akkad (now Abu Habbah, southwest of Baghdad, Iraq).

Babylonian astronomers had been observing the skies for centuries and had recorded their observations in astronomical diaries, astronomical almanacs, catalogs of stars, and other texts. Because there were many data available to Babylonian astronomers, their results could be fairly accurate.

Unlike Greek astronomers Kidinnu did not believe the Moon, other planets, and Sun traveled at constant velocities. As a result, he was able to arrive at good approximations for their movements, apparently by using complex methods and equations for calculating their irregular movements.

For the Sun, he determined that the apparent angular velocity is a minimum when the Earth is farthest from it. His system used steadily increasing and decreasing values for the planetary positions. For the length of the tropical year, Kidinnu used 365 days and six hours.

About 383 BC, Kidinnu obtained a more accurate value for the synodic month (the period between two full Moons). For the mean length of the synodic month he obtained 29.53059414 days, which is very close to the modern estimate of 29.530589 days.

A similarly accurate result was obtained for the length of the solar year, which Kidinnu calculated at 365 days, 5 hours, 44 minutes, and 12.52 seconds, instead of the now accepted 365 days, 5 hours. 48 minutes, 45.17 seconds.

Kidinnu is thought to have introduced the 19-year cycle known as the Metonic cycle into the Babylonian calendar in 383 BC. In this system, each year had 12 lunar months, and seven extra months were inserted at intervals during the 19-year period. This cycle, with the value for the mean lunar month, was later adopted for the Hebrew calendar and has remained in use until today. Kidinnu's work has been labeled System B, with System A being due to Nabu-rimanni (c.560-c.480 BC).

He may have discovered the precession of the equinoxes, which is the slow rotation of the Earth's axis that results in slight variations in the length of the year.

Kidinnu died in about 310 BC.

1. Kidinnu, Wikipedia, http://en.wikipedia.org/wiki/Kidinnu.
2. Lendering, Jona, Kidinnu and Babylonian astronomy, http://ircamera.as.arizona.edu/NatSci102/text/babylonian.

Heraclides (c.388-c.315 BC)

Hypothesized that the Earth rotates daily on its own axis

The Greek philosopher, Heraclides, was born at in Pontus, now Karadeniz Ereğli, Turkey, in about 388 BC. He was the son of a wealthy man and a descendant of one of the founders of Heracleia. Virtually nothing is known of his early life. In fact, very little is known about his life at all, with the exception of some of his cosmological observations.

In 363 or 364 he went to Athens to study under Plato (427-347 BC), and later Speusippus (407-339 BC), at the Academy. He was placed in charge of the Academy when Plato went to Sicily in 360 BC. He later studied under Aristotle (384-322 BC).

Heraclides was a most prolific writer, composing dialogues on ethics, natural science, literary criticism, music, rhetoric, and the history of philosophy. However, some of the works attributed to him may have been done so in error.

Heraclides contradicted the Platonic and Aristotelian doctrine that the Earth must stand still, hypothesizing that it rotates once daily on its own axis. He asserted that the universe is infinite, and speculated on the existence of other Earths in the stellar systems that appeared to be fixed stars. He tried to make his observations fit into the day's cosmological framework—that the Earth was the center of the universe and everything rotated around it. This view is called geocentric or Earth-centered.

Heraclides appears to have limited his theorizing to the planets closer to the Sun than Earth. Their motions are certainly easier to plot. Both of these planets, Venus and Mercury, appear only in the morning or the evening, and both are usually close to the Sun in the sky. Their motions across the sky are by far simpler than those of the outer planets.

Heraclides is thought to have believed that Venus and Mercury revolve around the Sun. This was a major departure from the thinking of the day, and, in fact, it would not be until about 2000 years later that this would again be suggested.

A punning on his name, Heraclides Pompicus, suggests he may have been a rather vain and pompous man.

Heraclides died in about 315 BC. His ideas do not seem to have taken hold during his lifetime. However, they did inspire Nicolaus Copernicus (1473-1543) , who in his book introducing the Sun-centered (heliocentric) universe cited Heraclides as one whose work supported his own.

1. Heraclides of Pontus, BookRags, from Encyclopedia of World Biography, www.bookrags.com/biography/heraclides-of-pontus.
2. Heraclides Ponticus, Wikipedia, http://en.wikipedia.org/wiki/Heraclides_Ponticus.
3. Heraklides of Pontus, BookRags, www.bookrags.com/research/heraklides-of-pontus-scit-01123.

Aristotle (384-322 BC)

Argued that the universe is spherical and finite

Aristotle was born in 384 BC at Stagirus, a Greek colony and seaport on the coast of Thrace. His father, Nichomachus, was court physician to King Amyntas of Macedonia. From age 17 to 37 Aristotle was a pupil of Plato in Athens.

Later, as a teacher at the Lyceum in Greece, Aristotle held that the universe was divided into two parts—the terrestrial region and the celestial region. In the realm of the Earth, Aristotle believed, as did the Greek philosopher Empedocles (c.490-430 BC), that all bodies were made of combinations of four substances—Earth, fire, air, and water. Heavy bodies, like one made of iron, consisted mostly of Earth, so they sought to return to the Earth when dropped. Less dense objects were thought to contain a larger admixture of the other elements, along with varying amounts of Earth. A feather, which contains a large amount of air, would have less attraction for the Earth, and therefore would fall more slowly. Smoke, containing even more air, would seek its own kind and rise.

Aristotle argued that the universe is spherical and finite. Spherical, because that is the most perfect shape; finite because it has a center (the center of the Earth), and a body with a center cannot be infinite. He believed that the Earth, too, is a sphere. He considered it to be relatively small compared to the stars, and in contrast to celestial bodies it was always at rest. He deduced that if the Earth were in motion, an observer on it would see the stars as moving. However, since this is not the case, the Earth must be at rest.

To prove that the Earth is a sphere, he argued that all earthly substances move towards the center, and therefore would eventually have to form a sphere. He also argued that if the Earth were not spherical, lunar eclipses would not show segments with a curved outline. Furthermore, when one travels northward or southward one does not see the same stars at night, nor do the stars occupy the same positions in the sky. In the case of the stars, Aristotle argued that they would have to be spherical, as this shape, which is the most perfect, allows them to retain their positions.

Aristotle's hierarchical model of the universe had a profound influence on medieval scholars, who modified it to correspond with Christian theology. Saint Thomas Aquinas, for example, re-interpreted the prime movers as angels. Aristotle died at Chalcis in Euboea in 322 BC.

1. Aristotle, Catholic Encyclopedia, www.newadvent.org/cathen/01713a.
2. Aristotle's Physics, http://aether.lbl.gov/www/classes/p10/aristotle-physics.
3. Fowler, Thomas, Aristotle's Astronomy, www.perseus.tufts.edu/GreekScience/Students/Tom/AristotleAstro.

Callipus (c.370-c.300 BC)

Refined the planetary theory of Eudoxus by adding additional spheres

Callipus, a Greek Astronomer and Mathematician, was born about 370 BC in Cyzicus, located in Hellespontine Phrygia on the southern shores of the Propontis (known today as the Sea of Marmara). He is best known for refining the planetary theory of Eudoxus (c.400-c.350 BC) by adding additional spheres. He also made accurate determinations of the lengths of the seasons and constructed a 76-year period to more accurately align the solar and lunar cycles.

Callipus studied with Polemarchus and followed him to Athens. He eventually came to live with Aristotle (384-322 BC), who encouraged him to devote his efforts to improving the Eudoxean system of concentric spheres. This system consisted of nested, concentric spheres rotating in such a way as to account, through compounded motions, for gross orbital appearances.

Callipus realized that Eudoxus's system required the Sun to move with an apparent constant velocity against the background of the fixed stars. This implied that the seasons were of equal length, which was contrary to common knowledge. Based on his own careful observations, Callipus accurately determined the lengths of the seasons. To account for his results, he found it necessary to refine the Eudoxean model by adding two more spheres for each of the lunar and solar models and one additional sphere each for the mechanisms of Mercury, Venus, and Mars. This brought the total number of spheres to 34. Unfortunately, all concentric-sphere models were incapable of explaining or reproducing the variation in the apparent diameters of the Sun and Moon and the requirement that the hippopede (horseshoe-curve) of retrograde motion repeat itself exactly from one orbit to the next.

By accurately determining the lengths of the seasons (94, 92, 89, and 90 days, respectively, from the time of the vernal equinox), Callipus reconciled the lunar and solar calendars. To bring the calendars into alignment, he combined four 19-year Metonic cycles, dropping one day from each. The resulting 76-year Callipic cycle provided a much more accurate measure for the year. It also became the reference standard by which later astronomers recorded their observations. The existence of this calendric standard made it possible to correct and correlate observations much more accurately. This in turn greatly contributed to the development of future astronomical theories. Callipus died about 300 BC.

1. Callipus, BookRags, www.bookrags.com/research/callipus-scit-01123.
2. Hatch, Robert A., Pre-classical and Classical Science, University of Florida, www.clas.ufl.edu/users/rhatch/his-sci-study-guide/0004_preandclassicalscience.

Aristarchus (310-230 BC)

The first to assert that the Earth revolves around the Sun

Aristarchus, a Greek mathematician and astronomer, was born in 310 BC. He was a student of Strato of Lampsacus (c.340-c.270 BC), who was head of Aristotle's Lyceum. However, it is not thought that Aristarchus studied with Strato in Athens, but rather in Alexandria. Strato became head of the Lyceum at Alexandria in 287 BC, and it is thought that Aristarchus studied with him there, starting his studies shortly after that date.

Aristarchus was the first person to present an explicit argument for a heliocentric model of the Solar System, placing the Sun, not the Earth, at the center of the known universe. He also placed the other known planets in correct order from the Sun. Aristarchus's idea was preceded by that of Heraclides of Pontus (*c*.388-315 BC), who proposed that the motions of Mercury and Venus were due to the fact that they revolved around the Sun, while the Sun revolved around the Earth.

Aristarchus's astronomical ideas were rejected in favor of the geocentric theories of Aristotle (384-322 BC) and Ptolemy (c.87-c.161) until they were successfully revived by Copernicus (1473-1543) and extensively developed and built upon by Johannes Kepler (1571-1630) and Isaac Newton (1643-1727) nearly 2000 years later.

Aristarchus's belief that the Earth revolves around the Sun is known only through the writings of Archimedes (c.287-c.212 BC). Aristarchus believed the stars to be very far away, and saw this as the reason why there was no visible parallax, that is an observed movement of the stars relative to each other as the Earth moved around the Sun.

In his only surviving work, *On the Dimensions and Distances of the Sun and Moon,* Aristarchus described a method for estimating the relative distances of the Sun and Moon from the Earth. He pointed out that the Moon and Sun have nearly equal apparent angular sizes, and therefore their diameters must be in proportion to their distances from Earth. He thus concluded that the diameter of the Sun was about 20 times larger than the diameter of the Moon; which, although wrong, follows logically from his data.

He also invented a sundial in the shape of a hemispherical bowl with a pointer to cast shadows placed in the middle of the bowl. Aristarchus died in 230 BC. The **Aristarchus crater** on the Moon was named in his honor.

1. Aristarchus of Sámos, Microsoft Encarta Online Encyclopedia, http://encarta.msn.com/encyclopedia_761568273/aristarchus_of_s%C3%A1mos, 2008.
2. Aristarchus of Samos, Wikipedia, http://en.wikipedia.org/wiki/Aristarchus_of_Samos.
3. O'Connor J. J. and E F Robertson, Anastarchus of Samos, MacTutor, www-history.mcs.st-andrews.ac.uk/Mathematicians/Aristarchus, April 1999.

Eratosthenes (276-194 BC)

Determined the circumference of the Earth

Eratosthenes, a Greek mathematician, poet, geographer, and astronomer, was born in Cyrene in modern-day Libya in 276 BC. He was educated at the academies of Athens, and became the chief librarian of the Great Library of Alexandria.

Eratosthenes wrote a comprehensive treatise about the world, called *Geography*. This was the first use of the word, which literally means "writing about the earth" in Greek. *Geography* also introduced the climatic concepts of torrid, temperate, and frigid zones.

He devised a system of latitude and longitude, calculated the length of the year as 365.25 days, and invented the leap day. He also created a map of the world based on the available geographical knowledge of the era. For his work, Eratosthenes is commonly called the father of geography.

Having heard of a deep well at Syene, near the Tropic of Cancer and modern Aswan, where sunlight only struck the bottom of the well on the summer solstice, Eratosthenes decided that he could determine the circumference of the Earth. To do so he needed two things—the distance between Syene and Alexandria and the angle of the shadow in Alexandria on the solstice. He knew the approximate distance between Syene and Alexandria as measured by camel-powered trade caravans. He then measured the angle of the shadow in Alexandria on the solstice. By taking the angle of the shadow (7°12') and dividing it into the 360 degrees of a circle he obtained the number 50. He then multiplied the distance between Alexandria and Syene by this number to determine the circumference of the Earth. Remarkably, he determined the circumference to be 25,000 miles, just 99 miles over the actual circumference at the equator of 24,901 miles.

Eratosthenes also measured the distance to the Sun and the distance to the Moon. He computed the distances using data obtained during lunar eclipses. He also measured the tilt of the Earth's axis with great accuracy, obtaining the value of 23° 51' 15".

In 193 BC, Eratosthenes became blind and, according to legend, a year later he starved himself to death. He never married.

1. Eratosthenes, Geography, About.com, http://geography.about.com/od/historyofgeography/a/eratosthenes.
2. Eratosthenes, Microsoft Encarta Online Encyclopedia, http://encarta.msn.com/encyclopedia_761557529/eratosthenes.
3. Eratosthenes, Wikipedia, http://en.wikipedia.org/wiki/Eratosthenes.
4. O'Connor J. J. and E F Robertson, Eratosthenes of Cyrene, MacTutor, www-history.mcs.st-andrews.ac.uk/Printonly/Eratosthenes, January 1999.

Apollonius (c.262-c.190 BC)

Used eccentric orbits to explain the apparent motion of the planets and the varying speed of the Moon

Apollonius of Perga, a Greek geometer and astronomer of the Alexandrian school, was born about 262 BC. Little is known of his life, but his works have had a very great influence on the development of mathematics. He is best known for his writings on conic sections. His innovative methodology and terminology influenced many later scholars, including Ptolemy (c.87-c.161), Francesco Maurolico (1494-1575), Isaac Newton (1643-1727), and René Descartes (1596-1650). It was Apollonius who gave the ellipse, the parabola, and the hyperbola the names by which they are known today. His treatise on Conics gained him the title of "The Great Geometer."

Apollonius's theorem demonstrates that two models are equivalent, given the right parameters. It is a theorem relating several elements in a triangle. It states that given a triangle ABC, if D is any point on BC such that it divides BC in the ratio $n{:}m$ (or $mBD = nDC$), then

$$mAB^2 + nAC^2 = mBD^2 + nDC^2 + (m+n)AD^2.$$

The first four books of his eight-volume treatise *Conica* consisted of an introduction and a statement of the state of mathematics provided by his predecessors. The last four books put forth his important work on conic sections, which is the foundation of much of the geometry still used today in astronomy, ballistic science, and rocketry.

Apollonius, in *Conics*, developed a method that is so similar to analytic geometry that his work is sometimes thought to have anticipated the work of Rene Descartes (1596-1650) by some 1800 years. His application of reference lines, a diameter, and a tangent is essentially no different than our modern use of a coordinate frame, where the distances measured along the diameter from the point of tangency are the abscissas, and the segments parallel to the tangent and intercepted between the axis and the curve are the ordinates.

He used eccentric orbits to explain the apparent motion of the planets and the varying speed of the Moon.

Apollonius died about 190 BC. The **Apollonius crater** on the Moon was named in his honor.

1. Apollonius of Perga | 262-190 BC | Greek mathematician, www.nahste.ac.uk/isaar/GB_0237_NAHSTE_P1095.
2. Appolonius of Perga, MacTutor, ://www-history.mcs.st-andrews.ac.uk/Biographies/Apollonius, 1999.
3. Appolonius of Perga, Wikipedia, http://en.wikipedia.org/wiki/Apollonius_of_Perga.

Hipparchus (c.190-c.120 BC)

May have produced of the first known catalog of stars

Hipparchus was born about 190 BC in Nicaea, Bithynia (now İznik, Turkey), but spent most of his life in Rhodes. He is said to have been the most important Greek astronomer of his time. He was extremely accurate in his research, a partial record of which was preserved in the *Almagest*, which is the scientific treatise by the Alexandrian astronomer Claudius Ptolemy (c.87-c.161).

Hipparchus's recorded observations span the years 147 to 127 BC. He used an instrument described by Ptolemy as a dioptra, which, attached to a stand, was a sighting tube or a rod with a sight at both ends. Fitted with protractors, it could be used to measure angles. Hipparchus made extensive observations of star positions, and is credited by some with the production of the first known catalog of stars.

He turned his attention to a wide variety of astronomical questions, including the length of the year, the determination of lunar distance, and the computation of lunar and solar eclipses. He developed theories for the Sun and Moon, concluding that they are represented by uniform circular motions.

By comparing his own celestial studies with those of earlier astronomers, Hipparchus discovered the precession of the equinoxes. His calculation of the tropical year, which is the length of the year measured by the Sun, was within 6.5 minutes of modern measurements.

He compiled the planetary observations to which he had access into a more useful arrangement, and demonstrated that the phenomena were not in agreement with the hypotheses of the astronomers of that time.

Hipparchus devised a method of locating geographic positions by means of latitudes and longitudes. He cataloged, charted, and calculated the brightness of perhaps as many as 1000 stars. He also compiled a table of trigonometric chords that became the basis for modern trigonometry.

Other works by Hipparchus, now lost, include an astronomical calendar, books on optics and arithmetic, a treatise entitled *On Objects Carried Down by their Weight*, geographical and astrological writings, and a catalog of his own work.

Hipparchus died about 120 BC.

1. Hipparchus, Microsoft Encarta Online Encyclopedia, http://encarta.msn.com/encyclopedia_761555438/hipparchus, 2008.
2. O'Connor, J. J. and E F Robertson, Hipparchus of Rhodes, MacTutor, www-groups.dcs.st-and.ac.uk/~history/Biographies/Hipparchus, April 1999.
3. Taub, Liba, Hipparchus, Department of History and Philosophy, University of Cambridge, www.hps.cam.ac.uk/starry/hipparchus, 1999.

Posidonius (c.135-c.50 BC)

Advanced the theory that the Sun emanated a vital force which permeated the world

Posidonius, a Greek philosopher, politician, astronomer, geographer, historian and teacher native to Apamea, Syria, was born about 135 BC. He settled about 95 BC in Rhodes, a maritime state which had a reputation for scientific research, and became a citizen. He served as an ambassador to Rome in 87-86 BC.

Posidonius made one or more journeys traveling throughout the Roman world, and even beyond its boundaries, to conduct scientific research. He traveled in Greece, Hispania, Africa, Italy, Sicily, Dalmatia, Gaul, Liguria, North Africa, and on the eastern shores of the Adriatic.

In Hispania, on the Atlantic coast at Gades, Posidonius studied the tides. He observed that the daily tides were connected with the orbit and the monthly tides with the cycles of the Moon, and he hypothesized about the connections of the yearly cycles of the tides with the equinoxes and solstices.

Some fragments of his writings on astronomy survive through the treatise by Cleomedes (c.10-c.70), *On the Circular Motions of the Celestial Bodies*. The first chapter of the second book appears to have been mostly copied from Posidonius.

Posidonius advanced the theory that the Sun emanated a vital force which permeated the world. He attempted to measure the distance and size of the Sun. In about 90 BC, he estimated the astronomical unit to be 9893 Earth radii, which would prove to be too small by half. The astronomical unit is based on the distance from the Earth to the Sun. In measuring the size of the Sun, however, he reached a figure larger and more accurate than those proposed by other Greek astronomers. He also calculated the size and distance of the Moon.

Some time not long after 100 BC, Posidonius became the head of the Stoic School in Rhodes. While in this position he also held political office in Rhodes.

Posidonius constructed an orrery, which is a mechanical device that illustrates the relative positions and motions of the planets and moons in the Solar System in a heliocentric model. Orreries are typically driven by a large clockwork mechanism. In modern ones, a globe represents the Sun at the center, with a planet at the end of each of the arms. Posidonius died about 50 BC.

1. O'Connor, J. J. and E F Robertson, Posidonius of Rhodes, MacTutor, www-groups.dcs.st-and.ac.uk/~history/Biographies/Posidonius, April 1999.
2. Posidonius, Wikipedia, http://en.wikipedia.org/wiki/Posidonius.

Claudius Ptolemy (c.87-c.161)

Believed that Earth is stationary and at the center of the universe

Claudius Ptolemy was born about 87 AD. He was an astronomer and mathematician whose astronomical theories and explanations dominated scientific thought until the 16th century. For most of his life he apparently lived and worked in Alexandria, Egypt at a time when Egypt was ruled by Romans and Alexandria was the center of widespread Greek culture.

The Ptolemaic system in astronomy is a theory of the order and action of the heavenly bodies. It held that Earth is stationary and at the center of the universe. Closest to Earth is the Moon, and beyond it, extending outward, are Mercury, Venus, and the Sun in a straight line, followed successively by Mars, Jupiter, Saturn, and the "fixed" stars. To explain the various observed motions of the planets, the Ptolemaic system described them as having small circular orbits called epicycles. The centers of the epicycles, on circular orbits around Earth, were called deferents. The motion of all spheres was said to be from west to east.

His astronomical book was called the *Megalê Syntaxis* (big explanation). It is a summary of all astronomical knowledge of his age. In it he gave mathematical explanations of various phenomena.

In the two books of *Planetary Hypotheses,* Ptolemy corrects some of the parameters of the *Megalê Syntaxis* (also known as *Almagest*), and suggested an improved model to explain planetary latitude. The Ptolemaic system was finally superseded in the 16th century by the Copernican system, named after Nicolaus Copernicus (1473-1543).

The *Apotelesmatica* was Ptolemy's attempt to place astrology on a sound scientific basis. At that time, astronomy and astrology were considered one subject.

Ptolemy advanced the study of trigonometry, and he applied his theories to the construction of astrolabes and sundials. An astrolabe is an astronomical computer for solving problems relating to time and the position of the Sun and stars in the sky. In geography, he charted the world as people of his time knew it. This work, which employed a system of longitude and latitude, influenced mapmakers for hundreds of years, although it suffered from a lack of reliable information.

Ptolemy also devoted a treatise, *Harmonica*, to music theory, and in *Optics* he explored the properties of light, especially refraction and reflection. He attempted to develop a mathematical theory of light's properties. Ptolemy died about 161.

1. Claudius Ptolemy, Answers.com, www.answers.com/topic/claudius-ptolemy.
2. Ptolemy, Microsoft Encarta Online Encyclopedia, http://encarta.msn.com/ encyclopedia_761562047/Ptolemy, 2008.

Aryabhata (476-550)

Ushered in the astronomical period that lasted until the year 1200

Aryabhata the Elder, a Hindu astronomer and mathematician, was born in 476 AD in Pataliputra (modern Patna, India). He was known to the Arabs as Arjehir, and his writings had considerable influence on Arabic science. He ushered in the astronomical period that lasted until the year 1200

Aryabhata is one of the first known to have used algebra. His writings include rules of arithmetic and plane and spherical trigonometry and solutions of quadratic equations. In India particularly, Aryabhata marked the end of the sacred or S'ulvasutra period during which mathematics was used primarily by priests for temple architecture.

In a time and place where people believed certain distant stars, called asuras, possessed powers capable of inflicting harm on Earth, Aryabhata took the first steps towards separating scientific explication from folklore and superstition. His *Aryabhatiya* (499) was the first major book on Hindu mathematics. It summarized knowledge of his predecessors. It consisted of a series of astronomical and mathematical rules and propositions, written in Sanskrit verse. While covering many aspects of arithmetic, algebra, and numerical notation, the majority of *Aryabhatiya* dealt with trigonometric tables and formulae for use in astronomy.

Aryabhata described a geocentric model of the Solar System in which the Sun and Moon are each carried by epicycles, which in turn revolve around the Earth. He held that the Earth rotates on its axis, and he gave the correct explanation of eclipses of the Sun and the Moon. He stated that the Moon and planets shine by reflected sunlight. He explained eclipses in terms of shadows cast by and falling on Earth. He described the orbits of the planets as ellipses. He correctly explained why equinoxes, solstices, and eclipses occur. These ideas were not accepted in Aryabhata's lifetime, but his mathematics set the foundation for developments in the Eastern and Western worlds for centuries to come.

Aryabhata calculated the rotation of the Earth, referenced to the fixed stars, as 23 hours, 56 minutes, and 4.1 seconds. The modern value is 23 hours, 56 minutes, and 4.091 seconds. Similarly, his value for the length of the year at 365 days, six hours, 12 minutes, and 30 seconds is an error of three minutes and 20 seconds over the length of a year.

Aryabhata died in 550.

1. Aryabhat, Wikipedia, http://en.wikipedia.org/wiki/Aryabhata.
2. Aryabhata, Microsoft Encarta Online Encyclopedia, http://encarta.msn.com/encyclopedia_761576604/aryabhata.
3. World of Mathematics on Aryabhata the Elder, BookRags, www.bookrags.com/biography/aryabhata-the-elder-wom

Brahmagupta (c.598-c.665)

Rejected the idea that the Earth was flat or hollow

Brahmagupta, a Hindu astronomer and mathematician, was born about 598 in Bhinmal city in the state of Rajasthan of northwest India. He became the head of the observatory at Ujjain—the foremost mathematical center of India at the time. He is said to have been the last and most accomplished of the ancient Indian astronomers.

His main work, *Brahmasphutasiddhanta* (The Opening of the Universe), written in 628, contains a good understanding of the mathematical role of zero, rules for manipulating both positive and negative numbers, a method for computing square roots, methods of solving linear and some quadratic equations, and rules for summing series.

Brahmagupta's theorem states that in a cyclic quadrilateral having perpendicular diagonals, the perpendicular to a side from the point of intersection of the diagonals always bisects the opposite side. A cyclic quadrilateral is a four-sided shape whose corners lie on a circle.

Brahmagupta's contributions to astronomy were equally ahead of their time. He applied algebraic methods to astronomical problems. He refuted the idea that the Moon is farther from the Earth than the Sun. He does this by explaining the illumination of the Moon by the Sun. He explains that since the Moon is closer to the Earth than the Sun, the degree of the illuminated part of the Moon depends on the relative positions of the Sun and the Moon, and this can be computed from the size of the angle between the two bodies.

Some of the important contributions made by Brahmagupta in astronomy are methods for calculating the position of heavenly bodies over time and their rising and setting, conjunctions (celestial bodies appearing near one another), and the calculation of solar and lunar eclipses. Brahmagupta criticized the view that the Earth was flat or hollow. Instead, he believed that the Earth and heaven were spherical and that the Earth is moving. About the Earth's gravity he said "Bodies fall towards the Earth as it is in the nature of the earth to attract bodies, just as it is in the nature of water to flow."

Brahmagupta's writings were taken to Baghdad, from where they influenced the development of the exact sciences in the Arab world. He died about 665.

1. Brahmagupta (c. AD 598–c. 665), The Internet Encyclopedia of Science, www.daviddarling.info/encyclopedia/B/Brahmagupta.
2. Brahmagupta, Wikipedia, http://en.wikipedia.org/wiki/Brahmagupta.
3. Brahmagupta (ca. 598-ca. 665), Eric Weinstein's World of Biography, Wolfram Research, http://scienceworld.wolfram.com/biography/Brahmagupta.

Albategnius (c.858-929)

Provided corrected values for many of the main parameters of planetary motion

Albategnius (Al-Battani) was born in or near Harran in northwestern Mesopotamia (modern Turkey) about 858. His father was thought to be a noted instrument maker. He is considered the greatest astronomer of the medieval Islamic world. Most of his research was conducted at al-Raqqa on the Euphrates River, where his family had moved when he was a youth.

Albategnius devised improved instruments and made accurate observations that allowed him to give corrected values for several astronomical constants, including the obliquity of the ecliptic (apparent motion of the sun in relation to the earth during a year) and the time of the equinoxes.

He is best known for his astronomical handbook *Kitab al-Zij*, which introduced new trigonometric methods for performing astronomical computations. *Kitab al-Zij* also includes a star catalog, as well as solar, lunar, and planetary tables, together with canons for their use. His observations revealed errors in the *Almagest* of Claudius Ptolemy (c.87-c.161) and allowed Albategnius to provide corrected values for many of the main parameters of planetary motion.

Albategnius's observations showed that the solar apogee, the point at which the Sun is farthest from Earth, had shifted from the position indicated by the *Almagest*. Among his other accomplishments, Albategnius rectified the Moon's mean motion in longitude, provided better estimates of the apparent solar and lunar diameters and their annual variation, and demonstrated the possibility of annular solar eclipses. He also developed a sophisticated method for determining the magnitude of lunar eclipses.

Albategnius's re-determination of the time of equinox allowed him to make an improved estimation of the tropical year, which he calculated as 365 days, five hours, 46 minutes, and 24 seconds—short by only two minutes and 22 seconds.

Kitab al-Zij proved influential in the development of European astronomy. The only surviving Latin version was made by Plato of Tivoli in about 1120. This version was first printed in Nuremberg in 1537 under the title *De motu stellarum* (On Stellar Motion). Albategnius died at Qasr al-Jiss in 929.

1. Albategnius (Al-Battani, Muhammad ibn Jabir) (c. 850-929), The Internet Encyclopedia of Science, ://www.daviddarling.info/encyclopedia/A/Albategnius.
2. Albategnius, Medieval History, About.com, http://historymedren.about.com/od/aentries/a/11_albategnius.
3. Al-Battani, BookRags, www.bookrags.com/research/al-battani-scit-021234.

Al-Sufi (903-986)

Documented the existence of the Andromeda galaxy

Abd al-Rahman Al-Sufi, a Persian nobleman and astronomer, was born on December 7, 903. He was also known in the West by the Latinized name Azophi. He carried out observations based on Greek work, especially the *Almagest* of Claudius Ptolemy (c.87-c.161). He contributed several corrections to Ptolemy's star list and did his own brightness and magnitude estimates, which frequently deviated from those in Ptolemy's work.

Al-Sufi's book *Kitab al-Kawatib al-Thabit al-Musawwar* (Book of Fixed Stars), published in about 964, includes a catalog of 1,018 stars, and gives their approximate positions, magnitudes, and colors. It contains Arabic star names that in corrupted form are still in use today and the earliest known reference to the Andromeda Galaxy.

Al-Sufi also recorded and named a southern celestial feature al-Baqar al-Abyad (the White Bull), which today we know as the Large Magellanic Cloud. He prepared an accurate map of the sky that became a standard work in the West for several centuries.

He had an enormous influence on the Arab astronomical studies of his time. He was a major translator into Arabic of the Hellenistic astronomy that had been centered in Alexandria, and he was the first to attempt to relate the Greek with the traditional Arabic star names and constellations.

Al-Sufi observed that the ecliptic plane is inclined with respect to the celestial equator, and more accurately calculated the length of the tropical year. The ecliptic is the annual path of the Sun in a circular path in the sky. For each constellation, he provided two drawings, one from the outside of a celestial globe and the other from the inside (as seen from the Earth).

He also wrote about the astrolabe, finding numerous additional uses for it. The astrolabe is an ancient astronomical computer for solving problems relating to time and the position of the Sun and stars in the sky.

Al-Sufi died on December 7, 986. A small mountainous ring on the Moon is named after him.

His observations were not known in Europe at the time of the invention of the telescope, so that the Andromeda Nebula M31 was independently rediscovered by Simon Marius (1573-1624) in 1612 with a moderate telescope. Al-Sufi died on May 25, 986 A.D.

1. Abd Al-Rahman Al-Sufi, Bookrags, www.bookrags.com/research/abd-al-rahman-al-sufi-scit-021234.
2. Abd-al-Rahman Al Sufi (December 7, 903 - May 25, 986 A.D.), SEDS, http://seds.org/Messier/xtra/Bios/alsufi.
3. Zahoor, A., 'Abd Al-Rahman Al-Sufi (Azophi), www.unhas.ac.id/~rhiza/saintis/sufi.

Abu'l Wafa (940-998)

Investigated the orbit of the Moon

Abu'l Wafa Muhammad Al-Buzjani, a Persian mathematician and astronomer, was born in 940 in Buzhgan, Nishapur in Iran. In 959, he moved to Iraq. He studied mathematics and worked principally in the field of trigonometry.

As an astronomer, he worked in a Baghdad observatory, where he created the first wall quadrant, a device used in observing the movement of heavenly bodies. In particular, he investigated the orbit of the Moon

Wafa constructed accurate trigonometric tables—to an accuracy of eight decimal places.

He was the first to describe geometrical constructions, possible only with a straight edge and a fixed compass that never alters its radius. He pioneered the use of the tangent function, apparently discovered the secant and cosecant functions, and compiled tables of sines and tangents at 15 arc-minute intervals—work done as part of an investigation into the orbit of the Moon.

Wafa established the trigonometric identities

$$\sin(a + b) = \sin(a)\cos(b) + \cos(a)\sin(b)$$
$$\cos(2a) = 1 - 2\sin^2(a)$$
$$\sin(2a) = 2\sin(a)\cos(a)$$

He also discovered the sine formula for spherical geometry.

Wafa wrote a number of books, one of which was for a practical use. It was on those geometric constructions which were necessary for a craftsman. It considered the design and testing of drafting instruments, the construction of right angles, approximate angle trisections, and constructions of parabolas. It also considered regular polygons and methods of inscribing them in and circumscribing them about given circles, inscribing of various polygons in given polygons, the division of figures such as plane polygons, and the division of spherical surfaces into regular spherical polygons. Wafa also wrote commentaries on the works of earlier mathematicians.

He died in 998 in Baghdad. The **Abul Wáfa crater** on the Moon is named for him.

1. Abu'l Wafa, The Internet Encyclopedia of Science, www.daviddarling.info/encyclopedia/A/Abul_Wafa.
2. Abū al-Wafā' al-Būzjānī, Wikipedia, http://en.wikipedia.org/wiki/Ab%C5%AB_al-Waf%C4%81'_al-B%C5%ABzj%C4%81n%C4%AB.
3. O'Connor, J. J. and E F Robertson, Abū al-Wafā' al-Būzjānī, MacTutor History of Mathematics, www-history.mcs.st-and.ac.uk/Printonly/Abu'l-Wafa, November, 1999.

Ali Al-Hazen (c.965-c.1040)

Gave a correct explanation of the apparent increase in size of the Sun and the Moon when near the horizon

Ali Al-Hazen (Abu Ali al Hassan ibn al Haitham) was born in about 965 in Basrah, Iraq, and received his education in Basrah and Baghdad. He spent most of his life in Spain, where he conducted research in optics, mathematics, physics, medicine, and development of scientific methods. He is considered the father of modern optics.

Al-Hazen studied the refraction of light rays through transparent media, such as air and water, and documented the laws of refraction. He carried out the first experiments on the dispersion of light into colors. He also was the first to described accurately the various parts of the eye, and gave a scientific explanation of the process of vision, contradicting the theories of Aristotle (384-322 BC), Euclid (c.325-270 BC), and Claudius Ptolemy (c.87-150) that the eye sends out visual rays to the object. Al Hazen believed that the rays originate in the object of vision and not in the eye. He also attempted to explain binocular vision.

Focusing on spherical and parabolic mirrors, Al-Hazen observed that the ratio between the angle of incidence and the angle of refraction does not remain constant. He also investigated the magnifying power of a lens. He developed a camera obscura to demonstrate that light and color from different candles pass through a single aperture in straight lines, without intermingling at the aperture. A camera obscura consists of small darkened rooms with light admitted through a single tiny hole. The result is an inverted image of the outside scene cast on the opposite wall.

His *Treasury of Optics* (first published in Latin in 1572) discusses lenses, plane and curved mirrors, and colors.

Al-Hazen gave a correct explanation of the apparent increase in size of the Sun and the Moon when near the horizon. He also studied atmospheric refraction, discovering that the twilight only ceases, or begins, when the Sun is 19 degrees below the horizon. From this, he deduced the height of the homogeneous atmosphere to be 55 miles. The homogeneous atmosphere is a hypothetical atmosphere in which the density is constant with height.

In mathematics, Al-Hazen developed analytical geometry by establishing linkage between algebra and geometry. He died about 1040.

1. Alhazen (Abu Ali al Hassan ibn al Haitham) (c. 965-c. 1040), The Internet Encyclopedia of Science, www.daviddarling.info/encyclopedia/A/Alhazen.
2. History of Optics, Wikipedia, http://en.wikipedia.org/wiki/History_of_optics.
3. Mirshhahi, Shahrokh, Al-Hazen, www.fravahr.org/spip.php?article80, June 4, 2003.
4. Zahoor, A, Abu Ali Hasan Ibn Al-Haitham (Alhazen) (965 - 1040 AD), www.geog.ucsb.edu/~jeff/115a/history/alhazen.

Al-Zarqali (1028-1087)

The first to determine the motion of the solar apogee

Abu Ishaq Ibrahim Ibn Yahya Al-Zarqali, a Spanish Arab, was born in 1028. He was a leading Arab mathematician who became known as the most prominent astronomer of his time. Al-Zarqali was known in the West as Arzachel. He carried out a series of observations at Toledo and compiled them and the work of others as the *Toledan Tables*. The *Tables* were responsible for invigorating the science of astronomy because it made possible the computation of planetary positions at any time based on observations. The *Tables* was translated into Latin in the twelfth century.

Al-Zarqali excelled at the construction of precision instruments for astronomical use. He developed a flat astrolabe, called a *Safihah.* An astrolabe is an instrument, now replaced by the sextant, which was once used to determine the altitude and position of the Sun or other celestial bodies. Al-Zarqali also built a water clock capable of determining the hours of the day and night and indicating the days of the lunar months.

He corrected Ptolemy's estimate of the length of the Mediterranean Sea from 62 degrees to the approximately correct value of 42 degrees. He was the first to show clearly that the motion of the solar apogee, which is when the Sun is furthest from Earth, amounts to 12.04 seconds per year relative to the stars. The actual value is now known to be 11.8 seconds per year.

In dealing with the complex model of Ptolemy for the planet Mercury, in which the center of the deferent moves on a secondary epicycle, Al-Zarqali concluded that the path of the center of the primary epicycle is not a circle, as it is for the other planets. Instead it is approximately oval. The deferent is the large circular orbit around which a planet was thought to orbit, in one or many epicycles. Epicycles are circular orbits within orbits that were used to (incorrectly) describe the orbits of objects in the Ptolemaic system.

Al-Zarqali's work was quoted a number of times in *De Revolutionibus Orbium Celestium* (On the Revolution of the Celestial Orbs) by Nicolaus Copernicus (1473-1543) in which he stated that the Sun, not the Earth, is the center of our Solar System. Al-Zarqali died in 1087. The Arzachel crater on the Moon is named after him.

1. Abu Ishaq Ibrahim Ibn Yahya Al-Zarqali, BookRags, ://www.bookrags.com/research/abu-ishaq-ibrahim-ibn-yahya-al-zarq-scit-021234.
2. Arzachel (Al-Zarqali, Abu Ishaq Ibrahim ibn Yahya) (1028-1087), The Internet Encyclopedia of Science, www.daviddarling.info/encyclopedia/A/Arzachel.
3. Zahoor, A., Abu Ishaq Ibrahim Ibn Yahya Al-Zarqali (Arzachel), www.unhas.ac.id/~rhiza/saintis/zarqali, 1997.

Omar Khayyam (1048-1131)

Helped reform the Moslem calendar

Omar Khayyam, a Persian mathematician, astronomer, and author of one of the world's best-known works of poetry, was born Ghiyath al-Din Abul Fateh Omar Ibn Ibrahim al-Khayyam in Nishapur, the capital of Khurasan, on May 18, 1048. Hs name means Omar the Tentmaker. It is possible that Omar or his father at one time worked in that trade.

The Vizier Nizam al-Mulk granted Khayyam a pension, which enabled him to devote himself to learning and research, especially in mathematics and astronomy. In 1074, Khayyam was invited to undertake astronomical research and commissioned to build an observatory in the city of Isfahan in collaboration with other astronomers.

As astronomer to the royal court, Khayyam was engaged with several other scientists to reform the Moslem calendar. Their work resulted in the adoption of a new era, called the Jalalian or the Seljuk. Omar's revision of the old Persian solar calendar was discontinued when Islamic orthodoxy gained power, but in 1925 it was again introduced in Iran. For comparison of Khayyam's accuracy, the length of one year at the end of the 19th century was 365.242196 days and today it is 365.242190 days.

Khayyam was one of the most notable mathematicians of his time. His series of astronomical tables is known as *Ziji Malikshahi*. Among his other mathematical writings is a work on algebra and a study *The Difficulties of Euclid's Definitions* (1077). His work in geometry was so far ahead of its time that it was not used again until René Descartes (1596-1650) built upon his theories in 17th century France.

Khayyam is most famous as the author of the *Rubáiyát*, which consists of about 1,000 stanzas reflecting upon nature and humanity.

The English poet and translator Edward FitzGerald (1809-1883) was the first to introduce the work of Khayyam to the West, through a version of 100 of the quatrains in 1859.

In the late 1960s, the quality of this translation was challenged by British poet Robert Graves (1895-1985). In 1968, in collaboration with Omar Ali-Shah (1922-2005), Graves produced *The Original Rubáiyát of Omar Khayyam*.

Khayyam died on December 4, 1131.

1. Omar Khayyam (1048-1131), Books and Writers, www.kirjasto.sci.fi/khayyam, 2003.
2. Omar Khayyam, Microsoft Encarta Online Encyclopedia, http://encarta.msn.com/encyclopedia_761556098/Omar_Khayyam, 2008.
3. Omar Khayyam, The Islamic World in 1600, The University of Calgary, www.ucalgary.ca/applied_history/tutor/islam/learning/khayyam, 1998.
4. Shahriari, Shahriar, Rubaiyat of Omar Khayyam, www.okonlife.com/poems, June 2, 2004.

Johannes Müller (1436-1476)

Calculated a 32-year period of celestial positions

Johannes Müller, a German astronomer and mathematician, was born at Königsberg (now Kaliningrad) on June 6, 1436. He was known as Regiomontanus, a Latinized form of his birthplace. At eleven years of age, he became a student at the university in Leipzig, Saxony. Three years later he continued his studies at the university in Vienna, Austria. He was appointed to the Arts Faculty of the latter institution in 1457.

Müller was a pupil of Georg von Pueurbach (1423-1461), with whom he worked at the University of Vienna on Latin translations of the *Almagest*, the influential work by the Greek geographer Claudius Ptolemy (c.87-c.161).

Following the death of Pueurbach, Müller moved in 1462 to Rome in search of better sources of Ptolemy's work. Later he returned to his native land, settling at Nuremberg, where he established an observatory under the patronage of Bernhard Walther (1430-1504). Here, he published *Ephemeredes*, which was astronomical tables showing the positions of celestial bodies, calculated for a 32-year period. These tables contained a description of the method of lunar distances, which made it possible to fix longitude at sea, providing the moon was visible.

Between 1467 and 1471, Müller worked in Hungary. With some help from the Hungarian court astronomer Martin Bylica (1433-1493) he compiled various astronomical and trigonometrical tables. He also constructed instruments and wrote treatises about them for King Matthias and the Archbishop of Gran. He produced a treatise on plane and spherical trigonometry, *De Triangulis Omnimodis,* and tables of sines and tangents known as *Tabulae Directionum*.

In 1475, Müller returned to Rome at the request of Pope Sixtus IV and was appointed Bishop of Regensburg, but died before taking office.

His contribution to the solving of triangles paved the way in succeeding centuries for the use of triangulation in which positions and distances can be calculated by trigonometry without having to be directly measured. Müller also made influential contributions to the science underlying cartography, which is the art and science of map-making. He died on July 6, 1476.

1. Regiomontanus, Microsoft Encarta Online Encyclopedia, http://au.encarta.msn.com/encyclopedia_781539354/Regiomontanus, 2008.
2. Mosely, Adam, Regiomontanus, Department of History and Philosophy of Science, University of Cambridge, www.hps.cam.ac.uk/starry/regiomontanus, 1999.
3. World of Mathematics on Regiomontanus, Bookrags, www.bookrags.com/biography/regiomontanus-wom.

Nicolaus Copernicus (1473-1543)

A major proponent of the heliocentric system

Nicolaus Copernicus was born on February 19, 1473 in Thorn (now Toruń), Poland to a family of merchants and municipal officials. Nicolaus was hardly ten years old when his father died. His uncle, Lucas, took charge of the children and gave the two boys a university education.

Copernicus entered the University of Kraków in 1491, studied liberal arts for four years without receiving a degree, and then in 1497 went to Italy to study medicine and law.

While studying canon law at the University of Bologna he lived at the home of a mathematics professor, Domenico Maria de Novara. Copernicus's geographical and astronomical interests were greatly stimulated by de Novara, an early critic of the accuracy of the geography of the second-century astronomer, Claudius Ptolemy (c.87-c.170).

Copernicus also went to Rome, where he observed a lunar eclipse and gave some lectures in astronomy and mathematics. He received a doctorate in canon law from Ferrara in 1503, and then returned to Poland to take up administrative duties.

Sometime between 1507 and 1515 he completed a short astronomical treatise, *De Hypothesibus Motuum Coelestium a se Constitutis Commentariolus* (known simply as *The Commentariolus*). In this work he laid down the principles of his new heliocentric (Sun-center) astronomy for which he is best known. The major premises of his theory are that the Earth rotates daily on its axis and revolves yearly around the Sun. He argued, furthermore, that the planets also encircle the Sun, and that the Earth precesses on its axis (wobbles like a top) as it rotates.

Another important feature of Copernican theory is that it allowed a new ordering of the planets according to their periods of revolution. In Copernicus's universe, unlike Ptolemy's, the greater the radius of a planet's orbit, the greater the time the planet takes to make one circuit around the Sun. The concept of a moving Earth was difficult to accept for most 16th-century readers.

The enunciation of the heliocentric theory by Copernicus marked the beginning of the scientific revolution and of a new view of a greatly enlarged universe.

Copernicus died at Frauenburg on May 24, 1543.

1. Copernicus, Nicolaus, Microsoft Encarta Online, Encyclopedia, www.encarta.co.uk/ encyclopedia_761571204/Copernicus_Nicolaus, 2008.
2. Knight, Kevin, Nicolaus Copernicus, New Advent, ://www.newadvent.org/cathen/04352b, 2008.
3. Copernicus, Nicolaus (1473-1543), Eric Weinstein's World of Biography, Wolfram Research, http://scienceworld.wolfram.com/biography/Copernicus.

Georg Rheticus (1514-1574)

Instrumental in spreading the work of Copernicus

Georg Joachim von Lauchen Rheticus (born Georg Joachim Iserin), a mathematician and astronomer, was born in Feldkirch, Austria on February 16, 1514. His father was the town doctor and also a government official. In 1528, his father was tried on a charge of sorcery, convicted, and beheaded. One of the legal requirements of such an execution was that the person's name could no longer be used, so Georg began calling himself Georg Joachim von Lauchen. He later took the additional name of Rheticus after the Roman province of Rhaetia in which he had been born.

In 1533, Rheticus entered the University of Wittenberg, and received his M.A. three years later. He was then appointed to teach mathematics and astronomy at the same University. Two years later, he left his position to study with some of the leading astronomers of the day, and in 1539 he arrived at Frauenburg in Ermland, where he spent about two years with Nicolaus Copernicus (1473-1543).

In September 1539, Rheticus went to Danzig, visiting the mayor there. The mayor gave him some financial assistance to help publish an introduction to Copernicus's work *Narratio Prima*. Copernicus could not have asked for a more erudite, elegant, and enthusiastic introduction of his new astronomy to the world of good letters. To this day, the *Narratio Prima* remains the best introduction to Copernicus's work.

In August 1541, Rheticus presented a copy of his work on a map of Prussia to Duke Albert of Prussia and the following day he sent him an instrument he had made to determine the length of the day. In October 1541, Rheticus returned to the University of Wittenberg, and there he was elected dean of the Faculty of Arts.

In early 1541, Rheticus published the trigonometrical sections of Copernicus's *De Revolutionibus,* adding work of his own that gave tables of sines and cosines (although he did not call them by these names). This was the first published table of cosines.

In 1542, Rheticus was appointed professor of higher mathematics at Leipzig. Eventually, he was made a member of the theological faculty there. However, a scandal forced him to leave Leipzig in April 1551. He was accused of having a homosexual affair with one of his students.

Rheticus died on December 4, 1574 in Kassa, Hungary (now Kosice).

1. Georg Joachim Rheticus, wikipedia, http://en.wikipedia.org/wiki/Georg_Joachim_Rheticus.
2. O'Connor, J. J. and E. F. Robertson, MacTutor, www-history.mcs.st-and.ac.uk/Printonly/Rheticus.
3. World of Mathematics on Georg Joachim Rheticus, BookRags, www.bookrags.com/biography/georg-joachim-rheticus-wom.

Leonard Digges (c.1520-c.1559)

Possibly co-invented the telescope with his son

Leonard Digges was born at Digges Court, near Canterbury, Kent about 1520. He was a member of the landed gentry of Kent. He studied at University College, Oxford, but never took his degree. He devoted his ample means and leisure to scientific pursuits. He was primarily known for his applied work in navigation, surveying, and ballistics. In 1553, he produced *A General Prognostication*, followed two years later by an enlarged edition called *A Prognostication of Right Good Effect,* and still another revision the next year with the title *Prognostication Everlasting*.

Pantometria, published posthumously by his son, Thomas Digges (c.1545-1595), in 1571, was a manual of practical mathematics that contained more advanced and up-to-date surveying techniques. Written in English rather than Latin, the works contributed greatly to a wider dissemination of the principles of astronomy, especially to those to whom Latin was unfamiliar.

These “Prognostications” contained a wealth of information. There were useful rules and tables for astronomy and astrology (which was not then separated from astronomy), calendar dates of moveable feasts for several years to come, tables of the Moon's motion, and a description of how to tell the time during day or night. The works also contained a short account of the universe according to the traditional system of Claudius Ptolemy (c.87-c.161), with tables of the dimensions of the planets and their orbits.

According to tradition, the telescope was invented in Holland around 1608 by Hans Lippershey (c.1570-1619). There is evidence, however, that it may have originated more than 30 years earlier in England, and that the inventors were Leonard and Thomas Digges. There is also reason to think that it was the Diggeses, and not Galileo, who first turned the telescope to the night sky, observing myriads of stars invisible to the naked eye.

Mary Tudor ascended the throne in 1553, restored Roman Catholicism as the state religion, and the next year was to be married to Phillip II of Spain. This triggered a rebellion by the men of Kent. The rebellion failed, and Leonard Digges, who had taken part in the insurrection, was condemned to death and his property confiscated. However, he received a reprieve. He died in England about 1559.

1. Digges, Leonard, The Galileo Project, http://galileo.rice.edu/Catalog/NewFiles/digges_leo.
2. Leonard Digges (scientist), Wikipedia, http://en.wikipedia.org/wiki/Leonard_Digges.
3. Leonard Digges, BookRags, ://www.bookrags.com/research/leonard-digges-scit-0312345.
4. Ronan, Colin A., Leonard and Thomas Digges, The Digges Telescope, www.chocky.demon.co.uk/oas/diggeshistory.

Christoph Clavius (c.1537-1612)

His textbooks were used for astronomical education for over three generations in Europe

Christopher Clavius, a German Jesuit mathematician and astronomer, was born in Bamberg, Bavaria, Germany on March 25, 1536 or 1538. Very little is known about his early life. He joined the Jesuit order in 1555, and attended the University of Coimbra in Portugal. Following this, he went to Italy and studied theology at the Jesuit Collegio Romano in Rome. In 1575 he became a full member of the order.

In 1570, Clavius wrote what was to become one of the most influential textbooks on astronomy of its day. In the Catholic world, this was the textbook for three generations of astronomers, including Galileo Galilei (1564-1642), and most particularly for Jesuit astronomers throughout the world. In later editions of his book, Clavius pronounced himself relatively favorably on the Copernican system as a mathematical model, but to the end of his life rejected its physical reality.

In 1579, he was assigned to compute the basis for a reformed calendar that would stop the slow process in which the Church's holidays were drifting relative to the seasons of the year. Using the Prussian Tables of Erasmus Reinhold (1511-1553), he proposed a calendar reform that was adopted in 1582 in Catholic countries by order of Pope Gregory XIII, and is now the Gregorian calendar used worldwide.

Clavius was chief astronomer at the Jesuit's Collegio Romano at the time of Galileo's first telescopic discoveries, and was still quite active despite his advancing age. He held strictly to the geocentric model of the Solar System, in which all the heavens rotate about the Earth. Though he opposed the heliocentric model of Nicolaus Copernicus (1473-1543), he recognized problems with the orthodox model. He was staunchly opposed to Galileo's notion of mountains on the Moon.

Within the Jesuit order, Clavius was almost single-handedly responsible for the adoption of a rigorous mathematics curriculum in an age where mathematics was often ridiculed by philosophers and theologians.

Clavius remained professor of mathematics at the Jesuit's Collegio Romano until his death on February 12, 1612. **Clavius crater** on the Moon and **Clavius Base**, a Moon base located in the crater, are named after him.

1. Christoph Clavius (1537-1612), High Altitude Observatory, www.hao.ucar.edu/Public/education/bios/clavius.
2. Christoph Clavius (1537-1612), http://physics.ship.edu/~mrc/pfs/110/inside_out/vu1/Galileo/people/clavius.
3. Clavius, Christoph (1537-1612), The Internet Encyclopedia of Science, ww.daviddarling.info/encyclopedia/C/Clavius_Christoph.

Giordano Bruno (c.1548-1600)

Early proponent of the idea of an infinite and homogeneous universe

Giordano Bruno, an Italian Renaissance philosopher and poet, was born about 1548 at Nola, near Naples. Originally named Filippo, he took the name Giordano when he joined the Dominicans, who trained him in Aristotelian philosophy and Thomistic theology. He was frank, outspoken and lacking in reticence. It wasn't long before he got into trouble. However, it wasn't his behavior but his opinions that got him into trouble. He fled the order in 1576 to avoid a trial on doctrinal charges and began the wandering that characterized his life.

Bruno visited Geneva, Toulouse, Paris, and then London, where he spent two years, from 1583 to 1585, under the protection of the French ambassador and in the circle of the English poet, Sir Philip Sidney (1554-1586). It was a most productive period, during which Bruno composed *Ash Wednesday Supper* (1584) and *On the Infinite Universe and Worlds* (1584), as well as the dialogue *On the Cause, Principle, and Unity* (1584).

In *Ash Wednesday*, he spread the Copernican doctrine, which was in variance with the teachings of Aristotle and the Church. In *On the Infinite Universe and Worlds*, he argued that the universe was infinite, that it contained an infinite number of worlds, and that these worlds were all inhabited by intelligent beings.

In another poetic dialogue, *Gli eroici furori* (1585), he praised a kind of Platonic love that joins the soul to God through wisdom. That same year, Bruno returned to Paris, and then went on to Marburg an der Lahn, Wittenberg, Prague, Helmstedt, and finally Frankfurt, where he arranged for the printing of his many writings.

At the invitation of a Venetian nobleman, Giovanni Mocenigo, Bruno returned to Italy as his private tutor. The infinite universe of which he wrote left no room for God. He preached a philosophy which made the mysteries of the virginity of Mary, of the crucifixion, and the mass meaningless. He thought of the Bible as a book which only the ignorant could take literally. In 1592, Mocenigo denounced Bruno to the Inquisition. He was turned over to the Roman authorities, and then imprisoned for some eight years while questioning proceeded on charges of blasphemy, immoral conduct, and heresy. Refusing to recant, Bruno was burned at the stake in Campo dei Fiori on February 17, 1600.

1. Giordano Bruno (1548-1600), The Galileo Project, http://galileo.rice.edu/chr/bruno.
2. Giordano Bruno, Microsoft Encarta Online Encyclopedia, http://encarta.msn.com/encyclopedia_761552368/giordano_bruno.
3. Kesser, John J., Giordano Bruno: The Forgotten Philosopher, www.infidels.org/library/historical/john_kessler/giordano_bruno.

Thomas Digges (c.1545-1595)

Possibly co-invented the telescope with his father

Thomas Digges, a British astronomer, was born about 1545 in Kent. He was the only child of Leonard Digges (c.1520-c.1559). After the death of his father, Thomas grew up under the guardianship of John Dee (1527-c.1608). In the world of print, Thomas Digges was an editor and author, and in the government he was an engineer, Member of Parliament, and military officer.

In 1571, Digges published *Leonard Digges's Pantometria*, an introductory geometry textbook. Before his death, his father had prepared the text, but left it to Thomas to publish.

Digges attempted to determine the parallax of the supernova observed by Tycho Brahe (1546-1601) in 1572, and in 1573 published *Alae seu Scalae Mathematicae* in which he concluded it had to be beyond the orbit of the Moon. This contradicted the accepted view of the universe, according to which no change could take place among the fixed stars.

In 1576, Digges published a new edition of his father's perpetual almanac, *A Prognostication Everlasting*. The original text, written by Leonard Digges for the third edition of 1556, was left unchanged, but Thomas added new material in several appendices. Contrary to the Ptolemaic cosmology of the original book by his father, the appendix featured a detailed discussion of the controversial and still poorly known Copernican heliocentric model of the universe. This was the first publication of that model in English, and was a milestone in the popularization of science. For the most part, the appendix was a loose translation into English of chapters from Copernicus's book *De Revolutionibus Orbium Coelestium*.

With his father, he may have pioneered the construction of the telescope and its use, although Hans Lippershey (c.1570-1619) is given official credit for the invention.

Digges served as a Member of Parliament for Wallingford and also had a military career as a Muster-Master General to the English forces from 1586 to 1594 during the war with the Spanish Netherlands. He was the father of Sir Dudley Digges (1583-1639), politician and statesman, and Leonard Digges (1588-1635), poet. Thomas Digges died in London on August 24, 1595.

1. Dictionary of Literary Biography on Thomas Digges, BookRags, www.bookrags.com/biography/thomas-digges-dlb.
2. Digges Thomas, The Galileo Project, http://galileo.rice.edu/Catalog/NewFiles/digges_tho.
3. J. J. O'Connor and E. F. Robertson, Thomas Digges, MacTutor, ://www-groups.dcs.st-and.ac.uk/~history/Biographies/Digges, 2002.

Tyco Brahe (1546-1601)

Made precise astronomical measurements of the Solar System

Tycho Brahe, a Danish astronomer, was born in Knudstrup in southern Sweden (then part of Denmark) on December 14, 1546. Tycho's father was from the Danish nobility. Tyco studied law and philosophy at the universities of Copenhagen and Leipzig.

Brahe succeeded in detecting serious errors in the standard astronomical tables and set about correcting them. He made precise, comprehensive astronomical measurements of the Solar System and more than 700 stars. And in 1572 he discovered a supernova in the constellation Cassiopeia. The data Brahe accumulated was superior to all other astronomical measurements made until the invention of the telescope.

In 1566, a part of the bridge of his nose was cut off by a rapier in a duel. Disfigured, he had a silver and gold piece attached in its place.

Frederick II, king of Denmark and Norway, provided Brahe with funds to construct and equip an astronomical observatory on the island of Hven (now Ven). In 1576, construction began on the castle of Uranienborg, where for many years the astronomer pursued his observations. After the death of Frederick II in 1588, Brahe's benefits were withdrawn by Frederick's successor, and eventually the astronomer was deprived of his observatory. In 1597, the Holy Roman Emperor, Rudolf II (1552-1612), gave him a pension and an estate near Prague, where a new observatory was to be built. However, Brahe died before his new observatory was completed.

Brahe sought a compromise by combining the Copernican theory of the universe with the old Ptolemaic system. In Brahe's system, the five known planets revolved around the Sun, which circled the Earth each year. The sphere of the stars revolved around the immobile Earth once a day.

Although Brahe's theory of planetary motion was flawed, the data he accumulated during his life played a crucial role in developing the correct description of planetary motion. Johannes Kepler (1571-1630), who was Brahe's assistant for a relatively short time before Brahe's death, used Brahe's data to formulate his three important laws of planetary motion. Brahe died, apparently of Mercury poisoning, on October 24, 1601 in Prague, Bohemia (now Czech Republic).

1. O'Connor, J. J. and E. F. Robertson, Tyco Brahe, MacTutor, www-groups.dcs.st-and.ac.uk/ ~history/Printonly/Brahe, 2003.
2. Tyco Brahe, Microsoft Encarta Online Encyclopedia, http://encarta.msn.com/ encyclopedia_761558742/tycho_brahe2008.
3. World of Scientific Discovery on Tycho Brahe, BookRags, www.bookrags.com/biography/ tycho-brahe-wsd.

Michael Maestlin (1550-1631)

The mentor of Johannes Kepler

Michael Maestlin, a German astronomer and mathematician, was born in Göppingen, Germany on September 30, 1550. His father was a merchant. Maestlin studied theology, mathematics, and astronomy at the Tübinger Stift in Tübingen, a town in Württemberg, and graduated in 1571. In 1576, he became a Lutheran deacon in Backnang near Stuttgart, and continued his studies there. He is best known for being the mentor of Johannes Kepler (1571-1630).

In 1580, Maestlin became a Professor of mathematics, first at the University of Heidelberg, where in 1582 he wrote a popular introduction to astronomy, and then at the University of Tübingen, where he taught for 47 years. At Tübingen, he replaced Petrus Apian (1495-1552), who had been dismissed for refusing to sign a Lutheran oath of religious allegiance.

Although Maestlin primarily initially taught the traditional geocentric Ptolemaic view of the Solar System, he was also one of the first to accept and teach the heliocentric Copernican view (Sun is the center of the Solar System). He corresponded with Kepler frequently and played a sizable part in Kepler's adoption of the Copernican system. The adoption of heliocentrism by Galileo Galilei (1564-1642) was also attributed to Maestlin.

The first known calculation of the golden ratio as a decimal of about 0.6180340 was written in 1597 by Maestlin in a letter to Kepler (If you divide a line into two parts so that the longer part divided by the smaller part is also equal to the whole length divided by the longer part then you will have the Golden Ratio. It is exactly equal to $(1+\sqrt{5})/2$.

Other things ascribed to Maestlin are cataloging the Pleiades cluster on December 24, 1579. The Pleiades is a conspicuous object in the night sky with a prominent place in ancient mythology. The cluster contains hundreds of stars, of which only a handful are commonly visible to the unaided eye. Eleven stars in the cluster were recorded by Maestlin, and possibly as many as fourteen were observed. The Occultation of Mars by Venus on October 3, 1590 was viewed by Maestlin at Heidelberg. In astronomy, occulation is the passage of a celestial body across a line between an observer and another celestial object.

Maestlin died at Tübingenon on October 20, 1631. The lunar crater, **Maestlin**, is named after him, as is the lunar rille (**Rimae Maestlin**) and the asteroid **11771-Maestlin**.

1. Golden Ratio, Math is Fun, www.mathsisfun.com/definitions/golden-ratio.
2. Maestlin, Micael, The Galileo Project, http://galileo.rice.edu/Catalog/NewFiles/maestlin.
3. Michael Maestlin, Wikipedia, http://en.wikipedia.org/wiki/Michael_Maestlin.

Petrus Plancius (1552-1622)

Created a celestial globe

Petrus Plancius (Pieter Platevoet), a Dutch astronomer, cartographer, and theologian, was born in Dranoutre, Flanders in 1552. He studied theology in Germany and England, and at the age of 24 became a minister in the Dutch Reformed Church.

In 1585, because of fear of religious prosecution by the Inquisition, he fled from Brussels to Amsterdam after Brussels fell into Spanish hands. There, he became cartographer to the Dutch East India Company. Fortunate enough to have access to nautical charts recently brought from Portugal, he was soon recognized as an expert on the shipping routes to India.

In his capacity as cartographer, he produced over 100 individual maps and charts. Apart from maps, he published journals and navigational guides and developed a new method for determining longitude.

In 1589, Plancius created a celestial globe using the sparse information available about southern celestial features—Crux, the southern cross; Triangulum Australe, the southern triangle; and the Magellanic Clouds, Nubecula Major and Minor.

In 1595, he asked Pieter Keyser (c.1540-1596), the chief pilot on the Hollandia, to make observations to fill in the blank area around the south celestial pole on European maps of the southern sky. Keyser's catalog of 135 stars arranged in 12 new constellations was delivered to Plancius, who inscribed the new constellations on a globe he prepared in 1598.

Plancius also introduced the Mercator projection for navigational maps. The Mercator projection is a cylindrical map projection presented by the Flemish geographer and cartographer Gerardus Mercator (1512-1594) in 1569. It became the standard map projection for nautical purposes because of its ability to represent lines of constant true bearing or true course, known as rhumb lines, as straight line segments.

In 1592, Plancius published his best known world map entitled *Nova et Exacta Terrarum Tabula Geographica et Hydrographica*. In 1613, he introduced the eight new constellations on a celestial globe published in Amsterdam by Pieter van der Keere (1570-1630).

Plancius died on May 15, 1622. The minor planet **10648 Plancius** commemorates his contributions in celestial and terrestrial cartography.

1. Mercator Projection, Wikipedia, http://en.wikipedia.org/wiki/Mercator_projection.
2. Petrus Plancius, Wikipedia, http://en.wikipedia.org/wiki/Petrus_Plancius.
3. Petrus Plancius~1552, Topical Stamps, http://sio.midco.net/dansmapstamps/plancius.
4. Plancius, Petrus (1552-1622), The Internet Encyclopedia of Science, www.daviddarling.info/encyclopedia/P/Plancius.

Thomas Harriot (c.1560-1621)

Made a map of the Moon, calculated the orbits of Jupiter's moons, and studied sunspots

Thomas Harriot, a British astronomer, mathematician, translator, and ethnographer, was born about 1560 in Oxford, England (Ethnography deals with scientific descriptions of specific human cultures). He attended St. Mary's Hall, Oxford University, receiving a B.A. in 1580.

For a time, he appears to have worked as a mathematics tutor in London before securing employment with Sir Walter Raleigh (1554-1618) in 1584. During the following year, Raleigh sent Harriot with a group of colonists to Roanoke Island off the coast of what is now North Carolina.

Harriot was responsible for studying the indigenous peoples on the Island, as well as the local vegetation, animal life, and other natural resources. He published an account of his findings in 1588.

He was one of the first mathematicians to use a number of now-commonplace symbols. He invented the signs for greater than (>) and less than (<), and was one of the first to adopt the plus and minus signs, as well as the equal sign. He used lowercase letters for variables, and was among the first to write an equation with the sum of all terms equal to zero.

Harriot studied the parabolic path of projectiles; determined the specific weights of materials; calculated the areas of spherical triangles, and thus confirmed that the Mercator projection preserves angles; independently discovered the sine law of refraction associated with Willebrord Snell (1580-1626); built telescopes; and in 1607, long before the birth of Edmund Halley (1656-1742), observed what came to be known as Halley's Comet. A Mercator projection is a mathematical method of showing a map of the globe on a flat surface.

He also made a map of the Moon, calculated the orbits of Jupiter's moons, studied sunspots and the Sun's rotation speed, and observed another comet.

In 1605, Harriot was briefly imprisoned, along with the Earl of Northumberland, as a result of the Gunpowder Plot. Harriot was quickly released, but the earl remained in the Tower of London until 1622.

Harriot succumbed to a cancer of the nose on July 2, 1621. He left behind a vast array of papers, many of them published as *Artis Analyticae Praxis*, a significant algebra text, in 1631. The observatory in the campus of the College of William and Mary is named in Harriot's honor

1. Thomas Harriot (1560-1621), The Galileo Project, http://galileo.rice.edu/sci/harriot.
2. Thomas Harriot, BookRags, ://www.bookrags.com/biography/thomas-harriot-scit-03123.
3. Thomas Harriot, Wikipedia, http://en.wikipedia.org/wiki/Thomas_Harriot.

Galileo Galilei (1564-1642)

Believed the Earth, rather than being the center of the Solar System, actually revolves around the Sun

Galileo Galilei was born in Pisa, Tuscany on February 15, 1564, the same year Michelangelo died. He initially studied medicine, but then changed to mathematics. He became interested in motion, and soon was in opposition with others who held to the 2000-year-old idea of Aristotle (384-322 BC) that heavy objects fall faster than lighter ones. To prove his point, the story goes, he dropped two objects of different weight from the Leaning Tower of Pisa and showed that they hit the ground at the same time. In other words, he experimented.

Galileo was uncomfortable with Aristotle's explanation that a force is necessary to keep an object in motion once it has been set in motion. He believed correctly that once an initial force is applied to an object the object remains in motion without additional force being necessary.

By 1609, Galileo had determined that the distance fallen by a body is proportional to the square of the elapsed time (the law of falling bodies) and that the trajectory of a projectile is a parabola.

He discovered the four largest satellites of Jupiter and observed sunspots. He observed Saturn and its rings, and was the first to report lunar mountains and craters. He also observed the Milky Way, previously believed to be nebulous, and found it to be a multitude of stars packed so densely that from Earth they appeared to be clouds. Galileo also observed the planet Neptune in 1612, but did not realize it was a planet.

Galileo ran afoul of the Church when he became an advocate of the theory of the Solar System advanced by Nicolaus Copernicus (1473-1543); that is, the Earth, rather than being the center of the Solar System, actually revolves around the Sun. He was warned by the church not to teach this view. After 15 years of restraint, Galileo published his belief. For this, he was arrested and forced to stand trial. He was found guilty and, under threat of torture, renounced the discoveries that supported Copernicus's views. He was sentenced to house arrest at his home at Arcetri for the remainder of his life. Nevertheless, he completed his studies on motion, and his writings were smuggled out of Italy and published in Holland. He died four years later in 1642.

In 1992, Pope John Paul II officially conceded that Galileo had been correct—the Earth is not stationary. It revolves around the Sun.

1. Galileo (1564-1642), Wolfram Research, http://scienceworld.wolfram.com/biography/Galileo
2. Galileo Galilei, Biographies/Images of Physicists and Astronomers, www.mlahanas.de/Physics/Bios/GalileoGalilei.
3. Galileo Galilei, Starry Messenger, www.hps.cam.ac.uk/starry/galileo.

David Fabricius (1564-1617)

Discovered the first variable star, noted sunspots that moved, and suggested that the Sun rotates on its axis

David Fabricius, a pastor and astronomer, was born in Esens, Frisia (part of modern-day Germany) on March 9, 1564. He served as Lutheran pastor for small towns in Frisia and dabbled in astronomy.

Fabricius discovered the first known periodic variable star, Mira, in August of 1596. At first, he believed it to be just another nova, as the whole concept of a recurring variable did not exist at the time. When he saw Mira brighten again in 1609, however, it became clear that a new kind of object had been discovered in the sky.

Two years later, his son Johannes Fabricius (1587-1616), returned from the university in the Netherlands with telescopes. On March 9, 1611 at dawn, Johannes directed the telescope at the rising Sun and saw several dark spots. He called his father, and together the two investigated this new phenomenon—sunspots. It was the first confirmed instance of their observation.

The pair soon invented camera obscura telescopy so as to save their eyes and get a better view of the solar disk. They observed that the spots moved. Over the next several months they tracked the spots as they moved across the Sun's face, and found that a dozen or so days after they had disappeared from the western edge of the Sun they reappeared on the eastern edge. They would then steadily move to the western edge, disappear, and then reappear at the east again after the passage of the same amount of time that it had taken for them to cross the disk in the first place. This suggested that the Sun rotated on its axis, which had been postulated before, but never backed up with evidence. Johannes then published *Maculis in Sole Observatis, et Apparente Earum cum Sole Conversione Narratio* (Narration on Spots Observed on the Sun and their Apparent Rotation with the Sun) in June of 1611. Because of the lack of a powerful patron who might have called the little book to the attention of influential people, it drew very little attention, and it was eclipsed by the publication of the work of Christoph Scheiner (1575-1650) in 1612.

After denouncing a local goose thief from the pulpit on May 7, 1617, the accused man struck Fabricius in the head with a shovel and killed him. A large crater on the Moon's southern hemisphere was named after Fabricius.

1. David (1564-1617) and Johannes (1587-1616) Fabricius, The Galileo Project, http://galileo.rice.edu/ sci/fabricius.
2. David Fabricius Biography, biographybase, www.biographybase.com/biography/ Fabricius_David.

Hans Lippershey (c.1570-1619)

Given credit for inventing the telescope

Hans Lippershey was born about 1570 in Wesel (Germany). He later migrated to Middelburg, the capital of Zeeland, in the southwestern most province of the Netherlands. Middelburg was a flourishing city, especially after the fall of Antwerp to the Spanish in 1585, which caused many of its Protestant inhabitants to flee north to the Netherlands.

Little is known about Lippershey except that he married in 1594 and officially became a citizen of his adopted country in 1602.

Most historians credit him with creating and disseminating designs for the first practical telescope. Lippershey, a spectacle-maker, is said to have come up with the idea for the telescope after watching two children in his eyeglass shop play with his lenses. Supposedly, the children held two lenses and, looking through both simultaneously, noticed that a weather vane on a nearby church steeple seemed to appear larger and clearer than visible by normal sight.

Lippershey proceeded to make a telescope, which he called a kijker or looker, by attaching lenses at the two ends of a tube. He applied for a patent in 1608, but because the device could not be kept a secret the patent was denied. The telescope Lippershey invented could only see three times farther than the naked eye.

Others came forth claiming to have invented the new device, including Jacob Metius (c.1571-1628) of Alkmaar, a city in the north of the Netherlands, and Sacharias Jansen (c.1585-c.1632), also of Middelburg. Still, Lippershey's application for a patent remains the earliest record of a telescope. Lippershey benefited financially from his invention through the Dutch government, which paid him to construct several telescopes for military use.

The first known mention of Lippershey's application for a patent for his invention appeared at the end of a September 10, 1608 diplomatic report on an embassy document to Holland from the Kingdom of Siam. The report was soon distributed across Europe, leading to the experiments by other scientists, including Galileo Galilei (1564-1642), who soon improved the device.

Lippershey remained in Middelburg until his death in September 1619. **Lippershey crater** on the Moon is named after him.

1. Hans Lippershey (d. 1619), The Galileo Project, http://galileo.rice.edu/sci/lipperhey.
2. Hans Lippeshey, Wikipedia, http://en.wikipedia.org/wiki/Hans_Lippershey.
3. World of Physics on Hans Lippershey, BookRags, www.bookrags.com/biography/hans-lippershey-wop.

Johannes Kepler (1571-1630)

Formulated three laws of planetary motion

Johannes Kepler, of German Nationality, was born on December 27, 1571 in Weil der Stadt, Württemburg, in the Holy Roman Empire. His parents were poor, but his evident intelligence earned him a scholarship to the University of Tübingen to study for the Lutheran ministry. There, he was introduced to the ideas of Nicolaus Copernicus (1473-1543).

In 1596, while a mathematics teacher in Graz, he wrote the first outspoken defense of the Copernican system, *Mysterium Cosmographicum.*

Kepler did not adhere to the Augsburg Confession (the primary confession of faith of the Lutheran Church), and refused to sign the Formula of Concord. Because of his refusal, he was excluded from the sacrament in the Lutheran church. He was forced to leave his teaching post at Graz due to the counter Reformation. He moved to Prague to work with the Danish astronomer, Tycho Brahe (1546-1601). Kepler inherited Brahe's post as Imperial Mathematician when Brahe died in 1601.

Using the precise data that Brahe had collected, in 1609 Kepler published *Astronomia Nova*, delineating his discoveries, which are now called Kepler's first two laws of planetary motion:

1. The orbit of every planet is an ellipse with the Sun at one of the foci. Thus, Kepler rejected the ancient Aristotelean, Ptolemaic, and Copernican belief in circular motion.
2. A line joining a planet and the Sun sweeps out equal areas during equal intervals of time as the planet travels along its orbit. With his law, Kepler rejected the Aristotelean astronomical theory that planets have uniform speed.

In 1612, Lutherans were forced out of Prague, so Kepler moved to Linz. Subsequently, he had to return to Württemburg, where he successfully defended his mother against charges of witchcraft. In 1619, he published *Harmonices Mundi* in which he describes his third law:

3. The squares of the orbital periods of planets are directly proportional to the cubes of the semi-major axes of their orbits. This means that the speed of a planet in a larger orbit is lower than in a smaller orbit.

Kepler died in Regensburg in 1630 while on a journey from his home in Sagan to collect a debt.

1. Field, J. V., Johannes Kepler, MacTutor, www-history.mcs.st-and.ac.uk/history/Printonly/Kepler.
2. Johannes Kepler: His Life, His Laws and Times, KeplerMission, http://kepler.nasa.gov/johannes.
3. Kepler's laws of planetary motion, Wikipedia, http://en.wikipedia.org/wiki/Kepler's_laws_of_planetary_motion.

Johann Bayer (1572-1625)

Published the first star atlas to cover the entire celestial sphere

Johann Bayer, a German astronomer and lawyer, was born in 1572 in Rain, Bavaria. He began his study of philosophy in Ingolstadt University in 1592, and moved later to Augsburg to begin work as a lawyer. He eventually became legal advisor to the Augsburg city council. He grew interested in astronomy during his time in Augsburg.

He is most famous for his star atlas *Uranometria*, published in 1603. It was the most comprehensive pre-telescopic star catalog to date. It introduced the nomenclature still in use for designating stars visible to the naked eye. It was the first atlas to represent the stars around the South Pole and to cover the entire celestial sphere. It contained over 2,000 stars, of which around 1,200 were taken from Tycho Brahe's (1546-1601) catalog. These were sorted into 49 constellation maps and two hemispheric charts.

When Bayer published *Uranometria*, he included 12 new southern asterisms. Asterisms are informal yet distinctive groupings of stars. A well-known example of a northern asterism is the Big Dipper in the constellation Ursa Major.

The significance of Bayer's work lies in his innovative method for naming stars within each constellation. The profusion of names for individual stars that resulted from the translation of Greek into various languages had proved most cumbersome and confusing. Bayer sought to reform this situation by systematically identifying each star precisely and succinctly. He assigned to each star in a constellation one of the 24 letters of the Greek alphabet. If a constellation had more than 24 stars, then additional characters were provided by the Latin alphabet. Bayer's new system of star designation has become known as the **Bayer designation**.

Bayer was not comfortable with the traditional heathen names assigned to the constellations. In the *Uranometria*, he therefore proposed (unsuccessfully) alternate names from the Bible. Constellations in the Northern Hemisphere were named for figures from the New Testament, while those in the Southern Hemisphere were given names from the Old Testament.

Bayer died on March 7, 1625. The **Bayer crater** on the Moon is named after him.

1. Bayer, Johann (1572-1625), The Internet Encyclopedia of Science, www.daviddarling.info/encyclopedia/B/Bayer.
2. Johann Bayer, BookRags, www.bookrags.com/research/johann-bayer-scit-031234.
3. Uranometria, Wikipedia, http://en.wikipedia.org/wiki/Uranometria.

Simon Marius (1573-1624)

Named Jupiter's four major moons

Simon Marius, a German astronomer, was born on January 10, 1573 in Gunzenhausen near Nuremberg, but most of his lifetime he spent in the city of Ansbach. His father was mayor of the city in 1576.

From 1586 to 1601, Marius studied off and on at the Markgrafschaft's Lutheran academy at Heilsbronn. During this period, he became interested in astronomy. His astronomical and meteorological observations began in 1594.

In 1596, Marius wrote a tract on the comet of that year, and in 1599 he published a set of astronomical tables. These efforts resulted in his appointment as mathematician of the Markgrafschaft of Ansbach in 1601. In that capacity, he printed prognostications each year until his death.

One of his first acts as the Markgrafschaft's mathematician was to travel to Prague to learn Tycho Brahe's observational techniques and instruments. Brahe died that year, and Marius's stay in Prague lasted only four months, but he did meet David Fabricius (1564-1617). He then went to Padua to study at the university there.

On his return from Italy, Marius settled in the city of Ansbach as court mathematician, and married Felicitas Lauer, the daughter of his publisher.

In 1614, Marius published *Mundus Lovialis*, describing the planet Jupiter and its moons. In it, he claimed to have discovered the planet's four major moons some days before Galileo (1564-1642). This led to a dispute with Galileo, who showed that Marius provided only one observation as early as Galileo's, and it matched Galileo's diagram for the same date as published in 1610. It is considered possible that Marius discovered the moons independently, but at least some days later than Galileo.

Regardless of priority, the mythological names by which these satellites are known today (Io, Europa, Ganymede, and Callisto) are those given them by Marius.

Marius also observed the Andromeda nebula, which had in fact already been known to Arab astronomers of the Middle Ages. Besides his annual prognostications, he published in his later years a book on the comets of 1618 and, posthumously, a book on Ptolemy's position circle.

Marius died in Ansbach after a brief illness on December 26, 1624. **Simon-Marius-Gymnasium Gunzenhausen**, was named after the astronomer

1. Simon Marius, Wikipedia, http://en.wikipedia.org/wiki/Simon_Marius.
2. Simon Marius, (1573-1624), The Galileo Project, http://galileo.rice.edu/sci/marius.
3. Simon Marius (January 20, 1573 - December 26, 1624), The Meisser Catalog, www.seds.org/messier/Xtra/Bios/marius.

Christoph Scheiner (1575-1650)

Devised techniques that greatly improved the accuracy of observed sunspot positions

Christoph Scheiner, a German astronomer, was born on July 25, 1575 in Wald at Mindelheim in Bavaria. He joined the Jesuit order in 1595, and started his studies in 1601 at Ingolstadt University. He taught mathematics there from 1610 to 1616. He moved to Innsbruck in 1616, and then lived in Rome from 1624 to 1633.

Early in his career, Scheiner became an expert on the mathematics of sundials, and also invented a pantograph (a device for copying and enlarging drawings). Upon hearing about Galileo's discoveries with the telescope, in 1610 Scheiner immediately set out to obtain good telescopes with which to scrutinize the heavens.

The controversy between Scheiner and Galileo (1564-1642) over priority in the discovery of sunspots was an important factor responsible for the degradation in the relationship between Galileo and Roman members the Jesuit Order. By his own account, Scheiner began observing sunspots in March or April 1611. The first published account of his observations are his three letters on solar spots dated November 11, 1611, addressed to Augsburg magistrate, Mark Wesler (1558-1614), and published in Augsburg in January 1612. These were followed by three more letters in September 1612, again published via Wesler. Scheiner was required by his ecclesiastic superiors to write under the pseudonym Appelles, to avoid possible embarrassment to the Jesuit order in the event that his findings were to prove spurious.

Scheiner's original opinion was that sunspots were small planets closely orbiting the Sun, a position convincingly refuted by Galileo in his own 1632 letters on solar spots. Unlike Galileo, Scheiner pursued sunspot observations on a continuous basis for more than 15 years. In the course of doing so, he devised techniques that greatly improved the accuracy of observed sunspot positions, and designed specialized solar observing instruments. Results of his observations were published in 1630 in his *Rosa Ursina,* a book that opened with a biting attack on Galileo, to which the latter was to reciprocate in his *Dialogues*.

Scheiner went on to publish books on atmospheric refraction and the optics of the eye, and in these works built on the optical achievements of Johannes Kepler (1571-1630). He died in Niesse on June 18, 1650.

1. Christoph Scheiner (1573-1650), The Galileo Project, http://galileo.rice.edu/sci/scheiner.
2. Christoph Scheiner (1575-1650), High Altitude Observatory, www.hao.ucar.edu/Public/education/bios/scheiner.
3. Christoph Scheiner, Wikipedia, http://en.wikipedia.org/wiki/Christoph_Scheiner.

Johannes Fabricius (1587-1615)

Observed movement of sunspots

Johannes Fabricius, a Frisian/German astronomer, was born on January 8, 1587 in Resterhafe (Friesland). He was the eldest son of David Fabricius (1564-1617) with whom he was a discoverer of sunspots independently of Galileo Galilei (1564-1642).

Fabricius returned from the university in the Netherlands with telescopes that he and his father turned on the Sun. Despite the difficulties of observing the Sun directly, they noted the existence of sunspots, the first confirmed instance of their observation. The pair soon invented camera obscura telescopy so as to save their eyes and get a better view of the solar disk. They observed that the spots moved. The spots would appear on the eastern edge of the disk, steadily move to the western edge, disappear, and then reappear at the east again after the passage of the same amount of time that it had taken for it to cross the disk in the first place.

Johannes wrote a tract on sunspots, *De Maculis in Sole Observatis, et Apparente earum cum Sole Conversione Narratio* (Narration on Spots Observed on the Sun and their Apparent Rotation with the Sun), the dedication of which was dated June 13, 1611. It was printed in Wittenberg (the site of the premier Lutheran university, where Johannes was apparently continuing his studies) in time for the autumn book fair in Frankfurt.

In the book, Johannes reviewed the observations made by him and his father, without giving times or dates or showing a picture of the spots, and then stated his opinion that they were on the Sun and that the Sun therefore probably rotated on its axis, a notion already suggested by Giordano Bruno (1548-1600) and Johannes Kepler (1571-1630).

Because of the lack of a powerful patron interested in scientific matters who might have called the little book to the attention of influential people, it drew very little attention.

Fabricius died in Marienhafe on March 19, 1616. A year later his father was killed when an irate peasant, whom he had accused of stealing a goose, hit him over the head with a shovel.

In 1895, a monument was erected to Johannes Fabricius's memory in the churchyard at Osteel where he was pastor from 1603 until 1617. the large **Fabricius crater** on the Moon's southern hemisphere is named after David Fabricius.

1. David (1564-1617) and Johannes (1587-1616) Fabricius, The Galileo Project, http://galileo.rice.edu/sci/fabricius.
2. Johannes Fabricius, Wikipedia, http://en.wikipedia.org/wiki/Johannes_Fabricius.
3. Johannes Fabricus, Connections, http://cnx.org/content/m11961/latest.

Giovanni Hodierna (1597-1660)

Described 40 nebulae he had observed

Giovanni Battista Hodierna was born in Ragusa, Sicily on April 13, 1597. He is thought to have grown up in a poor environment and to have been self-educated, at least in science. As a young man, he observed the three comets of 1618-1619 with a telescope of Galilean type and fixed magnification 20.

He became a Roman Catholic priest and was ordained at Syracuse, Sicily in 1622. From 1625 to 1636 he served as a priest in Ragusa and taught mathematics and astronomy at his hometown.

Hodierna was an enthusiastic follower of Galileo. In 1628, he wrote the *Nunzio del Secolo Cristallino*, an appraisal of Galileo's *Siderius Nuntius*. Hodierna was particular impressed by Galileo's resolution into stars of the Milky Way and the Nebulae.

In 1637, Hodierna followed Carlo and Gulio Tomasi, the Dukes of Montechiaro, to the newly founded Palma di Montechiaro. They gave him a house, a piece of land to live on, and funded his publications. He served them as a chaplain and parish priest. In 1644, he earned a doctorate in theology, in 1645 he was named archpriest, and in 1655 he was named court mathematician. In 1646 and 1653, he observed Saturn, created drawings showing the planet with its ring, and published *Protei Caelestis Vertigines seu Saturni Systema* in 1657.

In 1652, he observed eclipses of Jupiter's moons, as well as the passages of their shadows over the planet. In 1656, he published *Medicaeorum Ephemerides*, the first published table of the Galilean satellites based on an improved theory of the motion of Jupiter's moons. In his 1656 *De Admirandis Phasibus in Sole et Luna Visis*, Hodierna gave a treatise on the appearance of the Sun and the Moon, including sunspots and eclipses.

Hodierna thought there were profound differences between comets and nebulae. Because of the motion and changing appearance of comets, he thought them to be made up of a more terrestrial matter, while nebulae to be made up of stars. With finder charts and some sketches, he described 40 nebulae he had observed. He made the earliest surviving drawing of the Orion Nebula.

Hodierna died on April 6, 1660 in Palma di Montechiaro, Sicily. He was honored by the naming of **asteroid 21047 Hodierna** after him.

1. Giovanni Battista Hodierna (April 13, 1597 - April 6, 1660), The Messier Catalog, www.seds.org/ messier/xtra/Bios/hodierna.
2. Hodierna, Giovanni Battista (1597-1660), The Internet Encyclopedia of Science, www.daviddarling.info/ encyclopedia/H/Hodierna.

John Riccioli (1598-1671)

Co-created a nomenclature system for naming features on the Moon

John Baptist Riccioli, also known as Giovanni Battista Riccioli, was born in Ferrara, Italy on April 17, 1598. He received a Jesuit education, and eventually went on to study at the Jesuit college in Bologna, Italy. He completed a doctorate degree in theology at the age of 16.

His devotion to the Catholic Church led him to enter the Jesuit order in 1614. Jesuits serve the Pope directly and are called the "Foot soldiers of the Pope."

Riccioli and Francesco Grimaldi (1618-1663) were the first to study the acceleration rate of a free falling object, confirming that the distance of fall was proportional to the square of the time taken. Riccioli also improved on the law of the pendulum of Galileo Galilei (1564-1642). He went beyond the preliminary work of Galileo and succeeded in perfecting the pendulum as an instrument to measure time.

Since Riccioli was a Jesuit astronomer living in the Roman Empire, he had access to astronomical knowledge gathered by Jesuit students in India and China. Chinese and Indian astronomers tracked lunar and solar eclipses and made detailed maps of celestial objects. With this information, Grimaldi and Riccioli created the most detailed lunar map known at that time. It was the first map to name craters and mountains after scientists and prominent people instead of abstract concepts. Much of the nomenclature of lunar features still in use today is due to Riccioli and Grimaldi.

Riccioli published the map of the Moon in a work called *Almagestum Novum*. The purpose of the work was to discredit the Copernican theory of Heliocentrism because the Catholic Church at the time believed Earth was at the center of the Solar System.

Riccioli also observed Saturn, and was one of the first Europeans to note that Mizar was a double star.

Between 1644 and 1656, he was occupied by topographical measurements, working with Grimaldi to determine values for the circumference of Earth and the ratio of water to land.

Riccioli published tables of latitude and longitude for a great number of separate localities, correcting previous data and preparing the way for further developments in cartography. He died in Bologna on June 25, 1671.

1. Giovanni Battista Riccioli, Wikipedia, http://en.wikipedia.org/wiki/Giovanni_Battista_Riccioli.
2. John Baptist Riccioli, Fairfield University, www.faculty.fairfield.edu/jmac/sj/scientists/riccioli.
3. What does the John Baptist Riccioli biography have to do with the Moon?, Astronomy for Kinds Online, www.astronomy-for-kids-online.com/john-baptist-riccioli-biography.

17th Century Astronomers
1600 to 1699

Jakob Bartsch (1600-1633)

Published several star charts

Jakob Bartsch was born in 1600 in Lauban (Lubań) in Lusatia. He was taught how to use the astrolabe by Christopher Hauptfleisch, a librarian in **Breslau (Wrocław).** After studying astronomy and medicine at the University of Strassburg, he became a professor of mathematics at the same institution, where he became an assistant to Johannes Kepler (1571-1630) and helped Kepler with his calculations.

In 1624, Bartsch published several star charts entitled *Usus Astronomicus Planisphaerii Stellati*, which included a number of new constellations. These had been introduced around 1613 by Petrus Plancius (1552-1622) on a celestial globe published by Pieter van den Keere (c.1570-1630). The constellations included Camelopardalis, Crux, Monoceros, and Reticulum. Each map had two complimentary sheets showing the latitude and longitude and the constellations.

Camelopardalis has no mythology associated with its stars, being considered a modern constellation. The faintness of the constellation, and that of the nearby constellation Lynx, led to the early Greeks considering this area of the sky to be empty, and thus a desert.

Crux, commonly known as the Southern Cross, is the smallest of the 88 modern constellations, but nevertheless one of the most distinctive.

Monoceros, with only a few fourth magnitude stars, is a constellation not easily seen with the naked eye. It contains Plaskett's Star, which is a massive binary system whose combined mass is estimated to be that of almost 100 Suns put together.

Reticulum is one of the minor southern constellations. This constellation contains Zeta Reticuli, which is a pair of stars similar to the Sun. These stars gained some notoriety in ufology when they were alleged to be the home of an alien civilization.

Bartsch married Kepler's daughter, Susanna, on March 12, 1630.

After Kepler's death in 1630, Bartsch edited Kepler's posthumous work *Somnium*, which is a fantasy written between 1620 and 1630 in which a student of Tycho Brahe (1456-1601) is transported to the Moon by occult forces. Bartsch also helped gather money from Kepler's estate for his widow.

Borsch died in Lauban in 1633, shortly after being called to Strassburg as Professor of astronomy.

1. Bartsch, Jakob (1600-1633), The Internet Encyclopedia of Science, www.daviddarling.info/encyclopedia/B/Bartsch.
2. Jakob Bartsch, Wikipedia, http://en.wikipedia.org/wiki/Jakob_Bartsch.

Ismaël Bullialdus (1605-1694)

Proposed that the force of gravity follows an inverse-square law

Ismaël Bullialdus (Ismaël Boulliau) was born on September 28, 1605 in Loudun, Vienne, France. He was the first surviving son of Calvinists Susanna Motet and Ismaël Boulliau, a notary by profession and an amateur astronomer. In Loudun was an important circle of intellectuals of which Ismaël Boulliau Senior was part. The group met in a local hotel and provided a focus for other visiting scholars. Young Ismaël attended these meetings and became acquainted with many leading men of the day.

At age 21, Bullialdus converted to Catholicism, and by age 26 was ordained as a priest. In 1632, he moved to Paris, where he worked as a librarian for the Bibliothèque du Roi, traveling widely within Italy, Holland, and Germany to purchase books.

In 1657, he became secretary to the French ambassador to Holland, and then once again a librarian, and in 1666 moved to the Collège de Laon.

Bullialdus was an active supporter of Galileo Galilei (1564-1642) and Nicolaus Copernicus (1473-1543). He was one of the first to accept the elliptical orbits of Johannes Kepler (1571-1630).

Boulliau is best known for his astronomical and mathematical works. Chief among them is his *Astronomia philolaica*, (1645), in which he proposed that the force of gravity follows an inverse-square law:

> "As for the power by which the Sun seizes or holds the planets, and which, being corporeal, functions in the manner of hands, it is emitted in straight lines throughout the whole extent of the world, and like the species of the Sun, it turns with the body of the Sun; now, seeing that it is corporeal, it becomes weaker and attenuated at a greater distance or interval, and the ratio of its decrease in strength is the same as in the case of light, namely, the duplicate proportion, but inversely, of the distances that is, $1/d^2$."

Isaac Newton (1643-1727) later made this idea precise in his 1687 work, the *Principia*, which praised Bullialdus's work for his accurate tables.

Bullialdus established the period of the variable star Mira Ceti in 1667. He was one of the earliest members of the Royal Society of London, having been elected in 1667, seven years after its founding.

During the final five years of his life, he returned to the priesthood at the Abbey St. Victor in Paris, where he died on November 25, 1694 at the age of age 89. The Moon's **Bullialdus crater** is named in his honor.

1. Ismael Boulliau, BookRags, www.bookrags.com/research/ismael-boulliau-scit-031234.
2. Ismaël Bullialdus Wikipedia, http://en.wikipedia.org/wiki/Ismael_Bullialdus.
3. J. J. O'Connor and E. F. Robertson, MacTutor, www-gap.dcs.st-and.ac.uk/~history/Printonly/Boulliau.

Martin van den Hove (1605-1639)

Developed a method for determining the diameters of planets based on the measured visual angle

Martin van den Hove (Latinized as Martinus Hortensius), a Dutch astronomer and mathematician, was born in 1605 in Delft, South Holland, Netherlands. From 1625 to 1627, he studied at Leiden under Willebrord Snellius (1580- 1626) and Isaac Beeckman (1588-1637). He received further instruction from Snellius from 1628 to 1630 at Leiden and at Ghent. He apparently did not receive a degree.

In 1628, Van den Hove began studying under Philippe van Lansberge (1561-1632). He became an enthusiastic supporter of Lansberge, who by then was quite aged, and helped him complete his project to create new systematic observations to replace old insufficient data.

Van den Hove attacked many of the claims of Tycho Brahe (1546-1601) in the preface to his Latin translation of a work by Lansberge—*Commentationes in Motum Terrae Diurnum, & Annuum* (1630). The first Latin edition taught the probability of Earth's motion according to the Copernican theory.

Van den Hove regarded Lansberge, not Tycho Brahe, as the one who was restoring astronomy. He felt that "Only Lansberge held all ancient observations in esteem, whereas Tycho, Longomontanus, and Kepler tended to neglect them."

Van den Hove began lecturing on the mathematical sciences at the Amsterdam Atheneum in 1634. On assuming his new duties he delivered an inaugural speech, later published as *De Dignitate et Utilitate Matheseos* (On the Dignity and Utility of the Mathematical Sciences).

Van den Hove also lectured on optics at Amsterdam in 1635 and on navigation in 1637.

In 1638, he was made a member of the commission negotiating with Galileo on the determination of longitude by the method of Jupiter's moons.

Van den Hove developed a method for determining the diameters of planets based on the measured visual angle that his telescope embraced. He was made full professor "in the Copernican theory" in 1635.

Van den Hove was nominated professor at Leiden in 1639, but died shortly afterwards at Leiden on August 7, 1639. He was 34 years old. **Hortensius crater** on the Moon is named after him.

1. Hortensius, Martinus [Ortensius or Van den Hove, Maarten], The Galileo Project, http://galileo.rice.edu/Catalog/NewFiles/hortens.
2. Martin Van den Hove, BookRags, www.bookrags.com/wiki/Martin_van_den_Hove.
3. Martin Van den Hove, Wikipedia, http://en.wikipedia.org/wiki/Martin_van_den_Hove.

Johannes Hevelius (1611-1687)

Laid the foundation for the study of lunar topography

Johannes Hevelius, a Polish astronomer, was born on January 28, 1611 in Danzig (now Gdańsk) into a family of wealthy brewing merchants of Bohemian origin. He studied law at Leiden University in the Netherlands, and after traveling in Europe settled in Danzig in 1634. There he became a member of a local government council and also worked as a brewer in the family business. He later became mayor of Danzig.

In 1641, he began to construct an observatory in his home. His studies of the surface of the Moon and his discovery of the libration of the Moon in longitude, recorded in *Selenographia* (Lunar Topography (1647), are said to have laid the foundation for the study of lunar topography. Libration is a slight rocking motion that makes about 59 percent of the Moon's surface visible from Earth, though only 50 percent is visible at any one time.

Hevelius discovered four comets, in 1652, 1661, 1672, and 1677. These discoveries led to his thesis that such bodies revolve around the Sun in parabolic paths.

Hevelius's observatory, instruments, and books were maliciously destroyed by fire on September 26, 1679. The catastrophe is described in the preface to his *Annus Climactericus* (1685). He promptly repaired the damage, which enabled him to observe the great comet of December 1680.

He also observed sunspots, cataloged many of the stars, studied the phases of Saturn, and was one of the first to observe the transit of Mercury. A transit of Mercury across the Sun takes place when the planet comes between the Sun and the Earth. Mercury is seen as a small black dot moving across the face of the Sun.

In late 1683, in commemoration of the victory of Christian forces led by polish King John III Sobieski at the Battle of Vienna, Hevelius named a constellation Scutum Sobiescianum (Sobieski's Shield), now called Scutum. He had the book printed in his own house at lavish expense, and engraved many of the printing plates himself.

Hevelius died on his 76th birthday (January 28, 1687), and was buried in St. Catherine's Church in Danzig. He was the last astronomer of repute to carry out major observational work without a telescope. Moon crater **Hevelius** and **asteroid 5703 Hevelius** were named in his honor.

1. Johannes Hevelius, Wikipedia, http://en.wikipedia.org/wiki/Johannes_Hevelius.
2. Johan [Jan] Hevelius (January 28, 1611 - January 28, 1687), The Messier Catalog, http://seds.org/messier/Xtra/Bios/hevelius.
3. Johannes Hevelius, Microsoft Encarta Online Encyclopedia, http://encarta.msn.com/encyclopedia_761575883/johannes_hevelius, 2008.

Francesco Grimaldi (1618-1663)

Drew an accurate map of the Moon

Francesco Maria Grimaldi, an Italian physicist and astronomer, was born in Bologna on April 2, 1618. He entered the Society of Jesus in 1632, and after the usual course of studies spent 25 years as professor of mathematics and physics at the Jesuit college in Bologna.

From 1640 to 1650, he assisted Giovanni Riccioli (1598-1671) in his experiments on falling bodies and in his surveys in 1645 to determine the length of an arc of the meridian.

Grimaldi was also an observer of the Moon's surface and constructed a map, which was incorporated in Riccioli's *Almagestum Novum*. He named the elevations and depressions on the Moon after astronomers and physicists.

Grimaldi is best known for his work in optics, He made several discoveries of fundamental importance, but which were much in advance of his time, and their significance was not recognized until over a century later. The first of his discoveries with light involved the phenomenon of diffraction. He allowed a beam of sunlight to pass through a small aperture in a screen, and noticed that it was diffused in the form of a cone. The shadow of a body placed in the path of the beam was larger than that required by the rectilinear propagation of light. Careful observation also showed that the shadow was surrounded by colored fringes, similar ones being seen within the edges, especially in the case of narrow objects. He showed that the effect could not be due to reflection or refraction, and concluded that the light was bent out of its course in passing the edges of bodies. He called this phenomenon "diffraction," which means "breaking up."

Grimaldi also discovered that when sunlight entering a room through two small apertures is allowed to fall on a screen, the region illuminated by the two beams is darker than when illuminated by either of them separately. This is now known as the principle of interference.

He was also the first to observe the dispersion of the Sun's rays in passing through a prism. *Physicomathesis de Lumine, Coloribus, et Iride, Aliisque Annexis* was published after his death in 1665.

Grimaldi was one of the earliest physicists to suggest that light is wavelike in nature. He died in Bologna on December 28, 1663.

1. Francesco M Grimaldi, S.J., Fairfield University, www.faculty.fairfield.edu/jmac/sj/scientists/grimaldi.
2. Giovanni Battista Riccioli, Catholic Encyclopedia, www.newadvent.org/cathen/13040a.
3. Grimaldi, Francesco Maria (1618-1663), The Internet Encyclopedia of Science, www.daviddarling.info/encyclopedia/G/Grimaldi.

Jeremiah Horrocks (c.1618-1641)

Known as the father of British astronomy

Jeremiah Horrocks, a British astronomer and clergyman, was born about 1618 in Lower Lodge, Otterspool in Toxteth Park, near Liverpool, Lancashire. His father was a small farmer. Jeremiah was relatively poor during his entire brief life of 23 years.

He entered Cambridge University in 1632 at the age of 14, but left three years later without a degree due to the cost of continuing his studies. He became a tutor in Toxteth before being ordained to the curacy of Hoole, a poor hamlet near Preston, Lancashire in 1639. There, he continued to supplement his small stipend as a private tutor.

Horrocks read most of the astronomical treatises of his day, found the weaknesses in them, and was suggesting new lines of research by the age of 17.

He applied the laws of Johannes Kepler (1571-1630) on planetary motion, rightly predicting that the next transit of Venus across the Sun would occur on the afternoon of Sunday, November 24, 1639. He was the first to record this phenomenon, which he managed between church services.

Horrocks was the first to ascribe to the Moon an elliptical orbit, which Isaac Newton (1643-1727), in his *Philosophiae Naturalis Principia Mathematica* (Mathematical Principles of Natural Philosophy), later demonstrated to be due to gravity and acknowledged the contribution made by Horrocks.

He showed that the planets Jupiter and Saturn were giants, totally opposing the biblical view that our own planet must be the grandest in creation.

Horrocks studied the tides, hinted at the perturbing effect of the Sun on the Moon and observed the mutual perturbation of Jupiter and Saturn. In 1637, he greatly improved the work of Kepler and Johannes Hevelius (1611-1687) on calculating the solar parallax, a measure of the Earth's mean distance from the Sun, which he reduced to 14 minutes of arc.

He resigned his curacy in 1640, and died at Toxteth Park on January 3, 1641 of an unknown cause.

The achievements in his brief life were such that Sir John Herschel (1792-1871) called him "the pride and boast of British astronomy." His observations were published posthumously.

1. Horrocks, Jeremiah, Microsoft Encarta Online Encyclopedia, http://au.encarta.msn.com/encyclopedia_1461500548/Horrocks_Jeremiah.
2. Jeremiah Horrocks - Father of British Astronomy, BBC, www.bbc.co.uk/dna/h2g2/A2769041.
3. Jeremiah Horrocks, Wikipedia, http://en.wikipedia.org/wiki/Jeremiah_Horrocks.

Adrien Auzout (1622–1691)

Made contributions to the telescope

Adrien Auzout, the son of a concierge at the court of Rouen, was born on January 28, 1622 in Rouen, France. Facts about his basic education are unknown. He apparently did not attend a university. He carried out research mainly in the fields of astronomy and pneumatics. In 1664 and 1665, he argued in favor of comets having elliptical orbits, which was opposed to the view of his rival, Johannes Hevelius (1611-1687).

In 1666, Auzout was admitted to the Académie des Sciences in Paris, but he resigned after only two years. He engaged in a dispute with M. Perrault, a physician. Perrault produced a flawed translation of the work of Roman architect and engineer, Vitruvius (first century BC), which Auzout criticized severely. This may have been the cause for which he was forced to resign from the Académie. Auzout was a founding member of the Royal French Observatory.

He subsequently moved to Italy, where he remained the rest of his life, except for the period 1676 to 1685. There seems to be no information about his means of support during these final 20 years. He may have had personal resources on which he lived.

In 1647, Auzout showed, by placing a barometric tube within a vacuum, that the mercury did not remain at the same level, but completely descended into the vessel below. The ingenious project of creating a vacuum within a vacuum showed the role of the pressure of air in the barometric experiment.

Auzout and Jean Picard (1620-1682) made contributions to the telescope, including improving the use of the micrometer. Their micrometer consisted of two parallel hairs whose separation could be varied by a precision screw. The instrument allowed measurements of image sizes at a telescope's focal point.

In 1667-1968, he and Picard joined a telescope to a 38-inch quadrant and used it to determine positions on Earth. He wrote a memoir on the measurement of the Earth in which he advised the attachment of telescopes to surveying instruments.

His published works on astronomy, physics, and mathematics are reprinted in *Mémoires de l'Académie Royale des Sciences, Depuis 1666 Jusqu'à 1699* (1729).

Auzout died in Rome on May 23, 1691. The lunar crater **Auzout** was named in his honor.

1. Adrien Auzout (1622–1691), Vacui?, www.imss.fi.it/vuoto/eauzou.
2. Adrien Auzout, Wikipedia, http://es.wikipedia.org/wiki/Adrien_Auzout.
3. Auzout, Adrien, The Galileo Project, http://galileo.rice.edu/Catalog/NewFiles/auzout.

Giovanni Cassini (1625-1712)

Discovered four of Saturn's moons and the major division in its rings

Giovanni Domenico Cassini, an Italian mathematician, astronomer, engineer, and astrologer, was born in Perinaldo, near Nice, France on June 8, 1625. His first employment was in calculating astronomical tables for a nobleman who dabbled in astrology. Cassini was an astronomer at the Panzano Observatory from 1648 to 1669. He then became a professor of astronomy at the University of Bologna.

He established his reputation through his observations of the motions of comets and the apparent motion of the Sun. He used the most advanced telescopes of his day to observe the satellites of Jupiter and to draw up accurate tables of their movements. These enabled sailors to find their longitude by using the satellites as a "celestial clock."

Cassini discovered the seasonal changes on Mars and measured the period of rotation of both Mars and Saturn. Along with Robert Hooke (1635-1703), he is given credit for the discovery of the Great Red Spot on Jupiter.

In 1669, he went to Paris to help with setting up the new Paris Observatory, and later became its director. There, he discovered four satellites of Saturn and observed a gap in the planet's ring system.

Cassini measured the distance of Mars from the Earth using observations of parallax, which enables the distances of all the planets to be more accurately calculated.

He was a late and only partial convert to the Copernican system, and never accepted the theory of universal gravitation espoused by Isaac Newton (1642-1727). He rejected the idea of the finite speed of light, which his own observations of Jupiter's satellites had revealed.

Cassini went blind in 1710 and died in Paris on September 14, 1712. Three further generations of Cassinis headed the Paris Observatory after him.

Cassini has a number of things named for him, including the **Cassini crater** on the Moon, the **Cassini crater** on Mars, the **Cassini division of the rings of Saturn**, the **Cassini asteroid,** and **Cassini Regio**, which is a dark area on Iapetus. A space probe that the United States launched in 1997 to investigate Saturn was named after him

1. Cassini, Giovanni Domenico (Jean-Dominique), Microsoft Encarta Online Encyclopedia, http://au.encarta.msn.com/encyclopedia_781538677/Cassini_Giovanni_Domenico_(Jean-Dominique).
2. Giovanni Cassini, World Book at NASA, www.nasa.gov/worldbook/cassini_gio_worldbook.
3. Giovanni Domenico Cassini Wikipedia, http://en.wikipedia.org/wiki/Giovanni_Domenico_Cassini.

Christiaan Huygens (1629-1695)

Described the rings of Saturn and discovered Saturn's moon, Titan

Christiaan Huygens was born in the Hague Netherlands on April 14, 1629. He was the leading proponent of the wave theory of light, and in 1678 stated a geometrical method for finding, from the known shape of a wave front at some instant, the shape of the wave front at some later time. The principle is now known as **Huygens' principle.**

Huygens also made important contributions to mechanics, stating that in a collision between bodies neither body loses nor gains momentum and the center of gravity moves uniformly in a straight line. He also formulated the expression for centrifugal force.

He worked on developing more accurate clocks, suitable for naval navigation. In 1658, he published a book on the subject called *Horologium*. He worked with pendulum clocks and invented the clock with a spiral spring or cycloid pendulum, which enables a more precise time measurement. In 1675, he patented a pocket watch. After Blaise Pascal (1623-1662) encouraged him to do so, Huygens wrote the first book on probability theory, which he published in 1657.

In 1659, Huygens sketched a dark triangular patch on the surface of Mars, which eventually came to be known as Syrtis Major. Syrtis Major was the first permanent marking to be glimpsed on the surface of another planet. Huygens quickly used his discovery to demonstrate that the Martian day (the time it takes Mars to spin once around on its axis) is similar in length to the Earth's.

With his brother, Constantijn Huygens (1628-1697), he built tubeless telescopes, supported by cables, of very long focal length to reduce the problem of aberration. He also invented the **Huygenian eyepiece**.

Huygens accurately described the rings of Saturn and discovered Saturn's moon, Titan. In 1656, he observed and sketched the Orion Nebula. His drawing, the first such known of the Orion nebula, was published in *Systema Saturnium* in 1659. Using his modern telescope, he succeeded in subdividing the nebula into different stars. He also discovered several interstellar nebulae and some double stars.

He moved back to The Hague in 1681 after suffering a serious illness. He died there on July 8, 1695.

1. Christiaan Huygens (1629-1695), Eric Weissstein's World of Biography, Wolfram Research, http://scienceworld.wolfram.com/biography/Huygens.
2. Christiaan Huygens, Biographies/Images of Physicists and Astronomers, www.mlahanas.de/Physics/Bios/ChristiaanHuygens.
3. Huygens, Christiaan (1629-1695), The Internet Encyclopedia of Science, www.daviddarling.info/encyclopedia/H/HuygensC.

Jean Richer (1630-1696)

Determined that pendulums beat more slowly at the equator than at higher latitudes

Jean Richer, a French mathematician and astronomer, was born in 1630. The details of his early life have been lost. He was an assistant of Giovanni Cassini (1625-1712).

He became a member of the Académie Royale des Sciences in 1666 with the title of astronomer. By 1670, however, he had been given the title mathematician by the Académie. He spent most of his life after this time undertaking work for the Académie.

In 1670, Richer was sent by the Académie to La Rochelle to measure the heights of the tides there at both the spring and vernal equinoxes.

In 1671, he was sent on an expedition to Cayenne, French Guyana by the French Government. His first task there was to measure the parallax of Mars, and the observations were to be compared with that taken at other sites to compute the distance to the planet. Richer's observations of Mars were used by Cassini to make the first accurate estimate of the size of the Solar System.

Another of Richer's important tasks was to examine the periods of pendulums at different points on Earth. He examined the period of a pendulum while on the expedition to Cayenne, French Guyana and found that the pendulum beat more slowly than in Paris. From this, Richer deduced that gravity was weaker at Cayenne, so it must be further from the center of the Earth than was Paris.

His discovery initiated the famous controversy between Newtonians and Cartesians over Earth's shape. Newton (1642-1727) argued that his gravitational theory adequately accounted for Richer's results because it predicted Earth was an oblate spheroid—bulging equator and flattened poles. This contradicted the Cartesian view that Earth is a prolate spheroid—elongated along the polar axis.

His lunar and planetary observations corroborated the accuracy of existing astronomical tables. He also carried out extensive solar observations, and accurately determined the obliquity of the ecliptic and time of the solstices and equinoxes.

In 1673, Richer was given the title of Royal Engineer and undertook work on fortifications. He published his observations in his only written work *Observations Astronomiques et Physiques Faites en L'isle de Caienne* (1679). He died in Paris in 1696.

1. J. J. O'Connor and E. F. Robertson, Jean Richer, www-groups.dcs.st-and.ac.uk/~history/
2. Jean Richer, Answers.com, www.answers.com/topic/jean-richer?cat=technology.
3. Jean Richer, BookRags, www.bookrags.com/research/jean-richer-scit-031234.

Isaac Newton (1643-1727)

Showed how a universal force, gravity, applies to all objects in all parts of the universe.

Isaac Newton was born in Lincolnshire, near Grantham, England on December 25, 1643. He came from a family of modest yeoman farmers. His father died several months before he was born. He was educated at Trinity College, Cambridge. Afterward, he was appointed professor of mathematics at Cambridge, and held the post for 28 years.

Seeing an apple fall from a tree led Newton, at the age of 23, to consider the force of gravity extending to the Moon and beyond. He formulated the law of universal gravitation. The Moon he saw as falling around the Earth an exact amount to match the Earth's curvature. He showed how a universal force, gravity, applied to all objects in all parts of the universe.

In his law of universal gravitation, Newton stated that every mass attracts every other mass with a force "F" that, for any two masses, is directly proportional to the product of the masses and inversely proportional to the square of the distance separating them. Expressed mathematically as $F = Gm_1m_2/r^2$, where "G" is the gravitational constant, "m_1" and "m_2" are the two masses, and "r" is the distance between them.

In 1671, Newton designed a telescope using mirrors instead of lenses, the first reflector telescope. He also co-invented calculus independently of Gottfried Leibniz (1646-1716), formulated a theory of the nature of light, and demonstrated with prisms that white light is composed of all colors. In mechanics, he built on Galileo's work, formulating the three fundamental laws of motion. He also enunciated the principles of conservation of momentum and angular momentum.

Newton stated his ideas in several published works, two of which, *Philosophiae Naturalis Principia Mathematica* (Mathematical Principles of Natural Philosophy) in 1687 and *Opticks* (Optics) in 1704, are considered among the greatest scientific works ever produced.

In 1705, Newton was knighted by Queen Anne. He died at the age of 85 at Kensington, London on March 20, 1727. The SI unit of force, the **Newton** (N), is named for him.

1. Isaac Newton, Biographies/Images of Physicists and Astronomers, www.mlahanas.de/Physics/Bios/IsaacNewton.
2. Newton, Sir Isaac, Famous Physicists and Astronomers, www.phy.hr/~dpaar/fizicari/xnewton.
3. Sears,, Francis W., Mark W. Zemansky, and Hugh D. Young, College Physics, Seventh Edition, Addison Wesley, New York, 1991.
4. Wilkins, David R., Isaac Newton, Mathematicians of the Seventeenth and Eighteenth Centuries, www.maths.tcd.ie/pub/HistMath/People/Newton/RouseBall/RB_Newton.

Olaf Rømer (1644-1710)

Discovered that light travels at a finite speed

Olaf (Ole, Olaus) Christensen Rømer was born in Århus, Denmark on September 25, 1644. His father was a merchant and skipper. In 1662, Rømer graduated from the University of Copenhagen. He was then employed by the French government. Louis XIV made him teacher for the Dauphin (heir apparent of the throne of France). During this time, he also took part in the construction of the magnificent fountains at Versailles.

In 1681, Rømer returned to Denmark and was appointed professor of astronomy at the University of Copenhagen and royal mathematician. He was also an observer at the University Observatory at Rundetårn, where he used improved instruments of his own construction.

In his position as royal mathematician, Rømer in 1683 introduced the first national system for weights and measures in Denmark. He also developed one of the first temperature scales. Daniel Fahrenheit (1686–1736) visited him in 1708 and improved on the Rømer scale, the result being the Fahrenheit temperature scale still in use today in a few countries.

In 1705, Rømer was made Chief of the Copenhagen Police, a position he held until his death. As one of his first acts, he fired the entire force, convinced that the morale was alarmingly low. He invented the first street lights (oil lamps), and worked hard to control the beggars, poor people, unemployed, and prostitutes of Copenhagen.

Rømer joined the observatory of Uraniborg on the island of Hven, near Copenhagen, in 1671. Over a period of several months, he and Jean Picard (1620-1682) observed about 140 eclipses of Jupiter's moon, Io. By comparing the times of the eclipses, they calculated the difference in longitude of Paris to Uraniborg.

In 1672, Rømer went to Paris, where he continued observing the satellites of Jupiter. He noted that times between eclipses got shorter as Earth approached Jupiter, and longer as Earth moved farther away. It became obvious to Rømer that the speed of light was finite. However, he did not calculate a value for it.

Rømer died in Copenhagen, Denmark on September 19, 1710. In 1809, Jean Baptiste Delambre (1749-1822), using Rømer's approach, reported the time for light to travel from the Sun to the Earth as eight minutes and 12 seconds, yielding a value for the speed a little more than 300,000 km/second.

1. Olaus Roemer, Microsoft Encarta Online Encyclopedia, http://encarta.ca/encyclopedia_762512069/Olaus_Roemer.
2. Ole Rømer, Wikipedia, http://en.wikipedia.org/wiki/Ole_R%C3%B8mer.
3. Rømer, Olaus or Ole, Infoplease.com, www.infoplease.com/ce6/people/A0842330.

John Flamsteed (1646-1719)

Corrected the large number of errors in contemporary astronomical tables

John Flamsteed, a British astronomer, was born in Denby, England on August 19, 1646. His father was a wealthy businessman. At the age of 14, John developed a chronic rheumatic condition. From 1662 to 1669 he studied astronomy on his own without the help of teachers. Jesus College, Cambridge awarded him an M.A. in 1674.

Flamsteed accurately calculated the solar eclipses of 1666 and 1668. He was responsible for several of the earliest recorded sightings of the planet Uranus, which he mistook for a star. His 1690 sighting was the earliest known sighting of the planet.

He was ordained a deacon, and was preparing to take up a living in Derbyshire when he was invited to London in 1675 and appointed the first British Astronomer Royal.

When the Greenwich Observatory was founded in 1675, Flamsteed was made its first director. He lived at the Observatory until 1684 when he was appointed priest to the parish of Burstow, Surrey. He held that office, as well as that of Astronomer Royal, until his death.

Flamsteed's lunar observations furnished the data that his contemporary, Sir Isaac Newton (1642-1727), used to verify his theory of gravity.

In 1676, Flamsteed began a series of observations that, by exposing and correcting the large number of errors in contemporary astronomical tables, helped mark the beginning of modern, practical astronomy.

Flamsteed refused to publish his observations for a long time. Edmond Halley (1656-1742), despite Flamsteed's objections, finally published 400 copies of the work in 1712. Flamsteed obtained about 300 copies in 1715 and burned them publicly. It is in this unauthorized publication that the **Flamsteed Numbers** were assigned to the brighter stars of each constellation.

Flamsteed's own catalog of the fixed stars, *Historia Coelestis Britannica*, listing over 3000 stars, was larger than any previous star catalog. The catalog was published after his death in 1725.

Other notable work of Flamsteed included optics of telescopes and meteorological observations with barometers and thermometers, as well as longitude determination. He died in Greenwich on December 31, 1719.

1. John Flamsteed, Microsoft Encarta Online Encyclopedia, http://encarta.msn.com/encyclopedia_761565760/john_flamsteed.2008.
2. John Flamsteed, Wikipedia, http://en.wikipedia.org/wiki/John_Flamsteed.
3. O'Connor and E F Robertson, James Flamsteed, MacTutor, www-groups.dcs.st-and.ac.uk/~history/Printonly/Flamsteed.

Edmond Halley (1656-1742)

Showed that comets obey Newton's law of universal gravitation

Edmond Halley, a British mathematician and astronomer, was born in London, England on November 8, 1656. He was the son of a wealthy soap-maker. By the age of 18, he had found errors in authoritative tables on the positions of Jupiter and Saturn, and by age 19 had published a paper on the laws of Johannes Kepler (1571-1630).

In 1676, on St. Helena (an island west of Africa) Halley mapped the southern constellations, a task never before undertaken. He cataloged 341 stars before returning to England. He was awarded a master's degree from Oxford as well as election to the Royal Society.

Halley convinced his reticent friend, Isaac Newton (1643-1727), to publish his findings on universal gravitation. Using funds bequeathed to him by his father, Halley financed the publication of *Mathematical Principles of Natural Philosophy*, now considered one of the classic texts of modern scientific thought.

As comets streaked through the sky, they appeared ungoverned by Newton's law. Halley, however, believed that gravity did dictate their path, and that the rarity of their appearances was due to the vast length of their orbit. With the help of Newton, he compared the paths of comets that had appeared in 1531, 1607, and 1682. From this data, he determined that these seemingly separate comets were the same comet, and accurately predicted its reappearance in 1758. In 1705, Halley published his findings in *A Synopsis of the Astronomy of Comets*. Eventually, the comet that he predicted was named for him—**Halley's Comet**.

Halley undertook a lengthy study of solar eclipses and discovered that the so-called fixed stars actually moved with respect to each other. He also supported the theory that the universe is limitless and has no center.

Halley served as chief science adviser to Peter the Great (1672-1725) when the Russian tsar came to England in an attempt to integrate Western advances into his country's society. From 1698 to 1700, Halley commanded the Paramour, a Royal Navy ship, for a scientific expedition that studied the effects of Earth's magnetic field on magnetic needle compasses. In 1720, Halley succeeded John Flamsteed (1646-1719) as Astronomer Royal at Greenwich, a position that he held until his death on January 14, 1742.

1. Edmond Halley (1656 - 1742), BBC, www.bbc.co.uk/history/historic_figures/halley_edmond.
2. Edmond Halley, Microsoft Encarta Online Encyclopedia, http://encarta.msn.com/encyclopedia_761553871/edmund_halley, 2008.
3. World of Physics on Edmond Halley, BookRags, www.bookrags.com/biography/edmond-halley-wop.

Jacques Cassini (1677-1756)

Showed that the ancient belief in the unchanging sphere of the stars was incorrect

Jacques Cassini, a French-Italian astronomer, was born at the Paris Observatory on February 8, 1677. He was a son of Giovanni Cassini (1625-1712), the head of the Paris Observatory who was an academician and active in the cartographical projects of France.

Jacques entered the College Mazarin in 1691 and defended a thesis in optics at age 15. However, there is no record of any university degree.

He traveled with his father, making numerous geodesic measurements as well as several astronomical observations.

In 1694, Cassini, at the age of 17, was admitted to the Académie des Sciences and began to undertake scientific work on projects which his father was carrying out. He was elected in 1696 a fellow of the Royal Society of London, and became maître des comptes in 1706.

Cassini worked primarily in astronomy and cartography, but also did work in electricity and optics. He gave papers to the Academy on electricity, the recoil of firearms, barometers, and burning mirrors.

In astronomy, Cassini's primary interests were the study of planets and their satellites, the observation and theory of comets, and the tides. He presented a new method for determining longitudes by means of the eclipses of the stars and planets by the Moon.

He succeeded his father at the Paris Observatory in 1712. In 1813, he measured the arc of the meridian from Dunkirk to Perpignan, and published the results in a volume titled *Traité se la Grandeur et de la Figure de la Terre* in 1720. He published the first tables of the satellites of Saturn in 1716, and wrote *Eléments D'astronomie* in 1740.

Cassini supported the prediction of René Descartes (1596-1650) that Earth is a prolate spheroid (elongated polar axis) against the prediction of Isaac Newton (1642-1727) that it is an oblate spheroid (bulging equator and flattened poles). Expeditions to Peru (1734-1744) and Lapland (1736) later settled the debate decisively in favor of Newton.

He made important contributions with his study of the moons of Jupiter and Saturn and his study of the structure of Saturn's rings. He showed by direct measurement in 1738 that the ancient belief in the unchanging sphere of the stars was incorrect. Cassini died at Thury, near Clermont, on April 18, 1756.

1. Cassini, Jacques [Cassini The Galileo Project, ttp://galileo.rice.edu/Catalog/NewFiles/cassini_jac.
2. J. J. O'Connor and E. F. Robertson, Jacques Cassini, MacTutor, www-groups.dcs.st-and.ac.uk/~history/Biographies/Cassini_Jacques.
3. Jacques Cassini, www.bookrags.com/research/jacques-cassini-scit-041234

John Hadley (1682-1744)

Invented the octant, the precursor of the sextant

John Hadley, a British mathematician and inventor, was born on April 16, 1682 in Bloomsbury, London, England. His father, George, had an estate at Enfield Chase near East Barnet, Hertfordshire, which is now in London.

Hadley was elected a fellow of the Royal Society in 1717, and was elected Vice President in 1728. Although he had no need to earn a living, he did devote considerable time to looking after the family estates and was governor of Barnet grammar school from 1720.

His father died in 1729, and, as the eldest son, Hadley inherited the estate near East Barnet as well as other land owned by the family. This put him in a good financial position and allowed him to pursue his scientific interests.

He built, in collaboration with his brothers, the first effective Newtonian reflecting telescope during 1719-1720. It had a six-inch mirror.

Hadley is primarily known as the inventor of the octant, the precursor to the sextant, around 1730. At that time, one of the major problems was determining longitude at sea. The shipwreck of the fleet under the command of Sir Cloudesley Shovell on the Scilly Islands in 1707 led to Parliament putting up a large amount of money for any method to find longitude at sea to within one degree. This motivated Hadley to tackle the problem. As a result, he invented the reflecting octant.

The octant is used to measure the altitude of the Sun or other celestial objects above the horizon at sea. A mobile arm carrying a mirror and pivoting on a graduated arc provides a reflected image of the celestial body overlapping the image of the horizon, which is observed directly. If the position of the object on the sky and the time of the observation are known, it is easy for the user to calculate his latitude. The octant proved extremely valuable for navigation, and displaced the use of other instruments, such as the Davis quadrant. An American, Thomas Godfrey (1704-1749), independently invented the octant at approximately the same time as Hadley.

Hadley also improved the reflector telescope, building the first Gregorian telescope in 1721.

Hadley died on February 14, 1744 in East Barnet, Hertfordshire. **Mons Hadley** and **Rima Hadley** on the Moon are named after him.

1. Hadley, John H. (1682-1744), The Internet Encyclopedia of Science, www.daviddarling.info/encyclopedia/H/Hadley.
2. J. J. O'Connor and E. F. Robertson, John Hadley, MacTutor, www-gap.dcs.st-nd.ac.uk/~history/Biographies/ Hadley.
3. John Hadley, Wikipedia, http://en.wikipedia.org/wiki/John_Hadley.

James Bradley (1693-1762)

Discovered the angular discrepancy between the apparent position of a star and its true position

James Bradley, a British astronomer, was born in March 1693 in Sherborne, Gloucestershire, England. He was educated at Balliol College, Oxford University, where he received B.A. and M.A. degrees in 1714 and 1717, respectively.

In 1718, he was elected a fellow of the Royal Society, and the following year took orders on becoming vicar of Bridstow. At the age of 28, he became Savilian professor of astronomy at Oxford, resigning his religious orders.

In 1722, Bradley measured the diameter of Venus with a telescope over 212 feet in length.

In 1729, his theory of the aberration of the fixed stars, containing the important discovery of the aberration of light, was published. Aberration in astronomy is the angular discrepancy between the apparent position of a star and its true position. It arises from the motion of an observer relative to the path of the beam of light observed. Bradley arrived at almost precisely the modern value for the constant of aberration, about 20.5 seconds (the modern value being 20.47 seconds). He realized that aberration was due to the finite velocity of light.

Bradley also discovered the phenomenon of nutation (the vibration of the Earth's axis). He attributed this to the oscillation of the Earth's axis, caused by the changing direction of the gravitational pull of the Moon on the equatorial bulge.

At Greenwich, in 1742, he was appointed to succeed Edmund Halley (1656-1742) as Astronomer Royal. He applied for a set of instruments, including an 8-foot mural quadrant, completed for him in 1750. With the 8-foot mural quadrant he compiled a new catalog of star positions. It was published posthumously and involved some 60,000 observations.

The precise observations that Bradley made at the Greenwich Observatory were extremely useful, especially to the German astronomer Friedrich Bessel (1784-1846), who in 1818 published a catalog of star positions computed from Bradley's observations.

Bradley's health failed, and he retired to Chalford, Gloucestershire, where he died July 13, 1762.

1. Encyclopedia of World Biography on James Bradley, BookRags, http://www.bookrags.com/biography/james-bradley.
2. James Bradley, Microsoft Encarta Online Encyclopedia, http://encarta.msn.com/encyclopedia_761577559/james_bradley, 2008.
3. James Bradley, Wikipedia, http://en.wikipedia.org/wiki/James_Bradley.

John Bevis (1695-1771)

Discovered the Crab Nebula, M1

John Bevis, a British physician and amateur astronomer, was born on November 10, 1695 in Old Sarum in Wiltshire, England. He is best known as the discoverer of the Crab Nebula, M1, which he observed in 1731—27 years before independent rediscovery by Charles Messier (1730-1817).

Bevis in a letter of June 10, 1771 informed Messier of this, and Messier acknowledged the original discovery in the later publications of his catalog. The Crab Nebula is a crab-shaped, rapidly expanding cloud of gas in the constellation Taurus, containing a neutron-star pulsar. It is believed to be the remnants of the supernova of 1054

Bevis is reported to have observed an occultation by Venus on May 28, 1737. An occultation in astronomy is the passage of a celestial body between an observer and another celestial object. He also developed a prediction rule for eclipses of Jupiter's moons.

In 1738, he set up a private observatory at Stock Newington, North London. His *Uranographia Britannica* was to have consisted of 52 large plates of the sky with accompanying explanations. However, just before the work was to go to press in 1752 the publisher went bankrupt and the plates were sequestered. The atlas was published posthumously in 1786 from the plates that had been engraved earlier.

Bevis's atlas contains the nebulae M1, M11, M13, M22, M31, and M35, which might mean that he was the original discoverer of M35 also. Credit for discovering M35 is usually given to Phillip de Chéseaux (1718-1751).

Bevis is one of only two persons in Britain who are known to have observed comet Halley on its first predicted return in 1759. He observed it on May 1 and 2, 1759 after its perihelion.

From sometime in the 1760s until shortly before his death, Bevis was in correspondence with Messier, and translated some of Messier's earlier observing reports and publications for publication in the *Philosophical Transactions*. Bevis's *Atlas Celeste* also contains the star clusters M44 (Praesepe) and M45 (Pleiades), as well as southern Omega Centauri (NGC 5139) and NGC 6231.

In electricity, Bevis suggested the feature retained by all modern capacitors—two conductors separated by an insulating, or dielectric, layer. He died on November 6, 1771 from injuries received when falling from his telescope. He was 76 years old.

1. John Bevis, BookRags, www.bookrags.com/research/john-bevis-scit-041234.
2. John Bevis, The Messier Catalog, http://seds.org/messier/xtra/Bios/bevis.

18th Century Astronomers

1700 to 1799

Anders Celsius (1701-1744)

Confirmed Newton's belief that the shape of the Earth is an ellipsoid, flattened at the poles

Anders Celsius was born in Uppsala, Sweden on November 27, 1701. His father, Nils Celsius, was professor of astronomy at Uppsala University. Early on, Anders became engaged in the general problem of weights and measures, including temperature measurements. At that time there were a variety of thermometers, each based on a different scale.

As a student, he had assisted astronomy professor Erik Burman (1692-1729) in meteorology observations. In 1730, Celsius himself became professor of astronomy at Uppsala University. He remained there until 1744. From 1732 to 1735 he traveled, visiting notable observatories in Germany, Italy, and France. Soon after his return to Uppsala, he participated in an expedition to Torneå in the most northern part of Sweden. The expedition was headed by French astronomer, Pierre-Louis Maupertuis (1698-1759). Its aim was to measure the length of a degree along a meridian close to the pole and to compare the result with a similar expedition to Peru, near the equator. The two expeditions confirmed Newton's belief that the shape of the Earth is an ellipsoid, flattened at the poles.

The first real observatory at Uppsala University was the Celsius Observatory, founded through the efforts of Celsius in 1741.

In 1742, in a paper to the Royal Swedish Academy of Sciences, Celsius proposed the Celsius temperature scale. His thermometer had 100 divisions between the freezing point of water (100°C) and the boiling point of water (0°C). The scale was reversed by Carolus Linnaeus (1707-1778) in 1745 to its present form.

Celsius was the first to perform and publish careful experiments aimed at the definition of an international temperature scale on scientific grounds. In his paper *Observations of two Persistent Degrees on a Thermometer* he reported on experiments to check that the freezing point is independent of latitude and pressure. He determined the dependence of the boiling point of water on atmospheric pressure. He further determined a rule for determining the boiling point if the barometric pressure deviates from a given standard pressure.

Celsius died of tuberculosis in Uppsala on April 5, 1774. He was 43 years old. His grave is next to that of his grandfather, Magnus Celsius (1621-1679), in the Old Uppsala Church.

1. Anders Celsius, Biographies/Images of Physicists and Astronomers, www.mlahanas.de/Physics/Bios/AndersCelsius.
2. Anders Celsius, History, Uppsala Astronomical Observatory, www.astro.uu.se/history/Celsius_eng.
3. Beckman, Olaf, History of the Celsius Temperature Scale, www.astro.uu.se/history/celsius_scale.

John Dollond (1706-1761)

Patented an achromatic lens

John Dollond, a British optician and inventor, was born in Spitalfields, London, England on June 10, 1706. His father, a Huguenot refugee, was a silk-weaver. John followed in his father's trade, but found time to acquire knowledge of Latin, Greek, mathematics, physics, anatomy, and other subjects.

In 1752, Dollond abandoned silk-weaving and joined his eldest son, Peter Dollond (1730-1820), who in 1750 had started in business as a maker of optical instruments. John Dollond's reputation grew rapidly, and in 1761 he was appointed optician to the king.

In 1747, Leonhard Euler (1707-1783) suggested that achromatism might be obtained by the combination of glass and water lenses. Dollond disputed this possibility (1753), and subsequently began experiments to settle the question. An achromatic lens is desirable to produce images free of chromatic (color) aberrations.

Early in 1757, Dollond succeeded in producing refraction without color by the aid of glass and water lenses, and a few months later successfully obtained the same result with a combination of glasses of different qualities.

In 1758, he published an *Account of Some Experiments Concerning the Different Refrangibility of Light*, describing the experiments that led him to the discovery of a means of constructing achromatic lenses by the combination of crown and flint glasses. This allowed construction of telescopes free of color fringes.

For his achievement, the Royal Society of London awarded him the Copley Medal in 1758, and three years later elected him one of its fellows.

Dollond was the first person to patent the achromatic doublet. This was based on his own research in optics; however, he was not the first to make achromatic lenses. Optician George Bass, following the instructions of Chester Hall (1703-1771), made and sold such lenses as early as 1733.

His patent was disputed; however, the court found that the patent was valid. The patent remained valid until it expired in 1772.

Dollond also published two papers on an apparatus, the heliometer, for measuring small angles (1753, 1754). It was originally designed for measuring the variation of the Sun's diameter at different seasons of the year. Dollond died of a cerebrovascular accident (stroke) in London, England on November 30, 1761.

1. John Dollond, BookRags, ://www.bookrags.com/research/john-dollond-scit-041234.
2. John Dollond, Classic Encyclopedia, www.1911encyclopedia.org/John_Dollond.
3. John Dollond, Wikipedia, http://en.wikipedia.org/wiki/John_Dollond.

Roger Boscovich (1711-1787)

Developed the first geometric procedure for determining the equator of a rotating planet

Roger (Rudjer) Joseph Boscovich, a physicist, astronomer, mathematician, philosopher, diplomat, and poet, was born in Ragusa (today Dubrovnik, in Croatia) on May 18, 1711. He entered the novitiate of the Society of Jesus in Rome in 1725 and the Collegium Romanum in 1727. He devoted himself chiefly to mathematics and physics, and published his first scientific paper in 1736. He became professor of mathematics at the Collegium in 1740, and in 1744 he took his vows as a priest.

In 1750, Pope Benedict XIV commissioned him, with English Jesuit Christopher Maire (1697-1767), to measure an arc of the meridian through Rome. In 1754, Boscovich published a textbook, part of which was devoted to his theory of conic sections.

Boscovich's is best known for *Philosophiae Naturalis Theoria Redacta ad Unicam Legem Virium in Natura Existentium* (A theory of natural philosophy reduced to a single law of the actions existing in nature), published in 1758. In this work he presented an atomic theory on which he had been working for 15 years.

From 1759 on, he was engaged in extensive travels as far away as Constantinople. In 1760, he met Benjamin Franklin (1706-1790) and many other leading personalities in London and Cambridge, and he was elected the Royal Professor of mathematics at Pavia in 1765. A chair was created for him at Milan in 1769. There, he pursued studies at the Brera Observatory. In 1775, he was appointed director of naval optics for the French navy and went to Paris, where he was made a subject of France by Louis XV. He returned to Italy in 1783.

Boscovich did work on the accuracy of astronomical observations, the telescope, sunspots, eclipses, the determination of the Sun's rotation, the orbits of planets and comets, the aurora borealis, the transit of Mercury, the shape of Earth, and the variation of gravity. He developed the first geometric procedure for determining the equator of a rotating planet.

His last major publication was a five-volume work on optics and astronomy, *Opera Pertinentia ad Opticam et Astronomiam*, published in 1785. Boscovich died on February 13, 1787. During his last years he suffered from melancholia (depression).

1. Boscovich, Roger Joseph(1711–1787), Bookrags, www.bookrags.com/research/boscovich-roger-joseph-17111787-eoph.
2. World of Mathematics on Rudjer Boscovic, Bookrags, www.bookrags.com/biography/rudjer-boscovic-wom.

Nicolas de Lacaille (1713-1762)

Cataloged 10,000 southern stars

Nicolas Louis de Lacaille, a French astronomer, was born at Rumigny, in the Ardennes, on March 15, 1713. Nicholas spent his first years at Rumigny living in the house of his birth. He then studied the humanities at the college of Mantes-sur-Seine, northwest of Paris, until 1729. Afterward, he went to the college of Lisieux in Paris, where he studied history, antiques, mythology, and Latin poetry.

When his father, who held a post in the household of the duchess of Vendome, died, the duke of Bourbon paid for Nicolas's theological studies at the College de Lisieux in Paris. Afterward, Nicolas enrolled at the college of Navarre to obtain ordination for priesthood. In 1736, he graduated with an M.A. and a Bachelor of Theology degree, but some circumstance around his graduation caused him to turn away from formal theology.

Through the patronage of Jacques Cassini (1677-1756), Lacaille obtained employment, first in surveying the coast from Nantes to Bayonne and then in 1739 re-measuring the French arc of the meridian. This difficult operation, which occupied two years, corrected the incorrect result published by Cassini in 1718. For this work, Lacaille was rewarded by admission to the Academy and the appointment as mathematical professor in Mazarin College. He was also honored with a pyramid at Juvisy-sur-Orgea, a commune of the Essonne *département* in France.

He made an astronomical expedition to the Cape of Good Hope, resulting in determinations of the lunar and the solar parallax (the first measurement of a South African arc of the meridian) and the observation of 10,000 southern stars. This catalog, called *Coelum Australe Stelliferum*, was published posthumously in 1763. It introduced 14 new constellations, which have since become standard. He also calculated a table of eclipses for 1800 years.

On his return to Paris in 1754, Lacaille was distressed to find himself an object of public attention. He withdrew to Mazarin College, where he remained until his death on March 21, 1762. In honor of his contribution to the study of the southern hemisphere sky, a 60-cm telescope at Reunion Island is named **Lacaille Telescope**. **Lacaille crater** on the Moon and **asteroid 9135 Lacaille** are named after him.

1. Nicholas Louis de Lacaille (March 15, 1713 - March 21, 1762), The Messier Catalog, www.napoli.yurisnight.net/messier/xtra/Bios/lacaille.
2. Nicolas Louis De Lacaille, Classic Encyclopedia, LoveToKnow, www.1911encyclopedia.org.
3. Nicolas_Louis_De_Lacaille, Nicolas Louis De Lacaille, Wikipedia, http://en.wikipedia.org/wiki/Nicolas_Louis_de_Lacaille.

César Cassini (1714-1784)

The third in a line of Cassini astronomers

César François Cassini (Cassini III), a French astronomer and cartographer, was born in Thury-sous-Clermont, (Oise) on June 17, 1714. He was the second son of Jacques Cassini (1677-1756), the grandson of Giovanni Cassini (1625-1712), and would become the father of Dominique Cassini (1748-1825)—all astronomers.

César was brought up at the Paris Observatory, where his father had taken over as head about the time of his birth. It was in the Observatory that he was educated by his great uncle, Jacques Maraldi (1665-1729). In 1733, at the age of 19, he addressed the Académie des Sciences on the geodesic measurements he was carrying out with his father.

In 1735, he became a member of the French Academy of Sciences as a supernumerary adjunct astronomer, in 1741 as an adjunct astronomer, and in 1745 as a full member astronomer.

He succeeded his father at the Paris Observatory in 1756, and continued his surveying operations. In 1744, he began the construction of a great topographical map of France, which would become one of the landmarks in the history of cartography.

In 1748, Cassini, following in his father's footsteps, was appointed to the Chambre des Comptes, which was a financial court with administrative and legal duties relating to the King's accounts, and in particular to the land owned by the Crown. He was also appointed to the office of King's Counsel.

The post, Director of the Paris Observatory, was created for him in 1771 when the facility ceased being under the control of the French Academy of Sciences. The Directorship was to be a hereditary position, set up so that Cassini's son would succeed him, which he did in 1784.

Cassini's chief works were *La méridienne de l'Observatoire Royal de Paris* (1744), *Description Géometrique de la Terre* (1775), and *Description Géometrique de la France* (1784), which was completed by his son.

Cassini died of smallpox in Paris on September 4, 1784. Despite many observations made by Cassini in his role as head of the Paris Observatory, his work in astronomy is of relatively little importance. Although he is an acclaimed cartographer, in astronomy he is mainly remembered for his lineage.

1. César-François Cassini de Thury, MacTutor, www-groups.dcs.st-and.ac.uk/~history/Biographies/Cassini_de_Thury.
2. César-François Cassini de Thury, Wikipedia, http://en.wikipedia.org/wiki/C%C3%A9sar-Fran%C3%A7ois_Cassini_de_Thury.

John Winthrop (1714-1779)

Led Harvard University's first astronomical expedition

John Winthrop, an American educator and scientist, was born in Boston, Massachusetts on December 19, 1714. He was the great-great-grandson of Massachusetts Bay's first governor by the same name.

John graduated from Harvard in 1732 at the age of 18. He then studied science at home for six years, and at age 24 was named professor of mathematics and natural philosophy at Harvard. He helped liberalize the curriculum of Harvard College, and received English recognition as America's leading astronomer.

Winthrop's public lectures and demonstrations in physical science attracted wide attention, and the results of his extensive research were published by London's Royal Society.

His series of sunspot observations in 1739 were the first in Massachusetts, and necessitated close cooperation with both the Royal Society and Greenwich Observatory.

Winthrop noted transits of Mercury in 1740, 1743, and 1769, and accurately recorded the longitude of Cambridge, Massachusetts. Other studies included work on meteors (1755) and solar parallax and distance (1769). His study of the 1761 Venus transit in Newfoundland was Harvard's first astronomical expedition.

Winthrop established America's first laboratory of experimental physics in 1746, and his demonstrations on mechanics, heat, and light are thought to have influenced both Benjamin Franklin (1706-1790) and Benjamin Thompson (1753-1814).

In 1751, he inaugurated a new era in American mathematical study by introducing the elements of calculus at Harvard.

Winthrop's study of the New England earthquake of 1755 was a pioneering approach to seismology. His other interests included extensive research on magnetism, eclipses, and light aberrations.

He served as acting president of Harvard in 1769 and again in 1773; but both times he declined the offer of the full presidency. He served for a time as Massachusetts probate judge in Middlesex County and was a member of the governor's council.

During the American Revolution, Winthrop promoted the colonial cause, encouraged munitions production, and advised George Washington and other American leaders. He died in Cambridge on May 3, 1779.

1. Encyclopedia of World Biography on John Winthrop, BookRags, www.bookrags.com/biography/john-winthrop2.
2. John Winthrop (1714–1779), Wikipedia, http://en.wikipedia.org/wiki/John_Winthrop_(1714-1779).

Pierre Lemonnier (1715-1799)

Made 12 separate observations of Uranus before it was recognized as a planet

Pierre Charles Lemonnier was born in Paris on November 23, 1715. His father was professor of philosophy at the college d'Harcourt. In 1736, Pierre was chosen to accompany Pierre Maupertuis (1698-1759) and Alexis Clairault (1713-1765) on their geodesic expedition to Lapland.

In 1738, shortly after his return, Lemonnier explained to the Academy the advantages of determining right ascensions by the method of John Flamsteed (1648-1718). Right ascension is the astronomical term for one of the two coordinates of a point on the celestial sphere when using the equatorial coordinate system. The other coordinate is the declination.

His persistent recommendation of British methods and instruments contributed to the reform of French astronomy. King Louis XV of France furnished him with the means of procuring the best instruments, many made in Britain. He was admitted in 1739 to the Royal Society, and was one of the 144 original members of the Institute. He was the first to represent the effects of nutation in the solar tables, and introduced in 1741 the use of the transit-instrument at the Paris observatory. Nutation is the oscillatory movement of the axis of a rotating body.

Lemonnier visited England in 1748, and continued his journey to Scotland, where he observed the annular eclipse of July 25. He investigated the disturbances of Jupiter by Saturn, the results of which were confirmed by Leonhard Euler (1707-1783) in his prize essay of 1748.

He also made a series of lunar observations extending over 50 years, determined the locations of a great number of stars, and made 12 separate observations of Uranus and recorded it in his charts before it was recognized as a planet by William Herschel (1738-1822).

Lemonnier's work in physics included the Leyden jar experiment by which he established that water is one of the best electrical conductors and that the surface area, not the mass, of a conducting body determines its electrical charge. His research on electricity produced by storms confirmed the theories of Benjamin Franklin (1706-1790).

Lemonnier's temper and hasty speech resulted in many arguments and grudges. A stroke and paralysis late in 1791 ended his career. A second stroke took his life on May 31, 1799. **Lemonnier crater** on the Moon is named after him.

1. Pierre Charles Lemonnier, The Columbia Encyclopedia, Sixth Edition, www.encyclopedia.com/doc/1E1-LemonniP.
2. Pierre Le Monnier, Wikipedia, http://neohumanism.org/p/pi/pierre_lemonnier.

Antonio de Ulloa (1716-1795)

Participated in measuring a degree of the meridian at the equator

Antonio de Ulloa, a Spanish general, explorer, author, and astronomer, was born in Seville, Spain on January 12, 1716. He was the son of an economist. He entered the navy in 1733.

In 1735, he was appointed, with fellow Spaniard Jorge Juan y Santacilia (1713-1733), by the French Academy of Sciences to go on a geodesic mission to Peru to measure a degree of the meridian at the equator. The mission was led by Pierre Bouguer (1698).

Ulloa remained in Peru from 1736 to 1744, during which time the two Spaniards discovered the element platinum.

In 1745, having finished their scientific work, Ulloa and Santacilia prepared to return to Spain, traveling on different ships. Ulloa's ship was captured by the British, and he was taken as a prisoner to England. In a short time, through the influence of the president of the Royal Society of London, he was released and returned to Spain.

Back home, he was appointed to serve on various important scientific commissions. He is credited with the establishment of the first museum of natural history, the first metallurgical laboratory in Spain, and the observatory of Cadiz.

In 1758, Ulloa returned to South America as governor of Huancavelica in Peru and the general manager of the quicksilver mines there. He held that position until 1764. He went to New Orleans in 1766 to serve as the first Spanish governor of West Louisiana. However, he was expelled from Louisiana by a French-Creole uprising in 1768.

For the remainder of his life he served as a naval officer. In 1779, he became lieutenant-general of the naval forces.

As a result of his scientific work in Peru, he published *Relación Histórica del Viaje á la América Meridional* (1784), which contains a description of the greater part of South American geography and its inhabitants and natural history. In collaboration with Santacilia, he also wrote *Noticias Secretas de América* (1826), giving information about the early religious orders in Spanish America.

Ulloa's halo, a meteorological term for a fog-bow, is defined as a faint white, circular arc or complete ring of light that has a radius of 39 degrees and is centered on the antisolar point (a direction 180 degrees away from the Sun). Ulloa died at Isla de Leon, Cádiz, in 1795.

1. Antonio de Ulloa, New Advent, www.newadvent.org/cathen/15122b.
2. Antonio de Ulloa, Virtual American Biographies, http://famousamericans.net/antoniodeulloa.
3. Antonio de Ulloa, Wikipedia, http://en.wikipedia.org/wiki/Antonio_de_Ulloa.

Maximilian Hell (1720-1792)

Regularly published astronomical tables

Maximilian Hell, a Hungarian astronomer, was born Maximilian Höll in Štiavnické Bane (present-day Slovakia) on May 15, 1720. Höll later changed his surname to Hell. The third son of a second marriage of his father, he had 21 brothers and sisters. The place of birth of Maximilian's father is unknown. Hell considered himself a Hungarian.

Hell joined the Jesuit order in Trentschin in 1738. After some years as novice he was sent to Vienna for three years to study philosophy and mathematics. In 1745, he was invited to assist Joseph Franz (1704-1776), the astronomer at the Jesuit observatory in Vienna, with his observations.

After about a year and a half as teacher in Leutschau, he returned at the end of 1747 to Vienna to start his theological studies. In 1752, he was ordained priest and ordered to go to Klausenburg as a mathematics teacher and to build a new observatory there.

Hell became director of the Vienna Observatory in 1755. He regularly published the astronomical tables *Ephemerides Astronomicae ad Meridianum Vindobonemsem* (Ephemerides for the Meridian of Vienna). These were for several years the only astronomical tables other than those of the Paris observatory. The last volume edited by Hell was published in 1793.

At the invitation of the King of Denmark, Hell traveled to Vardø in the far north of Norway (then part of Denmark) to observe the 1769 transit of Venus. These observations were important, since the distance of the Sun could be calculated using data from the timing of the beginning and end of the transit.

He stayed in Norway for eight months, collecting non-astronomical scientific data about the arctic regions for a planned encyclopedia. The publication was delayed. There was some controversy about his trip. Some accused Hell posthumously of falsifying his astronomical results. However, Simon Newcomb (1835-1909) in Vienna carefully studied Hell's notebooks and exonerated him a century after his death.

Soon after the expedition, Hell was asked to build an observatory in Erlau for the Bishop Duke Carl Eszterhazy. Hell also had an interest in magnet therapy (the alleged healing power of magnets).

He fell ill with pneumonia in March 1792 and died in Vienna on April 14, 1792. He is buried in the cemetery at Enzersdorf. The **Hell crater** on the Moon is named after him.

1. Hell [Höll], Maximilian (1720 - 1792), List of Available Biographies, ww.plicht.de/chris/34hell.
2. Maximilian Hell, New Advent, http://home.newadvent.org/cathen/07211a.
3. Maximilian Hell, Wikipedia, http://en.wikipedia.org/wiki/Maximilian_Hell.

Tobias Mayer (1723-1762)

Developed lunar tables allowing accurate determination of longitude at sea

Tobias Mayer, a self taught German mathematician, cartographer, and astronomer, was born in Marbach, Württemberg, Germany on February 17, 1723. He was brought up at Esslingen in poor circumstances.

In 1746, he entered J.B. Homann's cartographic establishment at Nuremberg, where he introduced many improvements in mapmaking.

Mayer's first important astronomical work was a careful investigation of the libration of the Moon, which gained him fame, and led to his appointment as professor of economics and mathematics at Göttingen University in 1751. Libration of the Moon is small periodical changes in the position of the Moon's surface relatively to the Earth, which allows more than 50 percent of the Moon's surface to be seen over time.

In 1754, he became superintendent of the astronomy observatory at Göttingen University, a position he held until his death in 1762.

Mayer began calculating lunar and solar tables in 1753, and in 1755 he sent them to the British government. These tables allowed the determination of longitude at sea with an accuracy of half a degree. His work claimed the British government's prize for best such method.

His work was published after his death in 1770. Appended to the London edition of the solar and lunar tables were two short tracts. One tract was on determining longitude by lunar distances, together with a description of the reflecting circle, invented by Mayer in 1752. The reflecting circle is an octant extended to a full circle, designed for the purpose of determining longitude by measuring the distance between the Moon and a nearby star. The other tract was on a formula for atmospheric refraction, which applies a remarkably accurate correction for temperature.

Three years after his death, the British Board of Longitude awarded his widow 3,000 pounds in prize money for his longitude method.

Mayer also devised a method for fixing geographical coordinates independent of astronomical observations, invented a method for determining eclipses, investigated stellar proper motions, and produced an accurate map of the Moon's surface.

Mayer died in Göttingen, Germany on February 2, 1762. In 1881, Ernst Klinkerfues (1827-1884) published photo-lithographic reproductions of Mayer's local charts and general map of the moon.

1. J. J. O'Connor and E. F. Robertson, Tobias Mayor, MacTutor, www-groups.dcs.st-and.ac.uk/~history/Biographies/Mayer_Tobias, December 1996.
2. Johann Tobias Mayer, BookRags, www.bookrags.com/research/johann-tobias-mayer-scit-041234.
3. Tobias Mayer, Wikipedia, http://en.wikipedia.org/wiki/Tobias_Mayer.

John Michell (1724-1793)

The first to study the case of a heavenly object massive enough to prevent light from escaping

John Michell, a British theologian, geologist, and physicist, was born on December 25, 1724. He was educated at Queens' College, Cambridge, obtaining his M.A. in 1752 and Bachelor of Divinity degree in 1761. In 1762, he was appointed Woodwardian Professor of Geology at Cambridge, and in 1767 he became rector of Thornhill, West Yorkshire, a post he held the rest of his life. His work spanned a wide range of subjects from astronomy to optics, to gravitation, to geology.

In 1750, he published an article entitled *A Treatise of Artificial Magnets* in which he showed an easy and expeditious method of making magnets superior to the best natural ones (lodestones). In addition to a description of the method of magnetization, the work contained a variety of accurate magnetic observations, and explained the nature of magnetic induction.

He attempted to measure the radiation pressure of light by focusing sunlight onto one side of a compass needle. The experiment was not a success; the needle melted.

In a letter to Henry Cavendish (1731-1810), published in 1784, Michell discussed the effect of gravity on light. He is now credited with being the first to study the case of a heavenly object massive enough to prevent light from escaping. Such an object would not be directly visible, but could be identified by the motions of a companion star if it were part of a binary system. This concept was the predecessor of the modern idea of a black hole. Michell also suggested using a prism to measure the gravitational weakening of starlight due to the surface gravity of the source, a phenomenon now known as gravitational shift. By the time that Michell's paper was resurrected nearly two centuries later, these ideas had been reinvented by others.

He also invented and built a torsion balance for an experiment to measure Newton's gravitational constant "G", but did not live to put it to use. His apparatus passed to Cavendish, who performed the experiment in 1798.

Michell died at Thornhill on April 29, 1793. He is known as the father of seismology for his studies on earthquakes and vibrations within the Earth.

1. John Michell (1724 - April 21, 1793), SEDS, www.seds.org/messier/xtra/Bios/michell.
2. John Michell, Microsoft Encarta Online Encyclopedia , http://encarta.msn.com/encyclopedia_761594885/john_michell 2008.
3. John Michell, Wikipedia, http://en.wikipedia.org/wiki/John_Michell.

Guillaume Le Gentil (1725-1792)

Discovered several deep-sky objects

Guillaume Joseph Hyacinthe Jean-Baptiste Le Gentil de la Galaisière, a French astronomer, was born in Coutances, France on September 12, 1725. Initially, he intended to enter the Church, and studied to do so at the Collége de France. There he attended astronomy lectures, which caused him to become interested in that subject.

He became an assistant of Jacques Cassini (1677-1756) at the Paris Observatory and participated in Cassini's geodesic surveys. He then began observing deep-sky objects, and discovered M32 (the small bright companion of the Great Andromeda Galaxy) in 1747; M8 (a giant cloud of interstellar matter, which is currently undergoing vivid star formation); M36 (the first of three bright open clusters in the southern part of constellation Aurigaand); and M38 (one of the three Messier open clusters in the southern part of constellation Aurigain) in 1749. He presented these observations to the Paris Royal Academy of Sciences in 1749.

In 1758, Le Gentil summarized his observations of the Northern nebulae in a memoir presented to the French Royal Academy. His work was published in the volume of the Royal Academy for 1759, which was finally printed in 1765.

He made attempts to observe the transits of Venus in 1761 and again in 1769. The first transit found him stranded in the middle of the Indian Ocean, unable to make any useful observations. After spending four years in Mauritius and Madagascar and taking a side trip to the Philippines, he arrived in India, built an observatory at Pondicherry, and waited for the next transit, which was to occur on June 4, 1769. The weather clouded up on transit day, only to clear immediately after the long-awaited event had passed.

After a prolonged bout of dysentery and having his ship disabled in a hurricane off the Cape of Good Hope, he was forced to cross the Pyrenees on foot. Returning to France after an absence of more than 11 years, he learned that he had been declared dead, his estate looted, and its remains divided up among his heirs and creditors. He lived at the Paris Observatory after that.

Le Gentil died on October 22, 1792. He was honored by having the **Le Gentil crater** on the Moon named for him in 1961.

1. Guillaume Le Gentil, Wikipedia, http://en.wikipedia.org/wiki/Guillaume_Le_Gentil.
2. Guillaume-Joseph-Hyacinthe-Jean-Baptiste Le Gentil de la Galaziere (September 11, 1725 - October 22, 1792), The Messier Catalog, http://seds.lpl.arizona.edu/messier/xtra/Bios/legentil.
3. Le Gentil (de la Galaziere), Guillaume Joseph Hyacinthe Jean Baptiste (1725-1792), The Internet Encyclopedia of Science, www.daviddarling.info/encyclopedia/L/Le_Gentil.

Charles Mason (1728-1787)

Worked to perfect the Lunar Tables as a method of improving navigation at sea

Charles Mason, a British astronomer, was born at Weir Farm, England in April 1728. His father was a baker/miller. Charles's early career was spent at the Royal Greenwich Observatory near London, where he served as assistant astronomer from 1756 to 1760 under James Bradley (1693-1762), Astronomer Royal.

While employed at the Greenwich Observatory, Mason became familiar with the *Tables of the Moon* of Tobias Mayer (1723-1762). He then worked throughout his life to perfect the lunar tables as a method of improving navigation at sea. In 1787, Mason was awarded 750 pounds by the Board of Longitude for his work on the tables.

In 1760, he was assigned to travel to the island of Sumatra to observe the Transit of Venus as part of an international effort to record data that would enable scientists to determine the distance from the Earth to the Sun. Because of an attack by a French man-of-war, he did not reach his destination in time for the transit. Instead, he was forced to record his observations from the Cape of Good Hope.

In 1763, Mason and Jeremiah Dixon (1733-1779) were commissioned to survey the boundary-line between Pennsylvania and Maryland by the respective proprietors of these colonies. Within 36 miles of the entire distance to be determined, they had to suspend operations because of opposition by the Indians. The **Mason-Dixon Line** is a demarcation line between four U.S. states, forming part of the borders of Pennsylvania, Maryland, Delaware, and West Virginia.

Following the boundary survey in America, Mason returned to Greenwich, where he continued work on Mayer's lunar tables. He also contributed to the *Nautical Almanac*, working under Nevil Maskelyne (1732-1811), Astronomer Royal.

On September 27, 1786, Mason returned to Philadelphia with his wife, seven sons, and one daughter. He was very ill and confined to his bed. He died on October 26, 1787, and is buried at Christ Church Burial Ground in Philadelphia.

Mason crater on the Moon is named after him. Mason is one of the characters of Thomas Pynchon's 1997 novel *Mason & Dixon*. The song *Sailing to Philadelphia* from Mark Knopfler's album of the same name has strong references to Mason and Dixon.

1. Charles Mason, NNDB, www.nndb.com/people/528/000165033.
2. Charles Mason, Wikipedia, http://en.wikipedia.org/wiki/Charles_Mason.
3. Charles, Museum of History, www.famousamericans.net/charlesmason.

Johann Lambert (1728-1777)

Developed a theory of the generation of the universe

Johann Heinrich Lambert, a German mathematician, physicist, and astronomer, was born in Mülhausen (now Mulhouse, Alsace, France) on August 26, 1728. His father was a poor tailor, as his own father had been. When Johann was twelve years old he had to leave school to help his father. He continued to study in his spare time. At the age of 15, he worked as a clerk in an ironworks, and then gained a position in a newspaper office. In 1748, when he was 20 years old, the editor recommended him as a private tutor to a family, which gave him access to a good library and provided enough leisure time in which to explore it. He made his own astronomical instruments and delved deeply into mathematical and physical topics.

Lambert was the first to introduce hyperbolic functions into trigonometry, and in 1761 was credited with the first proof that π is irrational. He devised theorems regarding conic sections that made the calculation of the orbits of comets simpler. He also invented the first practical hygrometer and photometer. The **Lambert-Beer law** describes the way in which light is absorbed. In his *Cosmological Letters on the Arrangement of the Universe*, he coined the word "phenomenology." This signified the study of the way that objects appear to the human mind.

Lambert devised a formula for the relationship between the angles and the area of hyperbolic triangles. His first book, which was on the passage of light through various media, was published in The Hague in 1758. He also wrote a classic work on perspective, and contributed to geometrical optics.

In 1760, Leonhard Euler (1707-1783) recommended Lambert for the position of Professor of Astronomy at St. Petersburg Academy of Sciences to fill a vacancy which, due to a reorganization of the Academy and political changes, had remained unfilled for several years.

In 1761, Lambert hypothesized that the stars near the Sun were part of a group which traveled together through the Milky Way, and that there were many such groupings (star systems) throughout the galaxy. He also developed a theory of the generation of the universe that was similar to the nebular hypothesis of Immanuel Kant (1724–1804). Lambert published his own version of the nebular hypothesis of the origin of the Solar System. He died in Berlin, Prussia (today Germany) on September 25, 1777.

1. Johann Heinrich Lambert, MacTutor, www-groups.dcs.st-and.ac.uk/~history/Biographies/Lambert.
2. Johann Heinrich Lambert, Wikipedia, http://en.wikipedia.org/wiki/Johann_Heinrich_Lambert.

Charles Messier (1730-1817)

Invented comet hunting, a new discipline of astronomy

Charles Messier, a French astronomer, was born in Badonviller in the Lorraine region of France on June 26, 1730. In 1741, when Charles was 11 years old, his father died. Messier's interest in astronomy was stimulated by the appearance of a great six-tailed comet in 1744 when he was 14 years old, and by an annular solar eclipse visible from his hometown on July 25, 1748. In 1751, he went to Paris and entered the employ of Joseph Delisle (1688-1768), the astronomer of the French Navy. Messier was introduced to Delisle's observatory and instructed in using its instruments. His first job was copying a large map of China. His first documented astronomical observation was that of the transit of Mercury on May 6, 1753.

Messier discovered 12 comets and a 13th with his assistant Pierre Méchain (1744-1804). In 1774, Messier published his first astronomical catalog, *Catalogue des Nébuleuses et des Amas d'Étoiles* (Catalog of Nebulae and Star Clusters), consisting of deep sky objects, such as galaxies, planetary nebulae, open clusters, and globular clusters. The first version of his catalog contained 45 objects. By the time the final version of the catalog was published in 1781, the list of **Messier objects** had grown to 103. The purpose of the catalog was to help comet hunters and other astronomical observers to distinguish between permanent and transient objects in the sky.

Between 1721 and 1766, Messier discovered evidence of another seven deep-sky objects. These seven objects, M104 through M110, are accepted by many astronomers as official Messier objects. Because these objects were accessible to the relatively small aperture telescope used by Messier, they are among the most spectacular deep sky objects available to modern amateur astronomers using much better equipment.

Messier's work was interrupted by an accident, when he fell into the ice cellar about 25 feet deep. He was severely injured, and it took more than a year for him to recover.

He is credited with inventing comet hunting, a new discipline of astronomy. In 1815, he suffered a stroke, which left him partially paralyzed. He died on April 12, 1817. The **Messier crater** on the Moon and the **asteroid 7359 Messier** were named in his honor.

1. Charles Messier (June 26, 1730 - April 12, 1817), SEDS, www.seds.org/messier/xtra/history/biograph.
2. Charles Messier, Wikipedia, http://en.wikipedia.org/wiki/Charles_Messier.
3. Zander, Jon, The life and catalog of John Messier, Our Dark Skies, www.ourdarkskies.com/Messiers/cm_page1.

Benjamin Banneker (1731-1806)

America's first African-American scientist

Benjamin Banneker, an African-American mathematician and amateur astronomer, was born in Baltimore County, Maryland on November 9, 1731. He was the son of an African slave who had bought his freedom. Benjamin received little formal education, but enjoyed reading, and taught himself literature, history, and mathematics.

He worked as a tobacco planter for most of his life. In 1761, at the age of 30, he constructed a wooden clock, which operated successfully until the time of his death.

Banneker began his solo study of astronomy at age 58. The industrialist, George Ellicott, lent him several books on astronomy, as well as a telescope and drafting instruments. Without further guidance or assistance, Banneker taught himself the science of astronomy. He made calculations to predict solar and lunar eclipses and to compile an ephemeris for the *Benjamin Banneker's Almanac*, which was published from 1792 through 1797. An ephemeris is a table of values that gives the positions of astronomical objects in the sky at a given time or times. Because of the color of his skin he became known as the Sable Astronomer.

Banneker forwarded a manuscript copy of his calculations to Thomas Jefferson, then secretary of state, with a letter rebuking Jefferson for his proslavery views and urging the abolishment of slavery. Jefferson acknowledged Banneker's letter, and forwarded the manuscript to the Marquis de Condorcet (1743-1794), who was the secretary of the Académie des Sciences in Paris. The exchange of letters between Banneker and Jefferson was published as a separate pamphlet and given wide publicity at the time the first almanac was published.

As a result of Banneker's correspondence with Jefferson, in 1791 President George Washington appointed Banneker assistant to Major Andrew Ellicott as a part of the six-man team who surveyed the Territory of Columbia and planned Washington, D. C.

The last known issue of Banneker's almanacs appeared for the year 1797. However, he prepared ephemerides for each year until 1804. He died on October 9, 1806. In 1980, the U.S. Postal Service issued a postage stamp in his honor.

1. Encyclopedia of World Biography on Benjamin Banneker, BookRags, www.bookrags.com/biography/benjamin-banneker.
2. Who was Benjamin Banker?, The Banneker Center for Economic Justice, www.progress.org/banneker/bb.
3. World of Invention on Benjamin Banneker, BookRags, www.bookrags.com/biography/benjamin-banneker-woi.

David Rittenhouse (1732-1796)

Contributed to the best American calculations of solar parallax

David Rittenhouse, an American clockmaker, mathematician, and astronomer, was born in Roxborough, Pennsylvania on April 8, 1732. When he was 12 years old he inherited the mathematical library and tools of a deceased uncle, and without much instruction he learned to make clocks and scientific instruments, which became his occupation.

In 1763, Rittenhouse was commissioned by the Pennsylvania government to survey and determine the first part of what became the Mason-Dixon Line. In doing so, he used surveyor's instruments of his own making.

He built his own observatory at his father's farm in Norriton, outside of Philadelphia. He maintained detailed records of his observations and published a number of important works on astronomy, including a paper putting forth his solution for locating the place of a planet in its orbit. He calculated the transit of Venus in 1769, and contributed to the best American calculations of solar parallax. He later made a successful observation of Venus from his Norriton observatory.

Rittenhouse made orreries for what are now Princeton University and the University of Pennsylvania. The orreries show the solar and lunar eclipses and other astronomical phenomena for a period of 5,000 years, either forward or backward.

In 1770, Rittenhouse moved to Philadelphia, where he was elected to the provincial legislature in 1775. At the outbreak of the American Revolution he was a member of the Philadelphia Committee of Safety, becoming its president in 1776. He was also a member of the convention to form the Pennsylvania state constitution, and served as state treasurer from 1777 to 1789.

He was vice provost and professor of astronomy at the University of Pennsylvania from 1779 to 1782. He became the president of the American Philosophical Society in 1791, succeeding Benjamin Franklin (1706-1790).

Rittenhouse was the first director of the United States Mint, a capacity in which he served from 1792 to 1795.

He became one of the greatest American scientists of the 18th century, second only to Benjamin Franklin. He died on June 26, 1796.

1. David Rittenhouse (1732-1796), The University of Pennsylvania, www.archives.upenn.edu/histy/features/1700s/people/rittenhouse_david.
2. David Rittenhouse, Microsoft Encarta Online Encyclopedia, http://encarta.msn.com/encyclopedia_762512049/david_rittenhouse, 2008.
3. Encyclopedia of World Biography on David Rittenhousewww, BookRags, bookrags.com/biography/david-rittenhouse.

Jérôme Lalande (1732-1807)

Cataloged over 47,000 stars

Joseph Jérôme Lefrançais de Lalande, a French astronomer, was born at Bourg-en-Bresse (now in the département of Ain) on July 11, 1732. His parents sent him to Paris to study law, but as a result of lodging in the Hôtel Cluny where Joseph-Nicolas Delisle (1688-1768) had his observatory, he was drawn to astronomy.

Lalande's *Histoire Céleste Française* (1801) contained a catalog of over 47,000 stars, one of which, **Lalande 21185**, is now suspected of having extrasolar planets. His written works include the widely read *Traité D'astronomie* (1764) and *Bibliographie Astronomique* (1802).

In 1760, Lalande became professor of astronomy in the Collège de France, holding the post for 46 years. In 1768, he was appointed director of the Paris Observatory. He often included women in his work. He asked a well-known amateur astronomer, Nicole Lepaute (1723-1788), to work with noted astronomer Alexis Clairaut (1713-1765) on the prediction of the exact date of the next return of Halley's Comet.

Lepaute and Clairaut released their findings in September 1757. By Christmas of that year, the first sightings of the comet began. Their work was published in a paper by Clairaut, who initially gave full credit to Lepaute's efforts. Later, he retracted his statements and took full credit for himself. Lepaute's efforts, however, were recognized by Lalande, who included her in many other projects.

One of the first large-scale studies of lunar astronomy was undertaken at the Paris Observatory, and the chief investigator was an amateur, Mme du Pirey. Also, Lalande's niece by marriage, Marie de Lalande, (1768-1832) and her husband worked with Lalande. Women soon became known as "computers," and were eventually employed at observatories around the world.

Lalande's *Traite D'astronomie*, republished in three volumes under the title *Astronomie*, in 1792 includes passages that argue in favor of intelligent life throughout the universe. He reconciles his support for pluralism with Christian doctrine by arguing that the glory of God is magnified by the presence of life on other worlds. Later, however, he declared himself an atheist, stating "I have searched through the heavens, and nowhere have I found a trace of God." He died on April 4, 1807. **Lalande crater** on the Moon is named after him.

1. Jérôme Lalande Wikipedia, http://en.wikipedia.org/wiki/Joseph_J%C3%A9r%C3%B4me_Lefran%C3%A7ais_de_Lalande.
2. Lalande, Jérôme (1732-1807), The Internet Encyclopedia of Science, www.daviddarling.info/encyclopedia/L/LalandeJ.

Nevil Maskelyne (1732-1811)

Estimated the mean density of the Earth

Nevil Maskelyne, a British astronomer, was born in London on October 6, 1732. He was educated at Westminster School and Trinity College, Cambridge, where he graduated in 1754.

An ordained minister, he turned from the ministry to astronomy after observing a solar eclipse. His first contribution to astronomical literature was *A Proposal for Discovering the Annual Parallax of Sirius*, published in 1760.

The Royal Society sent him to British-controlled St. Helena, a small windswept island to the west of Africa, to observe the transit of Venus of June 6, 1761, but cloudy weather and faulty instruments prevented him from determining the solar parallax. A solar parallax is the angle subtended by the mean equatorial radius of the Earth at a distance of one astronomical unit (approximately 150 million kilometers).

In 1763, in the *British Mariner's Guide*, Maskelyne included the suggestion that to facilitate the finding of longitude at sea, lunar distances should be calculated beforehand for each year and published in a form accessible to navigators. In keeping with his previous suggestion, he induced the government to print his observations annually.

Subsequent volumes of his previous publication contained his observations of the transits of Venus (1761 and 1769), on the tides at St. Helena (1762), and on various astronomical phenomena at St. Helena and at Barbados (1764).

In 1765, he succeeded Nathaniel Bliss (1700-1764) as Astronomer Royal, a position he held from 1765 to 1811. Within the year, he had started the *Nautical Almanac*. It carried astronomical tables of the motions of the Sun, Moon, planets, and 36 important stars to help sailors navigate and determine longitude at sea.

While on St. Helena, Maskelyne calculated the gravitational constant relative to Greenwich by means of a seconds pendulum (two second period), and in 1774 he determined the absolute value of this constant from the deflection of a plumb line on the mountain of Schiehallion, Scotland. From this, he estimated the mean density of the Earth, which compares well with the modern computation.

Maskelyne died in Greenwich, London on February 9, 1811. He is buried in the churchyard of St. Mary the Virgin, Wiltshire, England.

1. Biography of Nevil Maskelyne, National Dictionary of Biography, Oxford, 1893.
2. Maskelyne, Nevil, Microsoft Encarta Online Encyclopedia, http://au.encarta.msn.com/encyclopedia_121503375/Maskelyne_Nevil, 2008.
3. Nevil Maskelyne, Wikipedia, http://en.wikipedia.org/wiki/Nevil_Maskelyne.

Charles Green (1735-1771)

Sent on an expedition to the Pacific Ocean to observe the transit of Venus

Charles Green, a British astronomer, was born in Yorkshire in 1735. He was the son of a farmer. Little is known about his childhood or education. In 1761, he was appointed assistant to the third Astronomer Royal, James Bradley (1693-1762), who was the successor to Edmund Halley (1656-1742) at the Royal Observatory, Greenwich. This position was described thus: "Nothing can exceed the tediousness of the life the assistant leads, excluded from all society, forlorn, he spends months in long wearisome computations."

In November 1763, Green and Nevil Maskelyne (1732-1811), who had observed the 1761 Transit of Venus in St. Helena, sailed for Barbados. Maskelyne had been appointed chaplain to HMS Princess Louisa and was to help Green with the astronomical work. Green had been instructed by the Board of Longitude to determine the longitude of the island by celestial observation in connection with the testing of the fourth marine chronometer of John Harrison (1693–1776).

Work completed, they returned home in the autumn of 1764, and on the death of Nathaniel Bliss (1700-1764) Maskelyne became Astronomer Royal, and Green was given a salary increase.

Green was assigned by the Royal Society in 1768 to the expedition sent to the Pacific Ocean to observe the transit of Venus aboard Captain James Cook's *Endeavour*. By the middle of the 18th Century the relative distances of the planets from the Sun were known, but the actual distances were still only estimates. If the distance of one planet from the Sun could be established accurately, the rest would follow. It was hoped that observation of the transit of Venus from places widely separated in latitude would provide this vital distance and so give scale to the Solar System.

In Tahiti, Green made observations and checked instruments ready for the Transit of Venus on June 3, 1769. The sky was cloudless, and they had every advantage they could desire in observing the whole passage of Venus over the Suns disk. Cook came to have a high regard for Green's ability as an astronomer and teacher of observational techniques.

Green died of a fever in Batavia on January 29, 1771 during the crew's voyage back to England.

1. Green, Charles (1734 - 1771), South Seas Corporation, http://southseas.nla.gov.au/biogs/P000377b.
2. Morris, Margaret, Man Without A Face - Charles Green, www.captaincooksociety.com/ccsu4143.
3. Paulding, Brenda, Charles Green Astronomer (1735-1771), www.captaincooksociety.com/ccsu4192.

Joseph Lagrange (1736-1813)

Studied the three-body problem for the Earth, Sun, and Moon

Joseph Louis Lagrange, although of French extraction, was born in Turin, Sardinia-Piedmont, Italy on January 25, 1736. The oldest of 11 children, only he and his youngest brother survived infancy. Lagrange was largely self-taught, and did not have the benefit of studying with leading mathematicians.

In 1754, at the age of 19, he was appointed professor of geometry in the Royal School of Artillery. The following year, he sent Leonhard Euler (1707-1783) a better solution than Euler's own for deriving the central equation in the calculus of variations. These solutions, and Lagrange's applications of them to celestial mechanics, were so monumental that by age 25 he was regarded by many of his contemporaries as the greatest living mathematician. In spite of his fame, Lagrange was always a shy and modest man.

With the aid of the Marquis de Saluces and the anatomist G. F. Cigna, Lagrange founded in 1758 a society which became the Turin Academy of Sciences.

In 1776, on the recommendation of Euler, Lagrange was chosen to succeed Euler as the director of the Berlin Academy. During his stay in Berlin, Euler's work covered many topics, including the stability of the Solar System, mechanics, dynamics, fluid mechanics, probability, the theory of numbers, and the foundations of calculus.

In astronomy, he studied the three-body problem for the Earth, Sun, and Moon (1764) and also for the movement of Jupiter's satellites (1766). In 1772, he found the special-case solutions to this problem that are now known as **Lagrangian points**.

In 1787, he moved to Paris. Napoleon was a great admirer of Lagrange, and showered him with honors—count, senator, and Legion of Honor.

One of Lagrange's most famous works is a memoir, *Mecanique Analytique*, in which he reduced the theory of mechanics to a few general formulas from which all other necessary equations could be derived.

Lagrange also established the theory of differential equations, and invented the method of solving differential equations known as variation of parameters. He died in Paris, France on April 10, 1813.

1. Joseph-Luis Lagrange, Biographies/Images of Physicists and Astronomers, www.mlahanas.de/Physics/Bios/JosephLouisLagrange.
2. Joseph-Luis Lagrange, Stetson, www.stetson.edu/~efriedma/periodictable/html/Lr.
3. Seikala, Nahla, Joseph-Louis Lagrange, Mathematicians, http://math.berkeley.edu/~robin/Lagrange.

Jean Bailly (1736-1793)

Wrote a five-volume history of astronomy

Jean Sylvain Bailly, a French astronomer, historian, and revolutionary leader, was born in Paris, France on September 15, 1736 into a family of minor courtiers. Originally intending to become a painter, he found that he preferred writing tragedies, until attracted to science by the influence of Nicolas de Lacaille (1713-1762). Bailly served as a president of the French National Assembly, and was the first mayor of Paris under the newly adopted system of the *Commune* after the storming of the Bastille. He is best known for his work on the history of astronomy.

In astronomy, he calculated an orbit for Halley's Comet when it appeared in 1759, and reduced Lacaille's observations of 515 zodiacal stars.

Bailly's *Essai sur la Theorie des Satellites de Jupiter* (Essay on the theory of the satellites of Jupiter, 1766) was an expansion of a memoir presented to the Academy in 1763. It was followed in 1771 by *Sur les Inegalites de la Lumiere des Satellites de Jupiter* (On the inequalities of light of the satellites of Jupiter).

He gained a high literary reputation for his writings on King Charles V of France, Nicolas de Lacaille, Jean-Baptiste Poquelin (stage name, Molière), Pierre Corneille, and Gottfried Leibniz, which were issued in collected form in 1770 and 1790. From then on, he devoted himself to the history of science.

Bailly was admitted to the Académie Française in 1784 and to the Académie des Inscriptions in 1785.

His writings on the history of science include *Histoire de L'astronomie Ancienne* (A history of ancient astronomy, 1775), *Histoire de L'astronomie Moderne* (A history of modern astronomy, 3 volumes, 1779-1782), *Lettres sur L'origine des Sciences* (Letters on the origin of the sciences, 1777), *Lettres sur L' Atlantide de Platon* (Letters on Plato's Atlantide, 1779), and *Traite de L'astronomie Indienne et Orientale* (A treatise on Indian and Oriental astronomy, 1787).

The dispersal by the National Guard under his orders to quell a riotous assembly in the Champ de Mars infuriated the populace and made him very unpopular. He was arrested and brought before the Revolutionary Tribunal at Paris on November 10, 1793. Two days later, on November 12, he was guillotined amid the insults of a howling mob.

1. Jean Sylvain Bailly, Wikipedia, http://en.wikipedia.org/wiki/Jean-Sylvain_Bailly.
2. Jean-Sylvain Bailly, Microsoft Encarta Online Encyclopedia, http://encarta.msn.com/encyclopedia_762510746/jean-sylvain_bailly, 2008.
3. Jean-Sylvain Bailly, NNDB, www.nndb.com/people/217/000100914.

William Herschel (1738-1822)

Discovered Uranus and its two largest moons

William Herschel was born in Hanover, Germany on November 15, 1738. At the time, the city belonged to England under the rule of George II. Herschel's father was a musician in the Hanoverian army, and William trained in music to enter the same profession. In 1757, he went to England, where he began working as an organist and music teacher. There, he became interested in astronomy.

Unable to find a telescope of a high enough resolution, he decided to grind his own lenses and to design his own instruments. With his first telescope completed (a 6-foot Gregorian reflector), he decided to conduct a systematic survey of the stars and planets.

Herschel's first major discovery took place in 1781—the existence of a new planet. Herschel named the planet Georgium Sidus (George's star) in honor of King George III. It eventually came to be known as Uranus, after the mythical father of Saturn. The discovery of Uranus caused a popular and scientific sensation, and George III appointed Herschel to the position of King's Astronomer, providing him with a small annuity that allowed him to pursue astronomy full time.

In 1783, he began to search for nebulae, and raised their known total from little more than 100 to 2,500. His work on nebulae led him to conclude that they might well be other Solar Systems in the universe.

Herschel also devoted much effort to measuring stellar distances, an essential element of determining the true size of the universe. He surveyed the sky for double stars, producing three catalogs over the next 40 years that listed 848 examples.

In addition to his work on stars, Herschel devoted significant effort to the Solar System. He studied the Sun, deducing that what we see is not a solid surface, but the diffuse, gaseous solar atmosphere. He also examined the nature of the infrared part of the spectrum in which the Sun radiates a significant portion of its energy. He then studied the Moon and the other planets, including Venus, Mars, Jupiter, and Saturn.

Herschel allowed too often his preconceived ideas about the presence of extraterrestrial life to color his observations or the inferences he drew from them. He died on August 25, 1822.

1. Encyclopedia of World Biography on William Herschel, Sir, BookRags, www.bookrags.com/biography/william-herschel-sir.
2. Friedrich Wilhelm (William) Herschel (November 15, 1738 - August 25, 1822), The Messier Catalog, www.seds.org/messier/xtra/Bios/wherschel.
3. World of Scientific Discovery on William Herschel, BookRags, www.bookrags.com/biography/william-herschel-wsd.

Pierre Méchain (1744-1804)

Discovered many new deep sky objects

Pierre François André Méchain, a French astronomer and surveyor, was born in Laon, France on August 16, 1744. His father was a ceiling designer and plasterer. Pierre studied math and physics, but due to financial difficulties left college. For some time, he worked as a tutor for two young boys, about 50 kilometers from Paris.

Méchain's talents in astronomy were noticed by Jérôme Lalande (1732-1807), for whom he became a proofreader. In 1774, Lalande secured a position for him with the Naval Depot of Maps at Versailles, where he worked through the 1770s, engaged in hydrographic work and coastline surveying. Hydrography is the measurement of physical characteristics of waters and marginal land. It was during this time that he met Charles Messier (1730-1817), who worked for the same department, but at the small observatory at Hôtel de Clugny.

His first astronomical work was on an occultation of Aldebaran (the brightest star in the constellation Taurus) by the Moon, which he presented as a memoir to the Academy of Sciences.

Méchain worked on maps of Northern Italy and Germany, but his most important mapping work was the determination of the southern part of an arc of the Earth's surface between Dunkirk and Barcelona, which he began in 1791. This measurement eventually became the basis of the metric system's unit of length, the meter.

Between 1779 and 1782 he discovered 30 deep sky objects, 26 of which were original findings. He communicated his observations to Messier, who usually checked their positions and added them to his catalog. Méchain also discovered five comets.

Méchain was interned in Barcelona after war broke out between France and Spain, and his property in Paris was confiscated during The Reign of Terror. When released, he went to live in Italy. He returned to France in 1795, and in 1799 became the director of the Paris Observatory.

Continuing doubts about his measurements of the Dunkirk-Barcelona arc led him to return to that work. This took him back to Spain in 1804, where he caught yellow fever and died in Castellon de la Plana on September 20, 1804. **Asteroid 21785 Méchain** was named for him.

1. O'Connor, J. J. and E. F Robertson, MacTutor, www-history.mcs.st-andrews.ac.uk/Biographies/Mechain.
2. Pierre François André Méchain (August 16, 1744 - September 20, 1804), SEDS, seds.lpl.arizona.edu/messier/xtra/history/pmechain.
3. Pierre François André Méchain, Wikipedia, http://en.wikipedia.org/wiki/Pierre_M%C3%A9chain.

Johann Schröter (1745-1816)

Produced the most detailed study of the Moon up to that time

Johann Hieronymus Schröter, a German astronomer, was born in Erfurt, Germany on August 30, 1745. He studied law at Göttingen University, during which time he developed an extracurricular interest in mathematics and astronomy. After graduation, he practiced law for 10 years.

Schröter's astronomical leanings were rekindled when he met William Herschel (1738-1822) in England, who was destined to become the foremost astronomer of the 18th century.

In 1772, Schröter set up a private observatory at Lilienthal, near Bremen. Financially secure through his appointment as district governor and judge, he carried out planetary and solar observations almost continuously for the rest of his life. His Lilienthal observatory was for a time home to the largest telescope in continental Europe.

Schröter was primarily interested in solar and planetary astronomy. He was the first, in 1787, to notice and comment on the solar surface feature now known as granulation. He also gave detailed descriptions of light bridges over the umbrae of sunspots.

He carried out many observations of Venus, and tried to determine its rotation period. His two volumes on lunar topography reached levels of detail that were to remain for many years unsurpassed.

In 1800, Schröter presided over the founding of the first astronomical society, with members distributed all over Europe. One of the foremost goals of the society was to organize a systematic observing program to detect a planet between the orbits of Mars and Jupiter.

In 1813, during the Napoleonic invasion of Northern Germany, retreating French imperial troops destroyed a good part of the town. Although Schröter's observatory and instruments were left untouched, many of his observational notebooks and unpublished scientific manuscripts were destroyed with the Lilienthal town hall.

Schröter was a pluralist who was convinced that every celestial body may be so arranged physically by the Almighty as to be filled with living creatures. He allowed too often his preconceived ideas about the presence of extraterrestrial life to color his observations or the inferences he drew from them.

He died on August 29, 1816.

1. Johann Hieronymus Schroeter (1745-1816), High Altitude Observatory, www.hao.ucar.edu/Public/education/bios/schroeter.
2. Johann Hieronymus Schröter, Wikipedia, http://en.wikipedia.org/wiki/Johann_Schr%C3%B6ter.
3. Schröter, Johann Hieronymous (1745-1816), The Internet Encyclopedia of Science, www.daviddarling.info/encyclopedia/S/Schroter.

Giuseppe Piazzi (1746-1826)

Discovered the dwarf planet Ceres in the Asteroid Belt

Giuseppe Piazzi, an Italian Astronomer, was born at Ponte in Valtellina, Italy in July 7, 1746. He became a Theatine monk, and then a professor of theology in Rome in 1779. He became professor of mathematics at the Academy of Palermo in 1780.

He taught philosophy for a time at Genoa and mathematics at the new University of Malta. In 1780, he assumed the chair of higher mathematics at the academy of Palermo. There, he soon obtained a grant from Prince Caramanico, Viceroy of Sicily, to build an observatory. To gain experience in Astronomy, he went to Paris in 1787 to study with Jérôme Lalande (1732-1807) and then to England in 1788 to work with Nevil Maskelyne (1732-1811) and the famous instrument-maker Jesse Ramsden (1735-1800).

Piazzi began observations at the new observatory in May 1791, and published the first reports in 1792. Soon, he began correcting errors in the estimation of the obliquity of the ecliptic, of the aberration of light, of the length of the tropical year, and of the parallax of the fixed stars.

In 1803, he published a list of 6784 stars, and in 1814 he published a second catalog containing 7646 stars. Both lists were awarded prizes by the Institute of France.

While looking for a small star mentioned in one of the earlier lists, Piazzi, on January 1, 1801, discovered the first known planetoid. Repeating the observation several nights in succession, he found that this object had shifted slightly. Believing it to be a comet, he announced its discovery as such.

Piazzi's measurements enabled Karl Gauss (1777-1855) to calculate the orbit and to determine that the object was actually a dwarf planet between Mars and Jupiter. Piazzi proposed the name of "Ceres Ferdinandea" in honor of his king. Over 600 of these so-called planetoids have since been located within the same space, the Asteroid Belt.

Piazzi also discovered that the star, 61 Cygni, has a large proper motion (change in position in the sky over time), which led Friedrich Bessel (1784-1846) to choose it as the object of his parallax studies.

Piazzi died at Naples on July 22, 1826. In 1923, the 1000th asteroid to be numbered was named **1000 Piazza** in his honor. More recently, a large feature on Ceres was named **Piazzi**.

1. Giuseppe Piazzi, New Advent, www.newadvent.org/cathen/12072d.
2. Giuseppe Piazzi, Wikipedia, http://en.wikipedia.org/wiki/Giuseppe_Piazzi.
3. Piazzi, Giuseppe (1746-1826), The Internet Encyclopedia of Science, www.daviddarling.info/encyclopedia/P/Piazzi.

Johann Bode (1747-1826)

Popularized an empirical mathematical rule giving the relative mean distances between the Sun and planets

Johann Elert Bode, a German astronomer, was born in Hamburg, Germany on January 19, 1747. As a youth, he suffered from an eye disease, and continued to have trouble with his eyes throughout his life. He never attended formal school, but was educated by his father, who was a merchant.

Bode was hired by the Berlin Academy of Sciences as a calculator. He also became director of the Berlin Observatory. He co-founded the German language ephemeris called the *Astronomisches Jahrbuch oder Ephemeriden* (Astronomical Yearbook and Ephemeris) in 1774. An ephemeris is a device that calculates the positions of astronomical objects.

He began to watch the night sky, looking for star clusters and nebulae. He found 20 of these in one year, three of which were new discoveries. In 1774, he discovered M81 and M82, and in 1775 he discovered M53. He also discovered an asterism (a cluster of stars smaller than a constellation), and made several other discoveries from 1777 to 1779. One of these was a comet.

Bode named the planet Uranus after the mythical father of Saturn, even though William Herschel (1738-1822) actually discovered the planet. Herschel had originally named the planet Georgium Sidus (George's Star) in honor of King George III.

He took great interest in Uranus. He collected various accounts of observations of the planet by astronomers worldwide and published them in *Astronomisches Jahrbuch*. Among others who witnessed the planet were John Flamsteed (1646-1719) in 1690 and Tobias Mayer (1723-1762) in 1756. In 1801, Bode published a popular star atlas called *Uranographia*, which contained several unique star formations.

Bode is best known for **Bode's law**, also called the **Titus-Bode law,** which is a mathematical rule about the relationship between the mean distances of each of the planets to the Sun. Johann Titus (1729-1796) discovered the rule 1766, but he was not given much credit at the time. Bode reformulated the law and made it popular in 1772 by publishing it.

A year after he retired as director of the Berlin Observatory, Bode died in Berlin on November 23, 1826.

1. Bode, Johann Elert (1747-1826), The Internet Encyclopedia of Science, www.daviddarling.info/encyclopedia/B/Bode.
2. Johann Elert Bode (January 19, 1747 - November 23, 1826), SEDS, www.seds.org/Messier/xtra/Bios/bode.
3. The Johann Bode Biography, Astronomy for the Kids Online, www.astronomy-for-kids-online.com/johann-bode-biography.

Dominique Cassini (1748-1845)

Helped test a new marine chronometer for determining longitude at sea

Dominique Cassini (Cassini IV), a French astronomer and cartographer, was born on June 30, 1748 in the Paris Observatory where his father, César Cassini (1714-1784), worked. Dominique was the grandson of Jacques Cassini (1677-1756) and the great-grandson of Giovanni Cassini (1625-1712), all astronomers.

Dominique received his early education in the Paris Observatory, and then attended the Collège du Plessis in Paris and the Collège Oratorien run by the Congregation of the Oratory at Juilly. Not wishing to become a priest, he studied physics, mathematics, and astronomy.

Cassini sailed on a scientific voyage in 1768, given the task of testing a new marine chronometer invented by Pierre Le Roy (1717-1785). If successful, the chronometer would allow ships to determine their longitude while at sea. The voyage took him to America, then to the coast of Africa, and finally back to home. In 1770, he published an account of his voyage in *Voyage Fait Par Ordre du Roi en 1768 Pour Éprouver les Montres Marines Inventées par M Le Roy*.

After the death of his father in 1784, Cassini assumed his role of director of the Paris Observatory. His first task was to complete the project his father had worked on for many years—a map of France. In 1790, Cassini presented the completed map to the National Assembly.

While he was completing his father's map, Cassini worked on another surveying project. In 1787, he was involved in a joint project with English scientists to determine the precise distance between the observatories at Greenwich and Paris. This would allow extremely useful scientific results to be obtained by combining data from the two observatories.

Because his position became intolerable due to the animosity of the National Assembly, Cassini resigned in 1793. He was thrown into prison in 1794, but was released after seven months. He then withdrew to Thury.

His *Mémoires Pour Servir à L'histoire de L'observatoire de Paris* (1810) included portions of an extensive work, the prospectus of which he had submitted to the Academy of Sciences in 1774. The volume included his eloges of several academicians and the autobiography of his great-grandfather, Giovanni Cassini.

Cassini died in Thury on October 18, 1845.

1. Jacques Dominique Cassini, NNDB, http://www.nndb.com/people/262/000095974.
2. Jean-Dominique, comte de Cassini, BookRags, www.bookrags.com/wik/ Dominique%2C_comte_de_Cassini.
3. O'Connor, J. J. and E. F. Robertson, Jean-Dominique Comte de Cassini, MacTutor, www-groups.dcs.st-and.ac.uk/~history/ Biographies/Cassini_Dominique.

Pierre Laplace (1749-1827)

Explained the long-term variations in the orbital speeds of Jupiter, Saturn, and the Moon

Pierre Simon Laplace, the son of a farmer, was born at Beaumont-en-Auge in Normandy on March 23, 1749. Little is known of his early life. Through the efforts of Jean D'Alembert (1717-1783), Laplace obtained a professorship of mathematics at the military school in Paris. His work would become pivotal to the development of mathematical astronomy.

In 1773, Laplace communicated the first of many memoirs to the Academy of Sciences in Paris. This contribution involved particular solutions of differential equations and the mean motions of the planets.

In 1787, he completed his demonstration of the stability of the Solar System, which had been in question. Over a period from 1799 to 1825 he published in five volumes his great work, the *Traité de la Mécanique Céleste*. His other major work, *Théorie Analytique des Probabilités*, was published in 1812.

Laplace discovered that two lunar perturbations were caused by the spheroidal shape of the Earth. He determined the solar distance from observations of the Moon, the masses of the satellites of Jupiter, and the period of revolution of the rings of Saturn.

In his *Exposition du système du monde*, he suggested that the Solar System had evolved from a quantity of incandescent gas. As the gas cooled, it contracted, and successive rings detached from the outer edge. These in turn cooled and condensed into planets, the Sun remaining as the central core.

Laplace derived the dynamical equations for the motion of the oceans caused by the attraction of the Sun and the Moon in a memoir of 1775. He developed a method of approximation for definite integrals with integrands containing factors of high powers, a type of equation frequently occurring in probability theory.

He formulated **Laplace's equation**, and invented the Laplace **transform.** The **Laplacian differential operator** is named after him.

Laplace is remembered as one of the greatest scientists of all time. He became a count of the First French Empire in 1806, and was named a marquis in 1817 after the Bourbon Restoration. He died in Paris on March 5, 1827.

1. Encyclopedia of World Biography on Laplace, Marquis de, BookRags, www.bookrags.com/biography/laplace-marquis-de.
2. World of Mathematics on Pierre Simon Laplace, BookRags, www.bookrags.com/biography/pierre-simon-laplace-wom.
3. World of Scientific Discovery on Pierre Simon Laplace, BookRags, www.bookrags.com/biography/pierre-simon-laplace-wsd.

Jean Delambre (1749-1822)

Calculated a table of the motions of Uranus

Jean Baptiste Joseph Delambre, a French mathematician and astronomer, was born at Amiens, France on September 19, 1749. Due to a childhood illness, he suffered from poor eyesight during his life. This caused him to score poorly on an examination, so he failed to gain a university scholarship. Since his parents were financially unable to send him to college, he remained in Paris and took a position as a tutor to the son of a nobleman in Compiègne. Delambre had to teach himself mathematics so that he could teach his pupil. As a result, he developed exceptional calculating skills.

Delambre's interest in Greek astronomy led him to learn about modern astronomy, and in about 1780 he read *Traité D'astronomie* by Jérôme Lalande (1732-1807). He began attending Lalande's astronomy lectures at the Collège de France, and became his best student. In 1783, Lalande hired him to help make observations for a new edition of *Traité D'astronomie.*

Lalande lent Delambre some equipment. The observational data he collected with it was incorporated into the third edition of *Traité D'astronomie*, which appeared in print in 1792.

In 1786, Delambre recorded a transit of Mercury across the Sun. At the time the transit was predicted, a cloud obscured the sky. Most astronomers gave up, but not Delambre, who did not believe the time predicted for the transit by Lalande was correct. When the cloud cleared 40 minutes after the transit was predicted to end, Delambre observed the transit taking place.

After the French Academy of Science announced that the Grand Prize for 1789 would be for the calculation of the precise orbit of Uranus, Delambre won the prize. From 1792 to 1799, he was occupied with the measurement of the arc of the meridian extending from Dunkirk to Barcelona, and published a detailed account of the operations in *Base du Système Métrique.*

In 1801, Napoléon Bonaparte took the presidency of the Academy of Sciences and appointed Delambre its Permanent Secretary for the Mathematical Sciences. In 1804, he was appointed director of the Paris Observatory. He was also professor of Astronomy at the Collège de France. He died in Paris on August 19, 1822. He was honored by having the **Delambre crater** on the Moon named for him.

1. Jean Baptiste Joseph Delambre, Wikipedia, http://en.wikipedia.org/wiki/Delambre.
2. Jean-Baptiste-Joseph Delambre NNDB, www.nndb.com/people/404/000097113.
3. O'Connor, J. J. and E. F. Robertson, Jean-Baptiste-Joseph Delambre's, MacTutor, www-groups.dcs.st-and.ac.uk/~history/Biographies/Delambre.

Caroline Herschel (1750-1848)

The first important woman astronomer

Caroline Lucretia Herschel was born in Hanover, Germany on March 16, 1750. Her father was a musician in the Hanoverian Guard. She did not receive a formal education. At the age of 17, her father died and her brother, William, who was 11 years her senior, invited her to live with him in Bath, England, where he was immersed in musical training and astronomy.

William increasingly needed her efficient, meticulous talents in copying his astronomy catalogs, tables, and papers. She also assisted him in grinding and polishing his telescope lenses. In 1781, William discovered Uranus. For this, he was appointed to the position of court astronomer to George III and was knighted. Caroline was appointed his assistant and given an annual stipend of 50 pounds. Herschel's appointment made her the first female in England honored with a government position.

She systematically collected data, trained herself in geometry, and learned formulas and logarithmic tables. She gained an understanding of the relationship of sidereal time (time measured by means of the stars) to solar time. Her record keeping was meticulous and systematic. The volume of her work was enormous.

When she was not engaged in other tasks, she too searched the night skies. In early 1783, she discovered the Andromeda and Cetus nebulae. By year's end, she had discovered 14 additional nebulae. She was also the first woman to discover a comet, and between 1789 and 1797 she discovered another seven comets.

Herschel calculated and cataloged nearly 2,500 nebulae. She also undertook the task of reorganizing the British catalog of John Flamsteed (1646-1719), which listed nearly 3,000 stars. Herschel's listings were divided into one-degree zones to allow William to use a more systematic method of searching the skies.

William died in 1822, and she spent the last years of her life in Hanover, organizing and cataloging the works of William's son, Sir John Herschel, who carried on his father's extensive work.

In 1828, the Royal Astronomical Society awarded Herschel a gold medal for her monumental works in science. Herschel never married. On January 9, 1848, she died at the age of 97.

1. Caroline Herschel, Wikipedia, http://en.wikipedia.org/wiki/Caroline_Herschel.
2. Encyclopedia of World Biography on Caroline Herschel, BookRags, www.bookrags.com/biography/caroline-hersche.
3. Herschel, Caroline Lucretia (1750-1848), BookRags, www.bookrags.com/research/herschel-caroline-lucretia-1750-184-woes-01.

Heinrich Olbers (1758-1840)

Originated the first satisfactory method for calculating the orbits of comets

Heinrich Wilhelm Matthäus Olbers, a German physicist, Physician, and astronomer, was born in Arbergen, near Bremen, Germany on October 11, 1758. His father was a minister. Heinrich studied to be a physician at Göttingen, and also pursued a mathematical course. After his graduation in 1780, he began practicing medicine in Bremen. At night, he dedicated his time to astronomical observation, making the upper storey of his home into an observatory. He reportedly only slept four hours a night so he could spend most of the night studying the skies.

In 1802, Olbers discovered and named the asteroid, Pallas. Five years later, in 1807, he discovered the asteroid, Vesta, which he allowed Carl Gauss (1777-1855) to name. Olbers proposed incorrectly that the asteroid belt, where these objects lay, was the remnants of a planet that had been destroyed. In 1815, Olbers discovered a periodic comet now named after him.

Olbers is probably best known for devising the first satisfactory method of calculating comet orbits. The treatise containing this important work was made public by Franz von Zach (1753-1832) under the title *Ueber die Leichteste und Bequemste Methode die Bahn eines Cometen zu Berechnen* in 1797. Olbers method is still used today.

He was chosen by his fellow-citizens to assist at the baptism of Napoleon II of France on June 9, 1811. He was a member of the corps législatif in Paris from 1812 to 1813. The Corps législatif was a part of the French legislature during the French Revolution and beyond.

Olbers' paradox, described by him in 1823 and reformulated in 1826 by Johann Bode (1747-1826), is the argument that the darkness of the night sky conflicts with the supposition of an infinite and eternal static universe. The explanation is that our universe is finite, both in time and place, and the total amount of matter and energy is far too small to light up the night sky.

Olbers died in Bremen, Germany at the age of 81 on March 2, 1840. The following celestial features are named for him: Periodic comet **13P/Olbers**, **asteroid 1002 Olbersia**, **Olbers crater** on the Moon, and **Olbers** (a 200 km-diameter dark albedo feature on Vesta's surface).

1. Heinrich Wilhelm Matthäus Olbers, Wikipedia, http://en.wikipedia.org/wiki/Heinrich_Wilhelm_Olbers.
2. Heinrich Wilhelm Olbers Biography - German physician, amateur astronomer, Anatomy for Kids Online, www.astronomy-for-kids-online.com/heinrich-wilhelm-olbers-biography.
3. Olbers' Paradox, Wikipedia, http://en.wikipedia.org/wiki/Olbers%27_paradox.
4. Wilhelm Olbers, NNDB, www.nndb.com/people/047/000102738.

Jean Pons (1761-1831)

The greatest comet discoverer of them all

Jean-Louis Pons was born into a poor family in Peyre, France on December 24, 1761. He received very little education. In 1789, at the age of 28, he became a porter and doorkeeper at the Marseille Observatory. There, he received some instruction on astronomy by the observers.

Pons learned fast, and soon he was allowed to do observations with the instruments. His favorite was a telescope with a three-degree-wide field of view. He had an extraordinary ability to remember the star fields he observed and to recognize changes in them.

Pons' discovered his first comet on July 11, 1801. Charles Messier (1730-1817) found it a day later. Pons's first discovery was Messier's last. From 1801 to 1827, Pons found a new comet almost every year.

His discovery of a comet in the early morning of February 9, 1808 was only confirmed recently, when it was shown to be the periodic comet, Grigg-Skjellerup, which returns every five years.

Pons's work was rewarded when he was promoted to Assistant Astronomer in 1813 and to Assistant Director in 1818. He discovered three comets that year. In 1819, he became director of a new observatory at Lucca in northern Italy.

Johann Encke (1791-1865) followed a suggestion by Pons, who suspected one of the three comets to be one already discovered by Encke in 1805, and calculated the elements of the orbit. The comet was found to have a period of 3.3 years, and Encke predicted it to return in 1822. This return was documented as predicted by Karl Ruemker (1788-1862). The comet is known today as Encke's Comet, but Encke himself always referred to it as Pons's Comet. Pons discovered two new comets in 1822.

Pons moved to the Florence Observatory in 1822. In 1825, the Grand Duke of Tuscany, Leopold II, invited him to become the director of the Florence Observatory. In Florence, he discovered seven more comets, the last in August 1827. Today, 26 comets bear his name, but he discovered or co-discovered up to 37.

From 1827 on, Pons's eyesight faded, and he had to stop observing in the first months of 1831. He died in Florence on October 14, 1831. Two comets, **7P/Pons-Winnecke** and **12P/Pons-Brooks**, bear his name. He was also honored by having a lunar crater named after him.

1. Jean-Louis Pons, Wikipedia, http://en.wikipedia.org/wiki/Jean-Louis_Pons.
2. Plicht Chris, Pons, Jean-Louis (1761-1831), www.plicht.de/chris/27pons.
3. Pons, Jean Louis (1761-1831), The Internet Encyclopedia of Science, www.daviddarling.info/encyclopedia/P/Pons.

John Goodricke (1764-1786)

Did important early work on variable stars

John Goodricke, a deaf amateur astronomer, was born in Groningen in the Netherlands on September 17, 1764. His father was British and his mother Dutch. He lived most of his life in England. He was named after his grandfather, Sir John Goodricke (1708-1789), the 5th Bart.

At an early age, John contracted scarlet fever and as a result became profoundly deaf. He never fully developed the power of speech, although he learned to read lips. Despite his handicap, his parents gave him a good education by sending him as a child to Thomas Braidwood's Academy, a school for the deaf in Edinburgh, and in 1778 to the Warrington Academy.

With the encouragement of his friend, Edward Pigott (1753-1825), Goodricke pursued astronomy. After observing ß Perseus, also known as Algol or the Demon star, he realized that the star was variable. Its variability, he suggested, might be due to it being an eclipsing binary. More than a 100 years later, astronomical instruments were developed sufficiently to confirm this. He presented a paper to the Royal Society regarding his observations of Algol, and in 1783, at the age of 19, received the Godfrey Copley medal from the society.

Goodricke was the first astronomer to discover the variable stars, beta Lyrae (an eclipsing binary) and delta Cephei (the first Cepheid). A Cepheid star is a star that expands and contracts. This finding would allow Edwin Hubble (1889-1953) to determine the distance of M31, the Andromeda galaxy

Goodricke looked for other variable stars, and found Sheliak, also known as beta Lyrae, in about 1783. Sheliak is a double star system with two giant stars close together so that they are constantly deforming each other and exchanging mass. He calculated the period of Sheliak to be 12 days and 20 hours.

He also discovered the variable star, Altais, better known as delta Cepheï. He calculated the period of this star to be 128 hours and 45 minutes with outstanding correctness.

Goodricke died on April 20, 1786 at the age of 22 after a bout with pneumonia. Shortly afterward, he was posthumously awarded membership of the Royal Society. **Goodricke College** at the University of York is named after him.

1. Goodricke, John (1764-1786), The Internet Encyclopedia of Science, www.daviddarling.info/encyclopedia/G/Goodricke.
2. John Goodricke and the Demon Star, Doncaster Astronomical Society, http://donastro.blogspot.com/ 2007/10/john-goodricke-and-demon-star.
3. John Goodricke, Wikipedia, http://en.wikipedia.org/wiki/John_Goodricke.
4. John Goodricke, www.surveyor.in-berlin.de/himmel/Bios/Goodricke-e.

William Wollaston (1766-1828)

Observed the dark Fraunhofer lines in the solar spectrum

William Hyde Wollaston, a British physician, chemist and physicist, was born in East Dereham, Norfolk, England on August 6, 1766. He was the son of priest-astronomer, Francis Wollaston (1737-1815). In 1793, William obtained a doctorate in medicine from Cambridge University. During his studies there he became interested in chemistry, crystallography, metallurgy, and physics.

In 1801, he left medicine to concentrate on his other interests. He is best known as a chemist. He developed the first physico-chemical method for processing platinum ore in practical quantities, and in the process of testing the device discovered the elements palladium in 1803 and rhodium in 1804.

Wollaston also performed important work in electricity. In 1801, he showed that electricity from friction was identical to that produced by voltaic piles. During the last years of his life he performed electrical experiments that paved the way to the eventual design of the electric motor.

Wollaston's claim to astronomical fame rests on his observations in 1802 of dark Fraunhofer lines in the solar spectrum, which eventually led to the discovery of the elements in the Sun. He noticed these while carrying out optical experiments aimed at determining refractive indices of various transparent substances. He did not attach great importance to his discovery, leaving it to Joseph von Fraunhofer (1787-1826) to rediscover and study the lines in great details 15 years later.

Wollaston invented the camera lucida (1807), the reflecting goniometer (1809), and the **Wollaston prism**. He also developed the **Wollaston's meniscus lens** in 1812. The lens was designed to improve the image projected by the camera obscura. By changing the shape of the lens, he was able to project a flatter image, eliminating much of the distortion that was a problem with many of that day's biconvex lenses.

Wollaston also served on a royal commission in 1819 that opposed adoption of the metric system, and one that created the imperial gallon.

He died in London on December 22, 1828, and was buried in Chislehurst, England. Also named for him are the **Wollaston Medal**, **Wollastonite** (a chain silicate mineral), **Wollaston crater**, and **Wollaston Lake** in Saskatchewan, Canada.

1. William Hyde Wollaston, Institute of Chemistry, http://chem.ch.huji.ac.il/history/wollaston.
2. William Hyde Wollaston, Wikipedia, http://en.wikipedia.org/wiki/William_Hyde_Wollaston.
3. Wollaston, William Hyde, The Internet Encyclopedia of Science, www.daviddarling.info/encyclopedia/W/Wollaston.

Alexis Bouvard (1767-1843)

Noticed that Uranus moved in a way that could not be completely explained as a result of the gravitational pulls of the then-known planets

Alexis Bouvard, a French astronomer, was born in Les Contamines-Montjoie Hamlet, France on June 27, 1767. He came from a poor background; in fact, he was born in a hut, and grew up learning the skills of a shepherd boy. There was little opportunity for schooling, so he was self taught. At age 18, he went to live in Paris, where he received mathematics lessons so that he could earn his living as a computer (a person who carried out numerical calculations).

Although he had little money to pay for an education, the Collège de France offered free courses, and Bouvard took these. He became fascinated by astronomy after a visit to the Paris Observatory. He realized that astronomy was an area in which his calculating skills could be particularly useful.

In 1794, Bouvard met Pierre Laplace (1749-1827), who was at that time working on *Méchanique Céleste*. Laplace recognized Bouvard's computing skills and hired him to carry out the complex calculations required for his theory. He then arranged for him to be offered a position in the Bureau de Longitudes in 1794, where he remained the rest of his life.

Bouvard's achievements include the discovery of eight comets and the compilation of astronomical tables of Jupiter, Saturn, and Uranus. While the former two tables were accurate, the latter showed substantial discrepancies with subsequent observations. To explain this, he hypothesized the existence of an eighth planet in the Solar System. The position of this planet, Neptune, was subsequently calculated from Bouvard's observations independently by John Adams (1819- 1892) and Urbain Le Verrier (1811-1877) after Bouvard's death.

In 1793, Bouvard became director of the Paris Observatory, following Dominique Cassini (1748-1845). He continued as a member of the Bureau of Longitudes. He was elected a member of the Academy of Science in 1803 and foreign member of the Royal Society in 1826.

He died in Paris on June 7, 1843. In Australia, a cape known as **Cape Bouvard** was named after him. **Bouvard** is also the name of a small Australian city south of Perth on the coast.

1. Alexis Bouvard, Wikipedia, http://en.wikipedia.org/wiki/Alexis_Bouvard.
2. Bouvard, Alexis (1767-1843), The Internet Encyclopedia of Science, www.daviddarling.info/encyclopedia/B/Bouvard.
3. O'Connor, J. J. and E. F. Robertson Alexis Bouvard, MacTutor, www-history.mcs.st-and.ac.uk/Printonly/Bouvard.

Franz Gruithuisen (1774-1852)

The first to suggest that craters on the Moon were caused by meteorite impacts

Franz von Paula Gruithuisen, a Bavarian physician and astronomer, was born on March 19, 1774. He taught medical students before becoming a professor of astronomy at the University of Munich in 1826.

During his period of medical studies and instruction, he was noted for his contributions to urology and lithotrity (process of crushing a calculus in the bladder into very small pieces so that it can be eliminated in the urine). He published his historic study on the subject in 1813 in the *Journal of Medicine and Surgery*. He developed crucial ideas on how to remove bladder stones trans-urethrally in a way that was safer and less likely to cause death. His instruments are regarded as the model on which subsequent devices are based.

In astronomy, Gruithuisen argued in favor of advanced life on the Moon and inner planets. On the subject of a lunar civilization, he supported the claims of Johann Schröter (1745-1816). He made multiple observations of the lunar surface and believed they supported his ideas. He even announced the discovery of a city in the rough terrain to the north of Schröter crater. This region contains a series of somewhat linear ridges that have a fishbone-like pattern, and, with the small refracting telescope he was using, could be perceived as resembling buildings complete with streets.

He wrote such papers as *Discovery of Many Distinct Traces of Lunar Inhabitants, Especially of One of Their Colossal Buildings* (1824), claiming to have seen roads, cities, and a star-shaped temple.

He proposed that jungles on Venus grew more rapidly than in Brazil due to the proximity of the planet to the Sun, and that as a consequence the inhabitants celebrated fire festivals, which were the cause of the bright caps on Venus. Such extraordinary inferences made him the object of ridicule by fellow astronomers, since his claims were readily refuted by those using more powerful instruments.

He was the first to suggest that craters on the Moon were caused by meteorite impacts instead of volcanic activity as was widely believed up until the 20th century. He died in 1852. The **Gruithuisen crater** on the Moon is named for him.

1. Franz von Gruithuisen, Wikipedia, http://en.wikipedia.org/wiki/Franz_von_Gruithuisen.
2. Gruithuisen, Franz von Paula (1774-1852), The Internet Encyclopedia of Science, www.daviddarling.info/encyclopedia/G/Gruithuisen.
3. Zajaczkowski, T, A.M. Zamann , and P. Rathert, Franz von Paula Gruithuisen (1774-1852): lithotrity pioneer and astronomer, World Journal of Urology, 2003 May;20(6):367-73.

Francis Baily (1774-1844)

Noted that as the Moon passes by the Sun during a solar eclipse the rugged lunar topography allows beads of sunlight to shine through in some places and not in others

Francis Baily, a British astronomer, was born at Newbury in Berkshire, England on April 28, 1774. His father was a banker. Francis sailed for America on October 21, 1796. The narrative of his experiences as a traveler is contained in *Journal of a Tour in Unsettled Parts of North America in 1796 and 1797.*

Baily entered the London Stock Exchange in 1799. The publication of *Tables for the Purchasing and Renewing of Leases* (1802), *The Doctrine of Interest and Annuities* (1808), and *The Doctrine of Life-Annuities and Assurances* (1810) earned him a high reputation as a writer on life-contingencies.

Having become interested in astronomy, he took a leading role in the foundation of the Royal Astronomical Society in 1820. For his preparation of the Astronomical Society's catalog of 2881 stars (*Memoirs Royal Astronomical Society II*), the Society awarded him its gold medal in 1827.

After amassing a fortune, Baily retired from business in 1825 to devote himself to astronomy. He initiated the reform of the *Nautical Almanac* in 1829, and he recommended to the British Association in 1837 the reduction of the catalogs of Jérôme Lalande (1732-1807) and Nicolas de Lacaille (1713-1762), which contained about 57,000 stars.

Baily superintended the compilation of the British Association's catalog of 8377 stars (published in 1845), and revised the catalogs of Tobias Mayer (1723-1762), Claudius Ptolemy(c.87-c.161), Ulugh Beg (c.1393-1449), Tycho Brahe (1546-1601), Edmund Halley (1656-1742), and Johannes Hevelius (1611-1687) in *Memoirs Royal Astronomical Society IV, XIII.*

On May 15, 1836, at Inch Bonney in Roxburghshire, he observed that as the Moon passed by the Sun during a solar eclipse, the rugged lunar topography allowed beads of sunlight to shine through in some places and not in others. The phenomenon, which depends upon the irregular shape of the moon's limb, was vividly described by him. This phenomenon has been named **Baily beads** in his honor.

Baily never married. He died in London on August 30, 1844. The lunar crater **Baily** was named in his honor.

1. Baily's Beads, Wikipedia, http://en.wikipedia.org/wiki/Baily's_beads.
2. Francis Baily (1774-1844), David Nash Ford's Royal Berkshire History, www.berkshirehistory.com/ bios/fbaily.

Mary Somerville (1780-1872)

One of the first two females to become honorary members of the Royal Astronomical Society

Mary Somerville, a Scottish mathematician, physical chemist, and astronomer, was born in Jedburgh, Scotland on December 26, 1780. Her maiden name was Fairfax. Her father was a vice admiral in the British Navy. Her education was spotty. At the age of 14 she became interested in algebra, and read *Elements of Geometry* by the Greek mathematician Euclid (c.325-c.270 BC).

In 1804, she married her cousin, Captain Samuel Greig, a member of the Russian navy. The marriage interrupted her studies because Samuel did not approve of women scholars. When he died suddenly in 1807, she resumed her reading of mathematics and science. In 1812, she married another cousin, Dr. William Somerville, inspector of the Army Medical Board, who encouraged her scientific interests. She mastered *Astronomy* by James Ferguson (1710-1776) and became a student of *Principia* by Isaac Newton (1643-1727).

When she and William moved to London in 1816, she became a part of the intellectual circles of that city. In 1826, she presented *The Magnetic Properties of the Violet Rays of the Solar Spectrum* to the Royal Society. The paper established her reputation as a scientist.

In 1831, she translated a complex treatise on physical astronomy, *Traité de mécanique céleste*, by the French astronomer Pierre Laplace (1749-1827. She published a separate volume the following year. Both books became standard classroom texts.

Her later publications included *On the Connection of the Physical Sciences* (1834), *Physical Geography* (1848), and her last work *On Molecular and Microscopic Science* (1869).

In 1834, the Royal Astronomical Society made her and German-born British astronomer Caroline Herschel (1750-1848) its first female honorary members.

In 1879, the University of Oxford named one of its first two colleges for women after her. Oxford also established the **Mary Somerville scholarship** for women in mathematics. Somerville spent the last 34 years of her life in Italy, where she and her husband moved for his health. She died at Naples on November 28, 1872.

1. Mary Somerville, Microsoft Encarta Online Encyclopedia, http://encarta.msn.com/encyclopedia_761582696/mary_somerville, 2008.
2. O'Connor, J. J. and E. F. Robertson, Mary Fairfax Greig Somerville, MacTutor, www-groups.dcs.st-and.ac.uk/~history/Biographies/Somerville.
3. Wood, Shane, Mary Fairfax Somerville, Biographies of Women Mathematicians, www.agnesscott.edu/lriddle/women/somer, April 1995.

Friedrich Bessel (1784-1846)

Obtained the distance of a star by parallax

Friedrich Wilhelm Bessel, a German mathematician and astronomer, was born in Minden, Germany on July 22, 1784. He became an accountant in Bremen, but his true interests were astronomy and mathematics. In 1806, at the age of 20, he recalculated the orbit of Halley's Comet, which was due to reappear in 1835. This so impressed astronomer Heinrich Olbers (1758-1840) that he helped Bessel obtain a post at the observatory.

Bessel produced a new star catalog of over 50,000 stars, and introduced improvements to astronomical calculations. He oversaw construction of Königsberg Observatory, and served as its director from 1813 until his death.

In 1838, Bessel announced he had obtained the parallax for a star, 61 Cygni. As the Earth orbits the Sun, our position, relative to any star, shifts by a maximum of 186 million miles—the diameter of the Earth's orbit. Thus, the apparent position of a star changes slightly during the year. The amount of observed shift is the parallax. Knowing an object's parallax, it is possible to calculate the distance to it from the Earth. Bessel had chosen Cygni because the star was relatively close, and the closer an object the greater its parallax. Bessel's calculations showed 61 Cygni to be about 10 light-years from Earth. This was the first time the term light-year was used. Up till that time, stars were believed to be much closer to the Earth. Since parallax could be obtained only from a moving Earth, Nicholas Copernicus's assertion that the Earth orbited around the Sun was further strengthened.

In 1841, Bessel noticed that displacements of the motion of the stars Sirius and Procyon could not be attributed to parallax—they wobbled. He concluded that there had to be invisible companions in orbit around each star. The gravitational tug of the companion would account for the observed wobble. He was later proven correct.

Bessel argued against the existence of life on the Moon at a time when several notable compatriots of his believed it possible. In particular, he noted the sharpness with which stars are occulted by the Moon, indicating the absence of a lunar atmosphere. He died in Königsberg on March 17, 1846. The largest crater in the Moon's Mare Serenitatis was named after him, as are **Bessel functions** in mathematics.

1. Bessel, Friedrich Wilhelm (1784-1846), The Internet Encyclopedia of Science, www.daviddarling.info/ encyclopedia/B/Bessel.
2. Encyclopedia of World Biography on Friedrich Wilhelm Bessel, BookRags, www.bookrags.com/ biography/friedrich-wilhelm-bessel.
3. World of Mathematics on Friedrich Wilhelm Bessel, BookRags, www.bookrags.com/biography/ friedrich-wilhelm-bessel-wom.

François Arago (1786-1853)

Determined that the Sun's limb is gaseous

François Jean Dominique Arago, a French mathematician, physicist, and astronomer, was born in Estagel, a small village near Perpignan, in the département of Pyrénées-Orientales, Catalan France on February 26, 1786. His father was the mayor of their small town and was eventually named cashier of the French Mint. Françoise was educated at the École Polytechnique in Paris.

In 1806, he traveled to southern Europe and North Africa to survey land to determine the arc of the Earth's meridian. Because the region where he was surveying was intolerant of the French, he spent a brief time as a prisoner in the French colony, Algiers. On his return to France in 1809, he was given a professorship of analytical geometry and geodesy at the École Polytechnique, and was elected to the French Academy of Sciences. He was allowed to lecture in the astronomy department as well. In his later years, he became director of the Paris Observatory.

During the early 1800s, scientists subscribed to one of two major theories concerning the nature of light—a stream of particles or a wave. In 1815, Arago began research with Augustin-Jean Fresnel (1788-1827) into the phenomenon of polarization. By the end of their work, the two scientists had ascertained the principles governing the polarization of light, and therefore shown that light behaved like a wave.

Arago worked on electromagnetism, showing that a coil of wire carrying a current could act as a magnet. He also found that a rotating copper disk could deflect a magnetic needle suspended above it.

During the solar eclipse of 1842, he examined polarized light from the chromosphere and corona and determined that the Sun's limb is gaseous. He proved the relationship between the aurora and variations in terrestrial magnetism, and suggested that his student Urbain Leverrier (1811-1877) investigate irregularities in the orbit of Uranus. After Neptune was discovered, he took part in the discussion regarding naming the planet.

Arago died in Paris on October 2, 1853. He was directly responsible for stimulating the scientific interests of a generation of French students, many of whom went on to attain prominence. Craters on Mars and the Moon and a ring of Neptune, are named after him, as well as the study association for Applied Physics at the University of Twente.

1. Arago, Dominique François Jean (1786-1853), The Internet Encyclopedia of Science, www.daviddarling.info/encyclopedia/A/Arago.
2. Dominique François Jean Arago, Institute of Chemistry, http://chem.ch.huji.ac.il/history/arago.
3. World of Scientific Discovery on Dominique-François-Jean Arago, BookRags, www.bookrags.com/ biography/dominique-francois-jean-arago-wsd

Joseph von Fraunhofer (1787-1826)

Founded stellar spectroscopy

Joseph von Fraunhofer, a German instrument maker and inventer, was born in Straubing, Bavaria on March 6, 1787. He became an orphan at the age of 12, and began working as an apprentice to a harsh glassmaker. When the workshop physically collapsed, Joseph was rescued by an operation led by Maximilian IV, who then provided Joseph with books and forced his employer to allow him time to study.

After eight months of studying, Fraunhofer went to work at the Optical Institute at Benediktbeuern, a Benedictine monastery devoted to glass making. There, he learned how to make the world's finest optical glass and invented very precise methods for measuring dispersion. In 1818, he became the director of the Institute. He instituted many improvements in the manufacture of optical glass, the grinding and polishing of lenses, and the construction of telescopes and other optical instruments.

In 1814, Fraunhofer invented the spectroscope, which is an optical instrument for producing spectral lines and measuring their wavelengths and intensities. Using his new invention, he discovered 574 dark lines appearing in the solar spectrum. These were later shown to be atomic absorption lines. The lines are still called **Fraunhofer lines** in his honor.

Using the spectroscope, Fraunhofer found that the spectra of Sirius and other first-magnitude stars differed from each other and from the Sun, thus founding stellar spectroscopy. He labeled the most prominent spectral lines with letters, establishing a nomenclature that is still used to this day.

Fraunhofer also invented the diffraction grating, comprised of 260 close parallel wires. He used his diffraction grating to accurately measure wavelengths of specific colors and dark lines in the solar spectrum. In doing so, he transformed spectroscopy from a qualitative art to a quantitative science. **Fraunhofer diffraction** is a form of wave diffraction which occurs when field waves are passed through an aperture or slit, causing only the size of an observed aperture image to change.

In 1817, Fraunhofer designed an achromatic objective lens. With minor modifications, his design is still in use today. In 1823, he was appointed director of the Physics Museum in Munich, and received the honorary title of Professor from the University of Erlangen. He died of tuberculosis in Munich, Germany on June 7, 1826 at the age of 39.

1. Fox, William, Catholic Encyclopedia, www.newadvent.org/cathen/06250a.
2. Fraunhofer, Joseph von, Microsoft Encarta Online Encyclopedia, http://uk.encarta.msn.com/encyclopedia_761575866/fraunhofer_joseph_von, 2008.
3. Joseph von Fraunhofer (1787-1826), High Altitude Observatory, www.hao.ucar.edu/Public/education/bios/ fraunhofer.

Fearon Fallows (1789-1831)

Built the first astronomical observatory in the southern hemisphere

Fearon Fallows, a British astronomer, was born in Cockermouth in Cumbria, England on July 4, 1789. His father was a weaver who was of a scholarly disposition and appreciated Fearon's quick intelligence and aptitude for learning.

Because of the generosity of the townspeople, Fearon was given the funds to attend St. John's College, Cambridge, where he studied mathematics, graduating in 1813. He obtained his M.A. in 1816, and went on to teach mathematics at Corpus Christi College, Cambridge. He also became an ordained priest in the Church of England.

On February 29, 1820 he was elected a fellow of the Royal Astronomical Society, and that same year he was appointed by the Admiralty to be the astronomer at the Cape of Good Hope. This involved picking a site and overseeing the construction of an observatory in what was then a British colony.

Before traveling to South Africa, Fallows married Mary Anne Hervey on January 1, 1821. He then worked on planning and developing the observatory, which would be the first astronomical observatory in the southern hemisphere.

The expedition embarked on May 4, 1821. It consisted of Fallows and Mary Anne, Sarah Bootle (their maid), assistant astronomer James Frayer, (who was also an instrument maker), and Frayer's sister Betsy.

When Fallows arrived, he only had two portable instruments and a clock. The instruments were a Circle and a Transit Instrument. When the observatory was built, he used a Jones Mural Circle and a Dolland Transit Circle. He cataloged over 300 stars from the observatory. The Royal Society published his *Catalogue of 273 Stars* in 1824.

He also served the Church of England in his time in South Africa. Fallows and all the observatory staff caught scarlet fever in 1830, and Fallows died of the illness in Simon's Town, South Africa on July 25, 1831 at the age of 43. He was buried in the Observatory grounds.

Mary Anne returned to England on September 13, 1831 and presented the Admiralty with all of Fallows papers and unpublished observations. The Observations were reduced by Royal Astronomer George Airy (1801-1892), and published in the *Memoirs of the Royal Astronomical Society* in 1851.

1. Atkinson, Stuart, Fearon Fallows, www.fallows.com/fearon_fallows.
2. Fallows, Fearon [MA; FRS] Reverend, Astronomical Society of South Africa, www.saao.ac.za/assa/html/his-astr-fallows_f.
3. Fearon Fallows, Wikipedia, http://en.wikipedia.org/wiki/Fearon_Fallows.

William Bond (1789-1859)

Co-discovered Saturn's moon, Hyperion

William Cranch Bond, an American astronomer, was born in Falmouth, Maine (now Portland) on September 9, 1789. When William was young, his father established himself as a clockmaker after a failed business venture. William received little formal education, but his father trained him in clock making. Aided by his penchant for engineering, William built his first clock when he was 15 years old. He eventually took over his father's business, becoming an expert clockmaker himself.

In 1806, when he was 17 years old, Bond saw a solar eclipse. Soon, he became an avid amateur astronomer. When he built his first house, he made its parlor an observatory, complete with an opening in the ceiling out of which his telescope could view the sky.

In 1815, Bond traveled to Europe, commissioned by Harvard University to gather information on European observatories. In 1839, He was allowed to move his personal astronomical equipment to Harvard and serve as its unpaid astronomical observer.

In 1843, a Sun-grazing comet aroused enough public interest in astronomy that Harvard was able to raise $25,730 towards the construction of a state-of-the-art observatory. Bond designed the building and the observing chair, and Harvard bought a 15-inch German-built refracting telescope, equal in size to the largest in the world at the time. The telescope was first put to use on June 24, 1847, when it was pointed to the Moon.

Bond and his son, George Bond (1825-1865), discovered Saturn's moon Hyperion. It was independently co-discovered at the same time by William Lassell (1799-1880) in Britain. Bond and son were the first to observe the then innermost ring of Saturn, termed the Crepe ring, when they pointed Harvard's telescope towards Saturn in 1850.

Working with John Whipple (1822-1891) the Bonds pioneered astro-photography, taking in 1850 the first daguerreo-type image of a star (Vega) ever taken from America. In all, the threesome took between 200 and 300 photos of celestial objects.

Bond died on January 29, 1859. A number of celestial objects were named in his honor. A few of them include **W. Bond crater** on the Moon, the **Bond-Lassell Dorsum** on Hyperion, and the **Bondia asteroid**.

1. Bond, William Cranch, The Columbia Encyclopedia, Sixth Edition,/www.bartleby.com/65/bo/Bond-Wil.
2. William Cranch Bond, Microsoft Encarta Online Encyclopedia, encarta.msn.com/encyclopedia_761589193/Bond_William_Cranch, 2008.
3. William Cranch Bond, Wikipedia, http://en.wikipedia.org/wiki/William_Cranch_Bond.

Samuel Schwabe (1789-1875)

Discovered that sunspots follow a roughly decade-long cycle

Samuel Heinrich Schwabe, a German pharmacist, botanist, and amateur astronomer, was born at Dessau, near Berlin, Germany on October 25, 1789. He began pharmaceutical studies at Berlin University, and in the course of which he became interested in astronomy and botany.

He returned to Dessau in 1812 to take over his family's pharmacy business, while pursuing astronomical and botanical researches on the side.

Schwabe obtained his first telescope in 1825, the result of a lottery win. By the following year, he had bought a more powerful telescope.

He originally aimed his observational work at discovering a possible planet orbiting within the orbit of Mercury, known as Vulcan. The problem was that the Sun had sunspots and these could easily be mistaken for the elusive planet. He decided to plot these sunspots so that he would not confuse them with the planet he was searching for.

Schwabe observed the Sun virtually every day that the weather permitted, and did so continuously for 42 years. In the process, he accumulated volumes of sunspot drawings.

Becoming increasingly absorbed with his astronomy work, he sold the family business in 1829.

Although not discovering a new planet, his 17 years of nearly continuous sunspot observations from 1826 to 1843 revealed a 10-year periodicity in the number of sunspots visible on the solar disk. This means that every 10 years the number of sunspots reaches its maximum, after which it starts to gradually decline.

Schwabe published his results in the journal *Astronomische Nachrichten*, but it attracted little attention until 1851 when his sunspot data was included by Alexander von Humboldt (1769-1859) in his *Kosmos, Volume III*, and used, together with other data, by Max Wolf (1863-1932) in 1857 to deduce a cycle of about 11 years.

In 1857, Schwabe was awarded the Gold Medal of the Royal Astronomical Society, and in 1868 he was elected to the Royal Society.

Schwabe is also credited with the first description and drawing, in 1831, of Jupiter's Great Red Spot. His work paved the way for future investigations in the fields of magnetism, weather, and organism growth rates. He died on April 11, 1875

1. Samuel Heinrich Schwabe (1789 - 1875), History of Science and Technology, The Open Door Web Site, www.saburchill.com/HOS/astronomy/026.
2. Schwabe, (Samuel) Heinrich (1789-1875), The Internet Encyclopedia of Science, www.daviddarling.info/ encyclopedia/S/Schwabe.

August Möbius (1790-1868)

Published two influential books on astronomy

August Ferdinand Möbius, a German mathematician and astronomer, was born in Schulpforta, Saxony-Anhalt, Germany on November 17, 1790. He was the only child of a dancing teacher and his wife. In 1809, August started classes at Leipzig University. At first, he believed he wanted to study law, but quickly turned to mathematics, astronomy, and physics.

In 1813, Möbius left Leipzig for Göttingen, having been chosen for a traveling fellowship. There, he spent half a year studying theoretical astronomy with Carl Gauss (1777-1855). The following year he received his doctorate and returned to Leipzig to serve as a junior astronomy professor. In 1816, having avoided an attempt to draft him into the Prussian Army, he also became an observer at the school's observatory. From 1818 to 1821, Möbius presided over the renovation and refurbishment of the outdated facility.

Over the next decade, he devoted himself to astronomy, publishing several important works on occultation phenomena, the path of Halley's Comet, and the basic laws of astronomy. *The Principles of Astronomy* came out in 1836. His most influential book on astronomy, however, was *The Elements of Celestial Mechanics* (1843).

In 1827, Möbius published *The Calculus of Centers of Gravity*, which introduced homogeneous coordinates into analytic geometry and discussed projective geometry. It was in this work that he first mentioned the configuration now known as the **Möbius net**.

During the 1830s, Möbius spent most of his time writing about statics, a branch of mechanics. *His Handbook on Statics*, which appeared in 1837, gave the area of a geometric treatment, which led to the study of systems of lines in space and the null system of planes and points.

He was made a full professor in 1844, and in 1848 the school made him director of the observatory. In 1858, Möbius entered a contest put on by the Paris Academy of Sciences for research on the geometrical theory of polyhedrons. His discovery came to be known as the **Möbius strip**. A rectangular, flat strip with a half twist, its ends connect to create a continuous, single-edged loop. Theoretically, it is a two-dimensional surface with only one side, but it can be constructed in three dimensions. Möbius died in Leipzig on September 26, 1868.

1. August Ferdinand Möbius, MacTutor, www-groups.dcs.st-and.ac.uk/~history/Biographie s/Mobius, January 1977.
2. August Ferdinand Möbius, Stetson, www.stetson.edu/~efriedma/periodictable/html/Mo.
3. World of Mathematics on August (Augustus) Ferdinand Möbius, BookRags, www.bookrags.com/biography/august-augustus-ferdinand-mobius-wom.

Johann Encke (1791-1865)

Developed the mathematical formula for calculating the orbits of short periodic comets

Johann Franz Encke, a German astronomer, was born in Hamburg on September 23, 1791. He studied mathematics and astronomy at Göttingen University with Carl Gauss (1777-1855). In 1812, he took a teaching job in Kassel, but in 1913 enlisted in the Hanseatic Legion for the liberation war against Napoleon. He became lieutenant of artillery in 1815. Afterward he became an observer at the observatory on the Seeberg.

At the end of 1818, Jean Pons (1761-1831) observed a weak comet, which had been seen earlier in 1786 by Pierre Mechain (1744-1804) and in 1795 by Caroline Herschel (1750-1848). In January 1819, Encke calculated the orbital time of the comet to be 3.3 years. Until then, all known comets had an orbiting time of 70 years or more, with the aphelion (point on the orbit of a celestial body that is farthest from the sun) far beyond the Uranian orbit. Halley's Comet, for instance, had an orbital time of about 76 years. The 3.3-orbit comet caused a sensation—the aphelion had to be within the Jovian orbit.

Encke sent his calculations as a note to Gauss, who published it, and Encke became famous as the discoverer of the short periodic comets. The comet was named **2P/Encke** after him.

In 1822, Encke became director of the observatory at Gotha, but in 1825 he was called to Berlin by the Prussian king, who ordered Encke to plan a new observatory in Berlin. The building was inaugurated in 1835, and Encke became director of this new observatory.

Subsequently, Encke was involved in the discovery and orbital parameter determination of other short periodic comets and planetoids. In 1837, he discovered a gap within the A ring of Saturn, now known as the **Encke gap**.

In 1838, his assistant, Johann Galle (1812-1910), discovered the dark, inner C ring of Saturn. In 1846, Galle also discovered the last big planet of the Solar System, Neptune, based on calculations by Urbain Leverrier (1811-1877). In 1849, Encke developed a method of determining an elliptic orbit from three observations.

Encke remained director of the Berlin observatory until his death in Spandau, Germany on August 26, 1865. Also named for him are **Encke crater** on the Moon and **asteroid 9134 Encke**.

1. Johann Franz Encke, NNDB, www.nndb.com/people/825/000095540.
2. Johann Franz Encke, The Discovery of the Short Periodic Comets, www.surveyor.in-berlin.de/himmel/Bios/Encke-e.
3. Johann Franz Encke, Wikipedia, http://en.wikipedia.org/wiki/Johann_Franz_Encke.

John Herschel (1792-1871)

Made observations of the stars in the southern hemisphere

John Frederick William Herschel, a British Astronomer, was born at Slough, England on March 7, 1792. His father was astronomer William Herschel (1738-1822), who had discovered Uranus. John was trained in mathematics at Cambridge, where he graduated in 1813. That same year, after publishing a mathematics paper, *On a Remarkable Application of Cotes's Theorem*, he was elected a fellow of the Royal Society of London.

Soon after graduation, Herschel and two classmates composed a textbook on calculus, which was aimed at introducing into England the more powerful mathematical methods that had been developed on the Continent during the preceding century.

Not until 1820 did he turn seriously to astronomy. After serving an apprenticeship to his father, he spent two years observing double stars. He published a catalog of double stars in the *Transactions of the Royal Society* in 1824.

After 1823, Herschel served as secretary of the Royal Society of London and president of the Royal Astronomical Society. He was knighted in 1831. The Paris Academy awarded him its Lalande Prize in 1825 and the Astronomical Society awarded him its Gold Medal the following year.

He did a systematic follow-up and extension of his father's surveys of double stars and nebulae. By 1833, he had finished with the northern hemisphere. He, therefore, moved his family to South Africa at the Cape of Good Hope for four years of observation of the southern skies. A chief feature in this endeavor was Herschel's inauguration of photometry, the precise measurement of stellar brightness. The survey, the first systematic observation of the sky in the southern hemisphere, yielded 2,307 nebulae and 2,102 double stars, which he listed, together with his father's discoveries, in the *General Catalogue of Nebulae and Clusters*

Returning to England in 1838, Herschel began research in chemistry. By 1839, he had invented methods of producing images on paper and glass, rather than metal, and had introduced "positive" and "negative" in the photographic context.

Herschel died in Collingwood, Kent, England on May 11, 1871, and was buried in West-minster Abbey. **Asteroid 2000 Herschel** was named after him.

1. Encyclopedia of World Biography on John Frederick William Herschel, Sir, BookRags, www.bookrags.com/biography/john-frederick-william-herschel-sir.
2. Herschel, John Frederick William (1792-1871), The Internet Encyclopedia of Science, www.daviddarling.info/encyclopedia/H/HerschelJ.
3. O'Connor, J. J. and E. F. Robertson, John Frederick William Herschel, MacTutor, www-history.mcs.st-and.ac.uk/Biographies/Herschel.

Friedrich Struve (1793-1864)

Extensively studied double stars

Friedrich Georg Wilhelm Struve, a German astronomer, was born in Altona (then in Denmark) on April 15, 1793. In 1808, he fled to Dorpat (now Tartu) in Estonia to avoid conscription into the German army. There, he began to attend the University of Dorpat where his brother, Karl, was a professor of philology, which is the science of the structure and development of languages. Friedrich studied philology also, and received his degree in 1810.

Influenced by physicist Georg Parrot (1767-1852), Struve became interested in sciences, especially astronomy. He began observing astronomical phenomena at the Dorpat Observatory in 1812. He was later appointed professor of mathematics and astronomy there.

In 1824, the observatory obtained the Fraunhofer equatorial telescope, which had a 9.6-inch achromatic objective lens, the largest of its time.

Struve's primary interest was binary stars. He published his discoveries on these in *Catalogus Novus* in 1827, *Mensurae Micrometricae* in 1837, and *Positiones Mediae* in 1852. His observations of double stars proved that Newton's law of gravitation also applies outside the Solar System.

In 1830, Czar Nicholas I funded an observatory near St. Petersburg, Russia. The observatory opened in 1839 with Struve as the director. The observatory had the best equipment in Europe. There, Struve continued his observations of double stars. Binary stars orbit around each other's center of mass and change position over time. Struve made measurements of these tiny position changes.

From 1824 to 1837, he measured over 2,714 double stars, and published his data in *Stellarum Duplicium et Dultiplicium Densurae Dicrometricae.*

Although Friedrich Bessel (1784-1846) was the first to measure a star's parallax, Struve was the first to measure Vega's parallax.

Between 1816 and 1855, he did geodetic surveying, trying to determine the exact size and shape of the Earth.

Because of failing health, Struve had to retire in 1861. He died in St. Petersburg on November 23, 1864. His son, Otto von Struve (1819-1905), was the director of the Pulkovo Observatory from 1858 to 1899.

1. Friedrich Georg Wilhelm Struve, Astronomy, http://pvastro0714.blogspot.com/2008/01/friedrich-georg-wilhelm-struve, January 11, 2008.
2. Struve, Friedrich Georg Wilhelm von (1793-1864), The Internet Encyclopedia of Science, www.daviddarling.info/encyclopedia/S/Struve_FGW.
3. Struve, The Columbia Encyclopedia, Sixth Edition, www.bartleby.com/65/st/Struve.

Johann von Mädler (1794-1874)

Co-published the most complete contemporary map of the Moon

Johann Heinrich von Mädler, a German astronomer, was born in Berlin on May 29, 1794. His interest in astronomy was sparked when he saw a comet as a child. He was orphaned at age 19 by an outbreak of typhus, and found himself responsible for raising three younger sisters. He studied astronomy and mathematics at the University of Berlin. While giving academic lessons as a private tutor in 1824, he met Wilhelm Beer, a wealthy banker.

In 1829, Beer decided to set up a private observatory and hired Mädler in 1830. They began producing drawings of Mars, which became the first true maps of that planet. They were the first to choose the Sinus Meridiani as the prime meridian for Mars maps. Sinus Meridiani is an albedo feature on Mars stretching east-west just south of that planet's equator. An albedo feature is a large area on the surface of a planet which shows a contrast in brightness or darkness with adjacent areas. Mädler and Beer made a preliminary determination for Mars's rotation period, which was off by almost 13 seconds. A later determination in 1837 was much improved, off by only 1.1 seconds.

They also produced the first exact map of the Moon, *Mappa Selenographica*, published in four volumes from 1834 to 1836. In 1837, they published a description of the Moon, *Der Mond Nach Seinen Kosmischen und Individuellen Verhältnissen*. Both were the best descriptions of the Moon for many decades, not superseded until the map of Johann Schmidt (1825-1884) in the 1870s. Beer and Mädler came to the conclusion that the features on the Moon do not change, and there is no atmosphere or water.

In 1836, Johann Encke (1791-1865) appointed Mädler an observer at the Berlin Observatory. He stayed there until 1840, when he was appointed director of the Dorpat Observatory in Estonia. There, he carried out meteorological as well as astronomical observations. He made calculations concerning the true length of the tropical year, with precision never attained before. In 1865, he retired and returned to Germany. He published many scientific works, among them the two-volume *History of Descriptive Astronomy* in 1873.

Mädler died on March 14, 1874. **Mädler crater** on the Moon and **Mädler crater** on Mars are named in his honor.

1. Hohann Madler, Megan's Astronomy Blog, http://pvastro0703.blogspot.com/2008/01/johann-madler, January 8, 2008.
2. Johann Heinrich von Mädler, Wikipedia, http://en.wikipedia.org/wiki/Johann_Heinrich_M%C3%A4dler.

Peter Hansen (1795-1874)

Contributed to the field of gravitational astronomy

Peter Andreas Hansen, a Danish astronomer, was born at Tønder, in the duchy of Schleswig Schleswig, Denmark on December 8, 1795. The son of a goldsmith, he learned the trade of a watchmaker at Flensburg, and practiced it at Berlin and Tønder from 1818 to 1820.

In 1820, he went to Copenhagen, where in 1821 he became the assistant of Heinrich Schumacher (1780-1850) at the new observatory of Altona. In 1925, he went to Gotha as director of the Seeberg observatory.

Hansen was primarily interested in the problems of gravitational astronomy. Research into the mutual perturbations of Jupiter and Saturn won him the prize of the Berlin Academy in 1830, and a memoir on comet disturbances was published by the Paris Academy in 1850.

In 1838, he published a revision of the lunar theory, entitled *Fundamenta Nova investigationis*, and in 1857 he published the improved *Tables of the Moon.* A theoretical discussion of the errors embodied in the previous tables was published in the *Abhandlungen* of the Saxon Academy of Sciences in 1862 to 1864.

Hansen twice visited England and was twice (1842 and 1860) awarded the gold medal of the Royal Astronomical Society. He presented to the Society in 1847 a paper on a long-period lunar inequality (*Memoirs Royal Astronomical Society, XVI, 465*), and in 1854 a paper on the Moon's shape, pointing out the mistaken hypothesis of its deformation by a huge elevation directed towards the Earth (*Memoirs Royal Astronomical Society XXIV, 29*).

He was awarded the Copley medal by the Royal Society in 1850. His *Solar Tables* appeared in 1854. He stated that the accepted distance of the Sun was too great by some millions of miles (*Notices Royal Astronomical Society, XV, 9)*, the error in the result of Johann Franz Encke (1791-1865), having become evident through Hansen's investigation of a lunar inequality.

In 1856, Hansen proposed a theory of the Moon, which included the possibility of an atmosphere and even of life on the far side. Speculation about life on the far side played a large role in Jules Verne's Moon-voyage fiction about the Moon.

Hansen supervised the construction of a new observatory at Gotha in 1857. He died on March 28, 1874 at the Gotha Observatory.

1. Beck, Daniel A., Life on the moon? A short history of the Hansen hypothesis, Annals of Science, Volume 41, Issue 5 September 1984 , pages 463 - 470
2. Peter Andreas Hansen, NNDB, www.nndb.com/people/029/000101723.

Wilhelm Beer (1797-1850)

Co-produced maps of the Moon and Mars

Wilhelm Wolff Beer, a banker and amateur astronomer, was born in Berlin, Prussia on January 4, 1797. He was the brother of Giacomo Meyerbeer (1791-1864), who was born Jakob Liebmann Beer, and poet Michael Beer (1800-1833). The four grand operas composed for Paris by Meyerbeer set a style that dominated the French lyric theater and exerted a powerful influence on opera production throughout Europe for a generation afterward. Michael Beer's most famous work is the play *The Pariah*.

Beer's fame derives from his hobby, astronomy. An amateur astronomer, Beer built a private observatory with a 9.5 cm refractor in Tiergarten, Berlin. Together with Johann Mädler (1794-1874) he produced the first exact map of the Moon (*Mappa Selenographica*), and in 1837 published a description of the Moon (*Der Mond Nach Seinen Kosmischen und Individuellen Verhältnissen*). Both remained the best descriptions of the Moon for many decades.

These works showed the Moon to be a world very unlike the Earth, and contradicted the pro-selenite claims of Franz Gruithuisen (1774-1852) and William Herschel (1738-1822). A selenite is a hypothetical lunar inhabitant. The name was coined by Johannes Hevelius (1611-1687) and later used by many others, both in philosophical discourse and fiction, including H. G. Wells in *First Men in the Moon* (1901).

Beer and Mädler's work helped persuade most professional astronomers that the Moon is uninhabited.

In 1830, Beer and Mädler created the first globe of the planet Mars, and in 1840 they made a map of Mars. They calculated the rotation period of Mars to be 24 hours, 37 minutes, and 22.7 seconds, only 0.1 seconds different from the actual period as it is known today.

In his last decade of life, he worked as a writer and politician. In 1849, he was elected as a Member of Parliament for the first chamber of the Prussian parliament.

In addition to astronomy and banking, Beer helped with the establishment of a railway in Prussia and promoted the Jewish community in Berlin. He died on March 27, 1850. **Beer crater** on Mars was named after him. It lies near Mädler crater. There is also a **Beer crater** on the Moon and an asteroid **1896 Beer** named for him.

1. Hevelius, Johannes (1611-1687), The Internet Encyclopedia of Science, www.daviddarling.info/encyclopedia/H/Hevelius.
2. Wilhelm Beer Biography, biographybase, www.biographybase.com/biography/Beer_Wilhelm.
3. Wilhelm Beer, BookRags, www.bookrags.com/wiki/Wilhelm_Beer.

Thomas Henderson (1798-1844)

The first person to measure the distance to Alpha Centauri

Thomas James Henderson, the first Astronomer Royal for Scotland, was born in Dundee, Scotland on December 28, 1798. He was the son of a tradesman. After high school, he trained as a lawyer. He then worked his way up through the profession as an assistant to a number of nobles. His major hobbies were astronomy and mathematics.

After developing a new method for using lunar occultation to measure longitude, he came to the attention of Thomas Young (1773-1829, superintendent of the Royal Navy's *Nautical Almanac*. Young helped Henderson enter the larger world of astronomical science, and on his death left a letter recommending to the Admiralty that Henderson take his place.

Henderson was passed over for that position, but the recommendation was enough to get him a position at the British observatory at the Cape of Good Hope in South Africa. There, he made a considerable number of stellar observations between April 1832 and May 1833.

He learned that the bright southern star, Alpha Centauri, had a large proper motion, and concluded that it might be a star "close" to Earth. Proper motion is a component of the space motion of a celestial body perpendicular to the line of sight, resulting in the change of a star's apparent position relative to that of other stars.

The big astronomical challenge of the 1830s was to be the first person to measure the distance to a star using parallax, a task which is easier the closer the star. After retiring back to the United Kingdom due to bad health, Henderson began analyzing his data and eventually came to the conclusion that Alpha Centauri was just slightly less than one parsec (3.25 light years) away.

Doubts about the accuracy of his instruments, however, kept him from publishing, and eventually he was beaten by Friedrich Bessel (1784-1846), who published a parallax of 10.4 light years for 61 Cygni in 1838. Henderson published his results in 1839, but was relegated to second place because of his delay.

In 1834, he was appointed the first Astronomer Royal for Scotland and given the chair of astronomy at the University of Edinburgh. From that time, he worked at the City Observatory, Edinburgh until his death from heart disease on November 23, 1844. He is buried in Greyfriars Kirkyard.

1. Thomas Henderson 1798-1844, Astronomical Society of Edinburgh Journal 38, www.astronomyedinburgh.org/publications/journals/38/hend.
2. Thomas Henderson, Significant Scots, ElectricScottland.com, www.electricscotland.com/hiStory/other/henderson_thomas.
3. Thomas James Henderson, BookRags, www.bookrags.com/wiki/Thomas_James_Henderson.

Friedrich Argelander (1799-1875)

Compiled a catalog of more than 300,000 stars over a 25-year period

Friedrich Wilhelm August Argelander, a Prussian astronomer, was born in Memel, East Prussia, Germany on March 22, 1799. He was the son of a Finnish father and German mother. He studied with Friedrich Bessel (1784-1846) at the University of Königsberg, obtaining his Ph.D. in 1822.

From 1823 to 1837, Argelander was the head of the Finnish observatory at Turku and then at Helsinki. He then moved to Bonn, Germany, where he developed a friendship with King Frederick William IV, who funded a new observatory at the University of Bonn. In 1837, Argelander was appointed to the chair of astronomy at that University. He remained there until his death.

He excelled in developing effective, simple, and fast methods for measuring star positions and magnitudes. He was the first astronomer to begin a careful study of variable stars. Only a handful of these stars were known when he began. A variable star is a star that undergoes significant variation in its luminosity. They are sometimes referred to as stars that are subjected to pulsations.

Argelander was responsible for introducing the modern system of identifying variable stars. He also is credited with establishing the study of variable stars as an independent branch of astronomy.

He is best known for publishing his *Bonner Durchmusterung*, which cataloged the position and brightness of more than 324,000 stars of magnitude greater than 9.5 from the North Pole to two degrees south of the equator. He later initiated the first *Astronomische Gesellschaft Katalog*.

Among his other achievements were the calculation of the motion of the Sun through space, the development of a system of magnitudes for describing the brightness of stars too dim to be seen by the naked eye, and the development of the system for naming stars that is still in use today.

In 1863, Argelander founded an international organization of astronomers named the Astronomische Gesellschaft. He won the Gold Medal of the Royal Astronomical Society in 1863. He died in Bonn, Germany on February 17, 1875. The **Argelander crater** on the Moon and **asteroid 1551 Argelander** are named for him. In 2006, the three astronomical institutes of Bonn University were merged and renamed the **Argelander-Institut für Astronomie**.

1. Argelander, Friedrich Wilhelm August (1799-1875), The Internet Encyclopedia of Science, www.daviddarling.info/encyclopedia/A/Argelander.
2. Friedrich Wilhelm August Argelander, Microsoft Encarta Online Encyclopedia, http://encarta.msn.com/encyclopedia_761573694/argelander_friedrich_wilhelm_august, 2008.
3. Friedrich Wilhelm August Argelander, NNDB, www.nndb.com/people/763/000096475.

William Lassell (1799-1880)

Discovered Neptune's moon Triton, Saturn's moon Hyperion, and Uranus's moons Ariel and Umbriel

William Lassell, a British amateur astronomer, was born in Bolton, England on June 18, 1799. He became a Liverpool Businessman, making his fortune in the brewing trade. A genius for instrument making and engineering, he is best known for creating the modern large reflecting telescope. Although the Newtonian reflector had been around for 150 years when Lassell started to cast his first mirrors around 1820, they suffered from their great wooden tubes, manhandled with ropes which made the long-term tracking of an object impossible.

When one is sweeping the sky for interesting deep-space objects, one could almost leave the telescope fixed and let the heavens do the moving, but if it was the planets and their satellites that were of interest, then mounts of wood and rope were severely limiting. Lassell, from the beginning, was a planetary astronomer.

He set up an observatory at Starfield, near Liverpool. There, he developed a reflecting telescope mounted in the equatorial plane, with smooth tracking. His home-made, 9-inch equatorial reflector was a revolutionary instrument. Everything was set on roller bearings and precision geared, so that nothing more than finger pressure was necessary to start and stop the instrument. With this instrument, in the early 1840s, Lassell did expert planetary work, which laid the foundations of his international reputation.

Lassell followed his 9-inch reflector with a 24-inch diameter mirror inside a 20-foot tube. In 1852, he took it to Malta, where the atmosphere was much clearer than in industrial England. Just 17 days after the discovery of Neptune by German astronomer Johann Galle (1812-1910), Lassell discovered the planet had a satellite, which was later named Triton.

In addition to Neptune's Triton, the same instrument revealed two new moons rotating around Uranus (Ariel and Umbriel) and Saturn's innermost rape ring. Lassell independently co-discovered Saturn's moon, Hyperion, with William Bond (1789-1859). In 1861, he built a 48-inch reflector and used it to observe and catalog hundreds of new nebulae. He died in Maidenhead, England on October 5, 1880. **Lassell crater** on the Moon, a crater on Mars, and a ring of Neptune are named in his honor.

1. Chapman, Allan, William Lassell (1799-1880) and the discovery of Triton, 1846, www.mikeoates.org/lassell/lassell_by_a_chapman.
2. Lassell, William (1799-1880), The Internet Encyclopedia of Science, www.daviddarling.info/encyclopedia/L/Lassell.
3. William Lassell, Wikipedia, http://en.wikipedia.org/wiki/William_Lassell.

19th Century Astronomers

1800 to 1899

William Parsons (1800-1867)

Built the world's largest telescope at the time

William Parsons (3rd Earl of Rosse), an astronomer and telescope maker, was born at Monkstown, County Cork, Ireland on June 17, 1800. He was the eldest son of the 2nd Earl of Rosse. Until his father's death, William was known as Lord Oxmantown. His youngest brother, Charles Parsons (1854-1931), is famous for his commercial development of the steam turbine.

William was educated at Trinity College, Dublin and Oxford University, where he graduated in 1822. He was a member of parliament for King's County, Ireland from 1822 until 1834, when he resigned to devote himself to science.

On inheriting his father's earldom in 1841, Parsons joined the House of Lords.

From 1827 Parsons devoted himself to the improvement of reflecting telescopes. His main aim was to build a telescope at least as large as that of William Herschel (1738-1822). As Herschel had left no details of how to grind large mirrors, Parsons had to rediscover the process.

It was not until 1839 that he completed a 3-inch mirror. This was followed by mirrors of 15 inches, 24 inches, and 36 inches. In 1842, he started work on his 72-inch mirror. The completed telescope weighed 8960 pounds. Its tube was over 50 feet long and was protected by two masonry piers 50 feet high and 23 feet apart. The telescope was supported by an elaborate system of platforms, chains, and pulleys. It was given the name, Leviathan. From 1845, it remained the world's largest telescope for the rest of the century.

From 1848 to 1878, Leviathan was, but with few interruptions, employed for observations of nebulae. Many previously unknown features in these objects were revealed by it, especially the similarity of annular and planetary nebulae, and the remarkable spiral configuration prevailing in many of the brighter nebulae.

A special study was made of the nebula of Orion. He named it the Crab Nebula, based on his feeling that it resembled a crab. He was the first to identify a spiral nebula, and went on to discover 15 of them. He was the first to discover binary and triple stars.

Parsons died on October 31, 1867.

1. William Parsons Rosse, third earl of. Answers.com, www.answers.com/topic/rosse-william-parsons-3d-earl-of?cat=technology.
2. William Parsons, 3rd Earl of Rosse (person), Everything, www.everything2.com/index.pl?node_id=1754889.
3. William Parsons, 3rd Earl of Rosse, Wikipedia, http://en.wikipedia.org/wiki/William_Parsons,_3rd_Earl_of_Rosse.

George Airy (1801-1892)

Determined the mean density of the Earth

George Biddell Airy, a British mathematician and astronomer, was born at Alnwick, Northumberland, England on July 27, 1801. In 1819, he entered Trinity College, Cambridge, where he paid a reduced fee but worked as a servant to make good the fee reduction. He graduated in 1823. In 1826, he was appointed Lucasian professor of mathematics. In 1828, he became Plumian professor of astronomy and director of the new Cambridge observatory. His work during this time was divided between mathematical physics and astronomy. The former was for the most part concerned with questions relating to the theory of light and a complete theory of the rainbow. In astronomy, he discovered, in correcting the elements of the solar tables of Jean Delambre (1749-1822), a new inequality in the motions of Venus and the Earth. The investigation represented the first specific improvement in the solar tables in England since the establishment of the theory of gravitation.

In 1826, the idea occurred to Airy of determining the mean density of the Earth by using pendulums placed at the top and bottom of a deep mine. The result showed that gravity at the bottom of the mine exceeded that at the top by 1/19286 of its amount. From this, he determined Earth's specific density of 6.566.

In 1835, Airy was appointed Astronomer Royal in succession to John Pond (1767-1836). The task of reducing the accumulated planetary observations made at Greenwich from 1750 to 1830 was already in progress under Airy's supervision when he became Astronomer Royal. Shortly afterward, he undertook the further task of reducing the enormous mass of observations of the Moon made at Greenwich during the same period under the direction, successively of James Bradley (1693-1762), Nathaniel Bliss (1700-1764), Nevil Maskelyne (1732–1811), and John Pond (1767-1836). As a result, no less than 8,000 lunar observations were rescued from oblivion.

In 1851, Airy established a new Prime Meridian at Greenwich. This line, the fourth Greenwich Meridian, became the definitive internationally recognized line in 1884. He died in London on January 2, 1892. The Martian crater **Airy** is named for him. Within that crater lies another smaller crater called **Airy-0** whose location defines the prime meridian of that planet. There is also a lunar crater **Airy**

1. Airy, George Biddell (1801-1892), The Internet Encyclopedia of Science, www.daviddarling.info/encyclopedia/A/Airy.
2. George Biddell Airy, MacTutor, www-groups.dcs.st-and.ac.uk/~history/Biographies/Airy.
3. George Biddell Airy, Wikipedia, http://en.wikipedia.org/wiki/George_Biddell_Airy.

Alvan Clark (1804-1887)

Made some of the world's largest and best lenses for refracting telescopes

Alvan Clark, an American astronomer and telescope maker, was born in Ashfield, Massachusetts on March 8, 1804. He was the descendant of a Cape Cod whaling family of English ancestry. Initially a portrait painter and calico engraver, at the age of 40 he become involved in telescope making. With his sons, George Bassett Clark (1827-1891) and Alvan Graham Clark (1832-1897), he established Alvan Clark & Sons at Cambridgeport, Massachusetts. It became the finest producer of telescope lenses. The first achromatic lenses made in the United States were produced there.

Using glass blanks made by Chance Brothers of Birmingham, England and Feil-Mantois of Paris, France, the firm ground lenses for refracting telescopes. The lenses included the largest in the world at the time—the 18.5-inch lens at Dearborn Observatory at the Old University of Chicago, the 26-inch lens at the United States Naval Observatory (Washington, D.C.), the 30-inch lens at Pulkovo Observatory (Saint-Petersburg), the 36-inch lens at Lick Observatory (California), and the 40-inch lens at Yerkes Observatory (Wisconsin). The 40-inch lens is still the largest refractor lens in the world.

The optical work of Clark & Sons was recognized as unsurpassed anywhere, and represented the first significant American contribution to astronomical instrument-making. Prior to this, American telescopes had never compared with those of European manufacture.

The younger of Clark's sons, Alvan, discovered a number of double stars, Sirius B (the dim companion of Sirius), and the first known white dwarf.

Clark died in Cambridge, Massachusetts on August 19, 1887. Craters on the Moon and on Mars are named in his honor.

Alvan Clark & Sons' assets were acquired by the Sprague-Hathaway Manufacturing Company in 1933, but continued to operate under the Clark name. In 1936, Sprague-Hathaway moved the Clark shop to a new location in West Somerville, Massachusetts, where manufacturing continued in association with the Perkin-Elmer Corporation, another maker of precision instruments.

Most of Clark's equipment was disposed of as scrap during World War II, and Sprague-Hathaway itself was liquidated in 1958.

1. Alvan Clark, NNDB, www.nndb.com/people/702/000167201.
2. Alvan Clark, Wikipedia, http://en.wikipedia.org/wiki/Alvan_Clark.
3. Clark, Alvan (1804-1887), The Internet Encyclopedia of Science, www.daviddarling.info/encyclopedia/C/Clark.

Urbain Leverrier (1811-1877)

Made theoretical investigations which led to the discovery of Neptune

Urbain Jean Joseph Leverrier, a French mathematical astronomer, was born at Saint-Lô in Normandy on March 11, 1811. He entered École Polytechnique with an early interest in chemistry. When a teaching post in astronomy fell vacant at the Polytechnique in 1837, Leverrier took it and thereby entered the discipline in which he was to spend the rest of his life.

He was primarily concerned with celestial mechanics, which is the mathematical analysis of the planetary motions. According to the principles of celestial mechanics, each planet moves around the Sun in an essentially elliptical orbit, with minor deviations due to attractions by the other planets. The computations involved are complicated, but the results are sufficient to provide predictions of considerable accuracy.

However, the planet Uranus did not obey the mathematics. Although it had been the subject of a great deal of study since its discovery in 1781 by William Herschel (1738-1822), attempts to reduce its motion to the rule had yet to meet with complete success. The error was small—one minute of arc—but astronomers were accustomed to accounting for angles less than one-tenth that size.

In 1845, Leverrier concluded the difficulty was probably due to the action of an unknown planet whose effects had not been taken into account. His calculations gave an estimate of the location of this planet. A young astronomer, John Adams (1819–1892), had done similar calculations shortly before Leverrier, but did not make them public.

On September 23, 1846, the planet, later named Neptune at Leverrier's suggestion, was discovered by German astronomer Johann Galle (1812-1910), the director of the Berlin Observatory.

Leverrier's other prediction was for a new innermost planet, which became known as Vulcan. Searches turned up nothing, and we now know that the irregularities in Mercury's orbit are an effect of general relativity.

Leverrier continued with exhaustive examinations and revisions of all the existing planetary theories. In addition, he served with distinction as director of the Paris Observatory, organized the French meteorological service, and worked for the inclusion of scientific instruction in the French educational system. He died in Paris on September 23, 1877.

1. Encyclopedia of World Biography on Urbain Jean Joseph Leverrier, www.bookRags.com/ biography/urbain-jean-joseph-leverrier
2. Leverrier, Urbain Jean Joseph (1811-1877), The Internet Encyclopedia of Science, www.daviddarling.info/ encyclopedia/L/Leverrier.
3. World of Scientific Discovery on Urbain Jean Le Verrier, BookRags, www.bookrags.com/ biography/urbain-jean-le-verrier-wsd.

John Draper (1811-1882)

Took the first photograph of the Moon

John William Draper, a British-American chemist and historian, was born in Saint Helens, Lancashire, England on May 5, 1811. His father was a Wesleyan clergyman, who was interested in scientific subjects. John studied at the University of London before settling in the United States in 1832. He received a medical degree in 1836 at the University of Pennsylvania in Philadelphia. That same year, he became a professor of chemistry and natural philosophy at Hampden-Sydney College in Virginia.

In 1839, he went to New York City, where he became a professor of chemistry at what would become New York University. He was a founder of the University's medical school in 1841. He remained there until 1882,

Draper made the second photographic portrait, taken in December 1839; the first astronomical photograph, a picture of the Moon in March 1840; the first microphotographs, engravings from which appeared in his *Human Physiology* (1856); and the first photograph of the diffraction spectrum in 1844.

Draper's law of photochemical absorption states that only absorbed rays produce chemical change. In 1847, he proved that all solid substances become incandescent at the same temperature, and that thereafter, with rising temperature, they emit rays of decreasing wavelength. He demonstrated in 1857 that the maximum luminosity and maximum heat in the spectrum coincide.

In 1876, he made a negative of the solar spectrum, and one of the spectrum of an incandescent gas on the same plate, with their edges in contact. These results and corroborative experiments led him to assume the presence of oxygen in the Sun.

Draper observed the solar eclipse of July 29, 1878, and obtained photographs of the corona. Later he photographed the great nebula of Orion, and in 1880 photographed the spectrum of Jupiter.

He helped found the American Chemical Society, and was elected its first president in 1876.

Draper's most famous book, *History of the Conflict between Religion and Science* (1874), was a vigorous argument against the persecution of scientists by religionists. He died in Hastings-on-Hudson, New York on January 4, 1882

1. John William Draper, Microsoft Encarta Online Encyclopedia, http://encarta.msn.com/encyclopedia_762511344/Draper_John_William, 2008.
2. John William Draper, NNDB, www.nndb.com/people/733/000167232.
3. John William Draper, Virtual American Biographies, www.famousamericans.net/johnwilliamdraper.

Johann Galle (1812-1910)

Discovered Neptune

Johann Gottfried Galle, a German astronomer, was born at Pabsthaus, in Germany on June 9, 1812. He started to work as an assistant to Johann Encke (1791-1865) in 1835 immediately following the completion of the Berlin observatory. In 1838, Galle discovered the crêpe ring of Saturn.

Galle's Ph.D. thesis, finished in 1845, was a reduction and critical discussion of the observations of Ole Rømer (1644-1710) of meridian transits of stars and planets on the days from October 20 to October 23, 1706.

He sent a copy of his thesis to Urbain Le Verrier (1811-1877), but only received an answer a year later, on September 18, 1846. In it, Le Verrier asked him to look at a certain region of sky to find a predicted new planet, which would explain the perturbations of Uranus. That same night, at the Berlin Observatory, on September 23, 1846, with the assistance of student Heinrich d'Arrest (1822-1875), Galle started observing. He was favored by having an unpublished copy of a new star chart covering the right part of the sky. He found a star that was not on the chart. A wait of 24 hours showed that it had moved against the background of the fixed stars. He confirmed that it was a planet over the next two nights. Galle became the first person to view the planet Neptune and know what he was looking at.

In 1851, he moved to Breslau (Wrocław) to become professor of astronomy and the director of the local observatory.

Throughout his career he studied comets, and in 1894 (with the help of his son, Andreas Galle, he published a list with 414 comets.

Galle also made an important contribution by determining the mean distance of the Sun from the Earth. Conventional means of determining the distance had leaned heavily on the two transits of Venus in each century. In practice, it turned out to be difficult to determine accurately the moment of first contact. Galle proposed instead, in 1872, that measuring the parallax of the planets would give a more reliable figure.

Galle lived long enough to receive the congratulations of the astronomical world on the 50th anniversary of the discovery of Neptune in 1896. He died in Potsdam, Germany on July 10, 1910. Two craters, one on the Moon and one on Mars, and a ring of Neptune have were named in his honor.

1. Johann Gottfried Galle, Answers.com, www.answers.com/topic/johann-gottfried-galle?cat=technology.
2. Johann Gottfried Galle, Wikipedia, http://en.wikipedia.org/wiki/Johann_Gottfried_Galle.

Anders Ångström (1814-1874)

One of the founders of the science of spectroscopy

Anders Jonas Ångström, a Swedish Physicist, was born at Lögdö, Medelpad, Sweden on August 13, 1814. He was educated at the University of Uppsala. In 1842, he went to the Stockholm Observatory to gain experience in practical astronomy work, and in the following year he was appointed keeper of the Uppsala Astronomical Observatory. In 1858, he succeeded Adolph Svanberg (1814-1874) in the chair of physics at Uppsala.

Ångström is best known for his work on spectroscopy. In his *Optiska Undersökningar*, published in 1853, he pointed out that an electric spark yields two superimposed spectra, one from the metal of the electrode and the other from the gas through which it passes. He deduced from the theory of resonance of Leonhard Euler (1707-1783) that an incandescent gas emits luminous rays of the same refrangibility as those that it can absorb.

After 1861, Ångström investigated the solar spectrum. In 1862, combining the spectroscope with photography, he showed that the Sun's atmosphere contains hydrogen, among other elements.

In 1867, he examined the spectrum of the aurora borealis, and detected and measured its characteristic bright line in its yellow green region.

In 1868, he published a map of the normal solar spectrum in *Recherches sur le Spectre Solaire*, including detailed measurements of more than 1000 spectral lines.

Ångström performed important work concerning heat conduction, and devised a method of measuring thermal conductivity, which he showed to be proportional to electrical conductivity. Becoming interested in terrestrial magnetism, he also carried out geomagnetical measurements in different places around Sweden.

Ångström died at Uppsala, Sweden on the June 21, 1874. The **angstrom** (Å) unit with which the wavelength of light is measured is named for him. One angstrom is 10^{-10} meter or 10^{-4} micron. The angstrom unit is used in crystallography as well as spectroscopy.

His son, Knut Ångström (1857-1910), became known for his research at Uppsala University on the radiation of heat from the Sun and its absorption by the Earth's atmosphere.

1. Anders Jonas Ångström, NNDB, www.nndb.com/people/929/000100629.
2. Anders Jonas Ångström, Wikipedia, http://en.wikipedia.org/wiki/Anders_Jonas_%C3%85ngstr%C3%B6m.
3. Cleveland, Cutler, J., Angström, Anders Jöns, The Encyclopedia of the Earth, www.eoearth.org/article/Angstr%C3%B6m,_Anders_J%C3%B6ns, December 1, 2006.

Daniel Kirkwood (1814-1895)

Discovered radial gaps in the main asteroid belt

Daniel Kirkwood, an American astronomer, was born in Harford County, Maryland on September 27, 1814. His father was a farmer. Daniel graduated in mathematics from York County Academy in York, Pennsylvania in 1838.

After teaching there for five years, he became principal of the Lancaster High School in Lancaster, Pennsylvania, and in 1848 he became principal of the Pottsville Academy in Pottsville, Pennsylvania.

In 1851, he accepted a position as Professor of Mathematics at Delaware College, and in 1854 he was promoted to its presidency. In 1856, he became Professor of Mathematics at Indiana University in Bloomington, Indiana. Kirkwood spent two years (1865 to 1867) at Jefferson College in Canonsburg, Pennsylvania.

He is best known for his study of the orbits of asteroids. In 1866, when arranging the 88 known asteroids by their distances from the Sun, he noted several gaps, now called **Kirkwood gaps**. He associated these gaps with orbital resonances with the orbit of Jupiter.

He was the first to suggest that meteor showers are debris from comets, and also the first to suggest that the Cassini Division in Saturn's rings is the result of a resonance with one of Saturn's moons.

In the same paper, he was also the first to suggest that meteor showers are debris from comets, and proposed the existence of a group of Sun-grazing comets. Kirkwood also identified a pattern relating the distances of the planets to their rotation periods, which is called **Kirkwood's Law**. This discovery earned Kirkwood an international reputation among astronomers. In 1891, at age 77, he became a non-resident lecturer in astronomy at Stanford University.

Kirkwood authored 129 publications, including three books. He died in Riverside, California on June 11, 1895, and is buried in Rose Hill Cemetery.

Kirkwood was a cousin of Iowa governor Samuel Kirkwood who became United States Secretary of the Interior. The asteroid 1951 AT was named **1578 Kirkwood** in his honor, as was the lunar impact crater **Kirkwood.** Indiana University's **Kirkwood Observatory** is named for him. **Kirkwood Avenue** in Bloomington, Indiana is also named for him.

1. Aley, Robert J., Biography of Daniel Kirkwood, Wikisource, http://en.wikisource.org/wiki/Biography_Of_Daniel_Kirkwood.
2. Daniel Kirkwood, Wikipedia, http://en.wikipedia.org/wiki/Daniel_Kirkwood.
3. Kirkwood, Daniel (1814-1895), The Internet Encyclopedia of Science, www.daviddarling.info/encyclopedia/K/Kirkwood.

Rudolf Wolf (1816-1893)

Calculated a period of 11.1 years for the sunspot cycle

Rudolf Wolf, a Swiss astronomer and mathematician, was born in Fällanden, near Zürich, Switzerland on July 7, 1816. He studied astronomy in Zürich, Vienna, and Berlin. After graduating, he moved to Bern to teach mathematics and physics. In 1847, he was appointed director of the small astronomical observatory at Bern.

In 1855, he moved back to Zürich, where he was appointed Professor of Astronomy at the University and at the Polytechnic school (now Eidgenössische Technische Hochschule), where he later became Director of the Observatory.

Wolf became interested in sunspots after observing a particularly large and spectacular sunspot group in 1847. He began his own telescopic observations and recording of sunspots, which he carried out continuously for the following 46 years.

Impressed with the sunspot cycle discovery of Heinrich Schwabe (1789-1875), he embarked on a program of reconstructing the variation in the number of sunspots as far back as possible. By 1868, Wolf had a sunspot-number reconstruction back to 1610. From this, he determined a sunspot cycle of 11.1 years. He was the first to note the possible existence in the sunspot record of a longer modulation period of about 55 years.

In 1852, Wolf was one of four people to notice independently the coincidence between the 11.1 years sunspot cycle and the cycle of geomagnetic activity. He and others also noted a similar correspondence between sunspot cycle and frequency of auroral activity. In 1848, he devised a way of quantifying sunspot activity. The **Wolf sunspot number**, as it is now called, remains in use today.

He was a prolific writer. His *Mathematics, Physics, Geodesy, and Astronomy* saw six editions between 1852 and 1893. His *History of Recent Astronomy* (1877) and his *Handbuch der Astronomie* (1893), were both extremely popular in the late 19th and early 20th centuries. He also contributed four volumes to the *Biographies of Swiss Men of Science* and two to the *Handbuch der Mathematik.*

Wolf's sunspot monitoring continued at the Zürich Observatory until 1979, when it was transferred to Brussels. He died in Zürich on December 6, 1893.

1. Rudolf Wolf (1816-1893), High Altitude Observatory, www.hao.ucar.edu/Public/education/bios/schwabe.
2. Rudolf Wolf, Wikipedia, http://en.wikipedia.org/wiki/Rudolf_Wolf.
3. Wolf, (Johann) Rudolf (1816-1893), The Internet Encyclopedia of Science, www.daviddarling.info/ encyclopedia/W/Wolf_Rudolf.

Angelo Secchi (1818-1878)

The first to make systematic use of spectroscopy in stellar classification

Angelo Secchi, an Italian astronomer and Jesuit priest, was born in Reggio, Italy on June 18, 1818. He joined the Jesuits at age 15, began his theological studies in 1844, and was ordained priest in 1847. In 1839, he began lecturing on physics and mathematics at the Jesuits' Collegio Romano.

In 1848, due to the political unrest in Italy that led to the expulsion of the Jesuit order, he traveled to England, then to Georgetown University near Washington, D.C. There, he taught the natural sciences. In 1852, he returned to Rome, where he founded a new observatory at the Collegio Romano.

Secchi was the first to classify stars according to their spectra. Between 1863 and 1867, he studied the spectra of some 4,000 stars, using a visual spectroscope on the Collegio Romano's observatory's telescope. His four-class scheme prevailed throughout much of the second half of the 19th century, and paved the way for all later classification schemes.

He studied solar prominences during eclipses, both visually and spectroscopically, and provided the first demonstrations that prominences are features belonging to the Sun. He was the first to suggest that the solar core is in a gaseous state, with the temperature steadily decreasing from the center to the surface. He also discovered **comet Secchi** (C/1853 E1).

Secchi wrote a number of astronomical books in Italian, some technical, others aimed at the general public, and one for children. His influential solar monograph *Le Soleil* was first published in Paris in 1870, with a later German translation. The second French edition was published in 1875.

In Italy, he presided for many years over the Accademia dei Nuovi Lincei, and founded the Societa degli Spettroscopisti Italiani, devoted to spectroscopic studies of the Sun.

Secchi saw his religion as being fully compatible with pluralism. In fact, he was such a committed pluralist that he often seemed to ignore the evidence of his own research.

In 1859, he became the first to use the term canali to describe two fine lines he had seen on the surface of Mars. He died in Rome on February 26, 1878. The **Secchi crater** on the Moon and a crater on Mars are also named after him.

1. Angelo Secchi, New Advent, www.newadvent.org/cathen/13669a.
2. Angelo Secchi, Wikipedia, http://en.wikipedia.org/wiki/Angelo_Secchi.
3. Secchi, Rev. Pietro Angelo (1818-1878), The Internet Encyclopedia of Science, www.daviddarling.info/ encyclopedia/S/Secchi.

John Adams (1819-1892)

Predicted the existence and position of Neptune

John Couch Adams, a British mathematician and astronomer, was born on June 5, 1819 in Laneast, Cornwall, England. His father was a poor tenant farmer. Adams graduated with a B.A. at Cambridge University in 1843. He was Lowndean Professor at the University of Cambridge for 33 years from 1859 to his death.

Adams' best known achievement was predicting the existence and position of Neptune, using only mathematics. The calculations were made to explain discrepancies with Uranus's orbit and the laws of Kepler and Newton. In October 1845, he gave his predicted position for Uranus to George Airy (1801-1892), the Astronomer Royal. However, Airy procrastinated for nine months until he heard of a similar claim by Urbain Leverrier (1811-1877).

In recognition of his discovery, Adams was eventually offered a knighthood and the post of Astronomer Royal after Airy, which he turned them down to remain at Cambridge as director of the Observatory.

The great meteor shower of November 1866 turned Adams' attention to the Leonids. Using a powerful and elaborate analysis, Adams ascertained that this cluster of meteors, which belongs to the Solar System, traverses an elongated ellipse in 33.25 years, and is subject to definite perturbations from the larger planets—Jupiter, Saturn and Uranus. He also gave tables of the positions of the moons of Jupiter.

He studied terrestrial magnetism, determining the Gaussian magnetic constants at every point on the Earth. He produced maps with contour lines of equal magnetic variation, which were published after his death.

After a long illness, Adams died at the Cambridge Observatory on January 21, 1892.

Three years after his death, he was honored with a memorial tablet in Westminster Abbey. The tablet is near the tablet that honors Sir Isaac Newton (1643-1727).

A crater on the Moon is jointly named after John Couch Adams, Walter Sydney Adams (1876-1956) and Charles Hitchcock Adams (1868-1951). Neptune's outermost known ring and **asteroid 1996 Adams** are also named after him. The **Adams Prize**, presented by the University of Cambridge, commemorates his prediction of the position of Neptune.

1. Adams, John Couch (1819-1892), The Internet Encyclopedia of Science, www.daviddarling.info/encyclopedia/A/AdamsJC.
2. John Couch Adams, Wikipedia, http://en.wikipedia.org/wiki/John_Couch_Adams.
3. John Couch Adams, Starchild, http://starchild.gsfc.nasa.gov/docs/StarChild/whos_who_level2/adams.

Jean Foucault (1819-1868)

Invented a pendulum to demonstrate the rotation of the Earth

Jean Bernard Léon Foucault, a French physicist, was born in Paris on September 18, 1819. His father was a publisher. After an education received chiefly at home, he studied medicine. However, he quickly abandoned it due to a fear of blood, switching to the physical sciences.

With French physicist Hippolyte Fizeau (1819-1896), he carried on a series of investigations on the intensity of the light of the Sun, the interference of infrared radiation, and the chromatic polarization of light.

Foucault is best known for his demonstration in 1851 of the diurnal motion of the Earth. To do this, he used a freely suspended, long and heavy pendulum in the Panthéon in Paris. The movement of the pendulum duplicated the rotation of the Earth on its axis, demonstrating clearly for the first time that the Earth rotates.

In the following year, he invented and named the gyroscope. In 1850, he showed that light travels more slowly in water than in air.

In 1855, he was made physicist at the imperial observatory at Paris. That year, he discovered that the force required for the rotation of a copper disc becomes greater when it is made to rotate with its rim between the poles of a magnet, the disc at the same time becoming heated by the eddy currents **(Foucault currents)** induced in the metal.

In 1857, he invented the **Foucault polarizer**. In the succeeding year, he devised a method of testing the mirror of a reflecting telescope to determine its shape—the **Foucault Test.** This allows the worker to tell if the mirror is perfectly spherical. Prior to Foucault's invention, testing reflecting telescope mirrors was a hit or miss proposition.

In 1862, with the revolving mirror of Charles Wheatstone (1802-1875) he determined the speed of light to be 298,000 km/s, only 0.6 percent off the currently accepted value.

In 1865, Foucault's papers on a modification of Watt's governor appeared in which he recommended making its period of revolution constant.

Foucault died of what was probably a rapidly developing case of multiple sclerosis in Paris on February 11, 1868. The **Foucault crater** on the Moon is named after him.

1. Foucault, Jean Bernard Léon (1819-1868), The Internet Encyclopedia of Science, www.daviddarling.info/encyclopedia/F/Foucault.
2. Jean Foucault, Microsoft Encarta Online Encyclopedia, http://encarta.msn.com/encyclopedia_761573292/foucault, 2008.
3. Léon Foucault, Wikipedia, http://en.wikipedia.org/wiki/L%C3%A9on_Foucault.
4. O'Connor, J. J. and E. F. Robertson, Jean Bernard Léon Foucault, MacTutor, www-history.mcs.st-andrews.ac.uk/Biographies/Foucault.

Charles Smyth (1819-1900)

Pioneered infrared astronomy

Charles Piazzi Smyth, a British astronomer, was born in Naples, Italy on January 3, 1819. He was the son of an amateur astronomer, who named him after his godfather, the Italian astronomer Giuseppe Piazzi (1746-1826). His father had met Piazzi's at Palermo. His father subsequently settled at Bedford and equipped an observatory there, at which Charles received his first lessons in astronomy.

At the age of 16, he became an assistant to Thomas Maclear (1794-1879) at the Cape of Good Hope, where he observed Halley's Comet and the great comet of 1843. He also took an active part in the verification and extension of the arc of the meridian of Nicholas de Lacaille (1713-1762).

In 1845, Smyth was appointed Astronomer Royal for Scotland and professor of astronomy at the University of Edinburgh. There, he continued the observations made by his predecessor, Thomas Henderson (1798-1844).

In 1853, he was responsible for installing the "time ball" on top of the Nelson Monument on Calton Hill to give a time signal to the ships at Leith. By 1861, this visual signal was augmented by the One O'clock Gun at Edinburgh Castle. The two signals were electrically connected by a long cable.

Obsessed by the pyramids of Egypt, Smyth took up the mystical pseudoscience of pyramidology. He resigned his fellowship of the Royal Society after they refused to publish his papers on this subject.

In 1856, Smyth founded the first high-altitude observatory on the site of what is now the Las Palmas Observatory in the Canary Islands.

In 1871-1872, he investigated the spectra of the aurora and zodiacal light. In 1877-1878, he constructed at Lisbon a map of the solar spectrum. He carried out further spectroscopic researches at Madeira in 1880 and at Winchester in 1884.

Smyth published *Three Cities in Russia* (1862), *Our Inheritance in the Great Pyramid* (1864), *Life and Work at the Great Pyramid* (1867), and a volume *On the Antiquity of Intellectual Man* (1868).

In 1888, Smyth resigned as Astronomer Royal for Scotland in protest of the chronic under-funding and age of the equipment at Calton Hill. He retired to Yorkshire, and devoted the rest of his life to photographing clouds. He died on February 21, 1900.

1. Charles Smyth, Classic Encyclopedia, www.1911encyclopedia.org/Charles_Smyth.
2. Smyth, Charles Piazzi (1819-1900), The Internet Encyclopedia of Science, www.daviddarling.info/ encyclopedia/S/Smyth.

Wilhelm Tempel (1821-1889)

Discovered numerous deep sky objects and comets

Ernst Wilhelm Leberecht Tempel, a German astronomer, was born in Nieder-Kunersdorf, near Löbau, in the kingdom of Saxony on December 4, 1821. His parents were poor, so he received a scanty education. When he was about 20 years old he went to Copenhagen, where he worked for about three years as a lithographer.

From Copenhagen, he went to Christiania, and then to Venice, Italy, where he became interested in astronomy. He purchased a 4-inch refractor telescope and began exploring the heavens. On April 2, 1859 he discovered his first comet.

In 1860, Tempel went to Marseilles, where he obtained employment at the Observatory. There, he discovered his second comet. That same year, he turned his attention to the minor planets, of which he discovered five from 1861 to 1868.

The Franco-Prussian War broke out in 1870. As a German, he was expelled from France by the Provisional Government in 1871. He went to Milan, Italy, where he became an assistant at the Brera Observatory. There he discovered three new comets.

Toward the end of 1874, he left Milan for the Arcetri Observatory, which is connected with the Reale Instituto di studi superiori of Florence. He now, for the first time, had use of larger instruments. He discovered only one more comet.

Tempel turned his attention to nebulae, and collected a considerable number of observations of them. He was the first to notice on October 19, 1859 the nebula around Merope in the Pleiades. He made accurate drawings of the more interesting nebulae, a pursuit for which his artistic skill and experience made him particularly fitted.

He was a prolific discoverer of comets, discovering, co-discovering, or rediscovering 26 in all, including **Comet 55P/Tempel-Tuttle** (now known to be the parent body of the Leonid meteor shower), and **9P/Tempel** (the target of the NASA probe Deep Impact in 2005).

Other periodic comets that bear his name include **10P/Tempel** and **11P/Tempel-Swift-LINEAR**. The asteroid **3808 Tempel** is named in his honor. **Tempel crater** on the Moon is also named after him.

He died on March 16, 1889.

1. Ernst Wilhelm Leberecht Tempel, Wikipedia, http://en.wikipedia.org/wiki/Ernst_Wilhelm_Leberecht_Tempel.
2. Ondra, Leos, Wilhelm Tempel's Biography, http://leo.astronomy.cz/tempel/tempel.
3. Tempel, (Ernst) Wilhelm (Leberecht) (1821-1889), The Internet Encyclopedia of Science, www.daviddarling.info/encyclopedia/T/Tempel.

Heinrich d'Arrest (1822-1875)

Assisted Johann Galle with the first observations of Neptune

Heinrich Louis d'Arrest, a Prussian astronomer, was born in Berlin on July 13, 1822. His name is sometimes given as Heinrich Ludwig d'Arrest. He studied Mathematics and Astronomy from 1839 in Berlin. While still a student, he was party to the search of Johann Galle 1812-1910) for Neptune. At the time, Galle was the chief assistant to Johann Encke (1791-1865).

On September 23, 1846, d'Arrest suggested to Galle that a recently drawn chart of the sky, in the region predicted by Urbain Le Verrier (1811-1877), could be compared with the current sky to seek the displacement characteristic of a planet, as opposed to a stationary star.

With Galle at the eyepiece and d'Arrest reading the chart, they scanned the sky and checked that each star seen was actually on the chart. Just a few minutes after their search began d'Arrest cried out "That star is not on the map!" and earned his place in history. Neptune had been discovered.

D'Arrest also discovered 342 *New General Catalog* (NGC) objects, mainly with a 28-cm refractor. NGC is an important catalog of nebulae, star clusters, and galaxies. The original version was published in Ireland in 1888 under the authorship of John Dreyer (1852-1926); it contained 7,840 northern-sky objects and was a revised and enlarged version of the General *Catalogue of Nebulae and Clusters* published in 1864 by John Herschel (1738-1822).

In 1852, d'Arrest became Professor of Astronomy at Leipzig University. His work at the Leipzig Observatory led him to the discovery of the comet named for him (**6P/d'Arrest**). He also studied asteroids and nebulae, and discovered asteroid 76 Freia,

In 1857, he moved to Copenhagen as Professor of Astronomy and director of the Observatory. He won the Gold Medal of the Royal Astronomical Society in 1875.

D'Arrest's published work, *Siderum Nebulosorum Observationes Hafniensis* (1876), contained 1942 nebula, 340 described for the first time. The work was done with the 11-inch refractor at Copenhagen Observatory

He died in Copenhagen, Denmark on June 14, 1875. **D'Arrest crater** on the Moon is also named after him, as well as a crater on the Martian satellite Phobos and the asteroid **9133 d'Arrest**.

1. Arrest, Heinrich Ludwig d' (1822-1875), The Internet Encyclopedia of Science, www.daviddarling.info/encyclopedia/A/Arrest.
2. Heinrich Louis d'Arrest, Wikipedia, http://en.wikipedia.org/wiki/Heinrich_d'Arrest.
3. Heinrich Ludwig [Louis] d'Arrest, www.klima-luft.de/steinicke/ngcic/persons/d-arrest.

Gustav Spörer (1822-1895)

Used observations of sunspots to determine the position of the Sun's equator

Friederich Wilhelm Gustav Spörer, a German Astronomer, was born in Berlin on October 23, 1822. Between 1840 and 1843 he studied mathematics and astronomy at the University of Berlin. He started his professional career as a schoolteacher, and only began his solar observations at age 36.

By the early 1860s, both Spörer and English astronomer Richard Carrington (1826-1875) had accumulated sunspot observations demonstrating the differential rotation of the Sun's surface and the gradual drift of sunspots toward the Sun's equator in the course of their cycle. The solar latitude at which new sunspots appear gradually decreases from 30° to 40° north or south of the solar equator at the beginning of a solar cycle to 5° to 10° at the end of the cycle.. Both Spörer and Carrington used observations of sunspots to determine the position of the Sun's equator (and hence the tilt of its axis) and to establish that it rotates at different rates at different latitudes. Spörer added to Carrington's observations of sunspot drift. The drift has been given the name **Spörer's Law**.

Spörer was invited in 1874 to join the Potsdam Astrophysical Observatory, which at the time was under construction. At Potsdam, he continued his solar observation. Through careful analyses of existing 17th century observational data, he compiled a list of the sunspots observed during that period. He was first to note a prolonged period of low sunspot activity from 1645 to 1715. This work attracted relatively little attention until it was publicized by Edward Maunder (1851-1928) as part of his own historical studies of 17th century sunspot observations. The period is known as the Maunder Minimum.

The **Spörer minimum** is a period of low sunspot activity that lasted from about 1420 to 1570. It was based on evidence such as carbon-14 in tree rings and records of auroral activity. It coincided with a period of low global temperatures known as the Little Ice Age. It was discovered by John Eddy (1931-), who named it after Spörer.

Spörer became chief observer at Potsdam in 1882. He retired from there in 1894. Despite having enjoyed perfect health throughout his life, he died suddenly on July 7, 1895.

1. Gustav Spörer (1822-1895), High Altitude Observatory, www.hao.ucar.edu/Public/education/bios/spoerer.
2. Gustav Spörer, Wikipedia, http://en.wikipedia.org/wiki/Gustav_Sp%C3%B6rer.
3. Spörer, Gustav Friedrich Wilhem (1822-1895), The Internet Encyclopedia of Science, www.daviddarling.info/encyclopedia/S/Sporer.

John Hind (1823-1895)

One of the early discoverers of asteroids

John Russell Hind, a British astronomer, was born in Nottingham, England on May 12, 1823. In 1840, at age 17, he went to London to serve an apprenticeship as a civil engineer, but through the help of Charles Wheatstone (1802-1875) he left engineering to accept a position at the Royal Greenwich Observatory under George Airy (1801-1892).

Hind is noted for discovering a number of asteroids and several variable stars, including Nova Ophiuchi 1848, U Geminorum, R Leporis (also known as **Hind's Crimson Star**), and a mysterious newcomer in the constellation of Taurus. He also discovered the variability of μ Cephei.

While scanning the sky through the Pleiades (a nearby, fairly loose open cluster of about 500 stars in the constellation Taurus) and in the direction of the Hyades (the nearest open cluster to the Sun, except for the Ursa Major Moving Cluster), he noticed a tenth magnitude star that was missing from the charts he was using. This object became known as the prototype T Tauri star.

Hind was sent as a member of the commission that was appointed to determine the exact longitude of Valentia Island, off the southwest coast of Ireland.

On his return, in June 1844, he succeeded William Dawes (1799-1868) as director of the private observatory of George Bishop (1785–1861) in Regents Park, London. Here he calculated the orbits and declination of more than 70 planets and comets, and between 1847 and 1854 discovered 10 asteroids.

Hind's naming of the asteroid 12 Victoria caused some controversy. At the time, asteroids were not supposed to be named after living persons. Hind somewhat disingenuously claimed that the name was not a reference to Queen Victoria, but the mythological figure Victoria.

In 1851, Hind was awarded the Lalande medal of the Academy of Sciences at Paris. In 1852, he was awarded the gold medal of the Astronomical Society of London and a pension of £200 a year from the British government. In 1853, he became Superintendent of the *Nautical Almanac*, a position he held until 1891.

Among his works are *Astronomical Vocabulary* (1852), *The Comets* (1852), *The Solar System* (1852), *Illustrated London Astronomy* (1853), *Elements of Algebra* (1855), and *Descriptive Treatise on Comets* (1857).

Hind died in Twickenham, London on December 23, 1895.

1. Hind, John Russell (1823-1895), The Internet Encyclopedia of Science, www.daviddarling.info/encyclopedia/H/Hind.
2. John Russell Hind, Wikipedia, http://en.wikipedia.org/wiki/John_Russell_Hind.

William Huggins (1824-1910)

Compared laboratory spectra with those of stars

William Huggins, a wealthy British amateur astronomer, was born in Stoke Newington, London, England, on February 7, 1824. Under no obligation to earn a living, he occupied his early years with the study of physics, chemistry, and physiology. He became intrigued with the concept of spectroscopy developed by Gustav Kirchhoff (1824-1887) and Robert Bunsen (1811-1899) and the fact that it was possible to analyze the light given off by a star. In 1856, he sold the family business, and with his friend William Miller (1817-1870) built his own private observatory at Tulse Hill, five miles outside London.

In 1863, Huggins announced that the spectra of stars revealed the same elements as those on the Earth. This disproved the hypothesis put forth by Aristotle (384-322 BC) over 2000 years earlier that the heavens were made of a unique material not found on the Earth.

Looking at the spectrum of a nebula in Draco, Huggins discovered that it was composed of glowing gases, not of stars as was suspected of all nebulae before his observation. His study of the spectrum of a nova in 1868 revealed that it was surrounded by hydrogen. The spectrum of a comet revealed carbon compounds. Huggins was also among the first to use photography to collect light. He used time exposures to enhance the visibility of the spectra of extremely dim objects. The photographic plate also allowed Huggins to make permanent records of his observations. Furthermore, he applied the Doppler effect to the analysis of his spectra. If a star is moving toward an observer, the wavelengths of light in its spectrum should be shifted toward the blue end of the spectrum; if moving away it should be stretched out toward the red end.

In 1868, he found that the star Sirius, the brightest star in the night sky, had a slight shift in one of the hydrogen lines toward the red end of the spectrum. He concluded that Sirius was moving away from the Earth, and that the amount of the red shift corresponded to its velocity. Analysis of the spectral shifts of objects would be fundamental for conclusions by Edwin Powell Hubble (1889-1953) about the structure of the universe in 1929. Huggins died in London on May 12, 1910. Named after him are **Huggins crater** on the Moon, a Crater on Mars, and **asteroid 2635 Huggins**.

1. World of Scientific Discovery on Sir William Huggins, BookRags, www.bookrags.com/biography/sir-william-huggins-wsd.
2. William Huggins, NNDB, www.nndb.com/people/389/000103080.
3. Encyclopedia of World Biography on William Huggins, Sirwww.bookrags.com/biography/william-huggins-sir

Jules Janssen (1824-1907)

Co-credited with the discovery of what would become known as helium

Pierre Jules César Janssen, a French astronomer, was born in Paris on February 22, 1824. An accident at a young age left him unable to walk, which prevented him from attending school. He began working as a bank clerk at age 16, studying mathematics in his spare time. He eventually entered the Sorbonne, graduating in 1852 with degrees in mathematics and physical sciences. He obtained his doctorate in 1860.

Remaining at his Alma Mata, he first worked on the faculty of medicine designing medical instruments, then he became professor of physics at the school of architecture. In 1876, he took the helm of the newly founded Observatoire de Meudon near Paris.

In 1862, impressed by the spectroscopic work of Gustav Kirchhoff (1824-1887) and Robert Bunsen (1811-1899), Janssen began studying the solar spectrum. His first important contribution was to show that some of the dark lines observed in the solar spectrum were caused by water vapor in the Earth's atmosphere. He also discovered the presence of the chromosphere, a gaseous envelope surrounding the Sun.

Following his observations of the 1868 solar eclipse in India, he suggested that some of the unknown spectral lines observed above the solar limb were due to a hitherto unknown chemical element. Joseph Lockyer (1836-1920) independently and simultaneously arrived at the same conclusion. Both men are now credited with the discovery of what would become known as helium.

Janssen traveled to the summits of the Mont Blanc in the French Alps, and to the Faulhorn in the Bernese Oberland to carry out spectroscopic observations above the bulk of the Earth's atmosphere. He also made extensive use of hot air balloons for the same purpose, including a particularly daring flight to escape the 1870 siege of Paris during the Franco-Prussian war, and a failed attempt to observe a solar eclipse at Oran, Algeria.

He was also an early pioneer in the use of photography in solar physics. His extensive series of photographs of the solar surface were published in 1904 in his *Atlas de Photographies Solaires*. The quality of Janssen's solar photographs was to remain unsurpassed for nearly half a century. Janssen died in Meudonon on December 23, 1907.

1. Janssen, (Pierre) Jules César (1824-1907), The Internet Encyclopedia of Science, www.daviddarling.info/encyclopedia/J/Janssen.
2. Pierre Jannsen, Microsoft Encarta Online Encyclopedia, http://ca.encarta.msn.com/encyclopedia_762508746/Pierre_Janssen, 2008.
3. Pierre Janssen, Wikipedia, http://en.wikipedia.org/wiki/Pierre_Janssen.

Gustav Kirchhoff (1824-1887)

Co-Invented the spectroscope

Gustav Kirchhoff, a German Physicist, was born in Königsberg, East Prussia on March 12, 1824. His father was a lawyer. He entered the University of Königsberg at the age of 18, and received his doctorate in 1847. In 1845, while still a student he formulated his circuit laws, which are now ubiquitous in electrical engineering. He completed this study as a seminar exercise; it later became his doctoral dissertation.

In 1848, he became a teacher in Berlin, and two years later he obtained the post of associate professor at Breslau. It was there that he first met Robert Bunsen (1811-1899). By 1854 both Kirchhoff and Bunsen were working together in Heidelberg.

Kirchhoff, one afternoon in the summer of 1859, looked at the interaction of sunlight and the light of table salt burning in the flame of a Bunsen burner and decided that there must be a relation. The next day, he came up with **Kirchhoff's law of radiation**, which states that the powers of emission and the powers of absorption for rays of the same wavelength are constant for all bodies at the same temperature.

Kirchhoff's now famous paper, written with Bunsen and published in 1859, stated that the dark lines, now known as Fraunhofer lines, of the solar spectrum (not caused by the terrestrial atmosphere) arise from the presence in the glowing solar atmosphere of those substances which in a flame produce bright lines in the same position.

In the early 1860s, Kirchhoff produced the first detailed map of the solar spectrum. His pioneering work in spectroscopy permitted investigation of the chemical composition of stars.

He also contributed to the fundamental understanding of the emission of black-body radiation by heated objects. He coined the term "black body" radiation in 1862. He also determined that electricity travels at the speed of light. Two sets of independent concepts in both circuit theory and thermal emission are named **Kirchhoff's laws** after him.

Kirchhoff occupied the chair of theoretical physics at the University of Berlin from 1875.

His failing health forced him to prematurely retire in 1886. He died on October 17, 1887. He was buried in the St. Matthäus Kirchhoff Cemetery in Schöneberg, Berlin.

1. Encyclopedia of World Biography on Gustav Robert Kirchhoff, BookRags, www.bookrags.com/biography/gustav-robert-kirchhoff.
2. Gustav Kirchhoff, Wikipedia, http://en.wikipedia.org/wiki/Gustav_Kirchhoff.
3. World of Scientific Discovery on Gustav Robert Kirchhoff, BookRags, www.bookrags.com/biography/gustav-robert-kirchhoff-wsd.

Benjamin Gould (1824-1896)

Discovered a partial ring of stars about 3000 light years across

Benjamin Apthorp Gould, an American astronomer, was born in Boston, Massachusetts on September 27, 1824. After graduating from Harvard College in 1844, he studied mathematics and astronomy under Carl Gauss (1777-1855) at Göttingen, Germany. Gould was the first American to receive a Ph.D. in astronomy. After receiving his degree, he toured European observatories asking for advice on what he could do to further astronomy as a professional science in the United States. He was advised to start a professional journal modeled after the then world's leading astronomical publication, the *Astronomische Nachrichten*.

He returned to America in 1848. From 1852 to 1867 he was in charge of the longitude department of the United States Coast Survey. He developed and organized the service, was one of the first to determine longitudes by telegraphic means, and employed the Atlantic cable in 1866 to establish accurate longitude-relations between Europe and America.

In 1849, Gould returned to Cambridge and started the *Astronomical Journal*, which he published until the outbreak of the Civil War in 1861. That same year, he undertook the task of preparing for publication the records of astronomical observations made at the U.S. Naval Observatory since 1850.

In 1868, on behalf of the Argentine republic, he established a national observatory at Córdoba and became its director. There, he extensively mapped the southern hemisphere skies using newly developed photometric methods. The need of astronomers for good weather prediction spurred Gould to collaborate with Argentine colleagues to develop the Argentine National Weather Service, the first in South America.

Gould put together a zone-catalog of 73,160 stars and a general catalog compiled from meridian observations of 32,448 stars. On June 1, 1884, he made the last definite sighting of the great comet of 1882.

In 1885, he returned to Massachusetts, where he restarted the *Astronomical Journal* and worked on the 1,000 photographic plates of star clusters he had brought back with him from Cordoba.

Astronomers continue to investigate the astrophysics of a large scale feature of the Milky Way to which he called their attention in 1877, and which is named for him, the **Gould Belt.** He died at Cambridge, Massachusetts on November 26, 1896. A crater on the Moon is named after him.

1. Benjamin Apthorp Gould, Wikipedia, http://en.wikipedia.org/wiki/Benjamin_Gould.
2. Gould, Benjamin Apthorp (1824-1896), The Internet Encyclopedia of Science, www.daviddarling.info/encyclopedia/G/Gould_Benjamin.

George Bond (1825-1865)

Independently discovered Saturn's eighth satellite, Hyperion

George Phillips Bond, an American astronomer and son of William Cranch Bond (1789-1859), was born in Dorchester, Massachusetts on May 20, 1825. He earned his B.A. from Harvard University in 1845. He then started work as assistant astronomer at the Harvard College Observatory, and earned his M.A. from Harvard in 1853.

With his father, he established the Harvard College Observatory as one of the most important centers of astronomical research in the United States.

Bond became a pioneer of photographic astronomy. He demonstrated photographic astronomy's use in mapping the sky and in measuring the comparative brightness and exact location of celestial objects. He also used photography to study the Moon's surface, and was the first person to photograph a binary star (a pair of stars revolving around a common center of mass).

While working with his father, Bond discovered Hyperion, the eighth satellite of Saturn. Hyperion was found independently in the same year by William Lassell (1799-1880).

Bonds observations led him to reject the generally accepted theory that Saturn's rings were of solid structure. However, his hypothesis of their being in a fluid state was soon discarded. Bond also discovered Saturn's Crêpe ring, the dim ring that is closer to Saturn than the planet's two brighter rings.

Bond became director of the Harvard College Observatory after his father's death in 1859. During the Civil War, the observatory lost virtually all its funding, and the facility could no longer be heated nor the leaking roof repaired, but Bond continued taking notes and photos of the skies, even as his chronic tuberculosis worsened.

He discovered numerous comets and calculated their orbits. He studied Saturn and the Orion Nebula. Also a cartographer, Bond surveyed and drew the first detailed maps of New Hampshire's White Mountains, where **Mount Bond** is named in his honor.

Bond was awarded the Royal Astronomical Society's Gold Medal in 1865, but the medal did not arrive until several days after his death in Cambridge on February 17, 1865.

1. Bond, William Cranch (1789-1859) and George Phillips (1825-1865), The Internet Encyclopedia of Science, www.daviddarling.info/encyclopedia/B/BondW.
2. George Phillips Bond, Microsoft Encarta Online Encyclopedia, http://encarta.msn.com/encyclopedia_761589192/Bond_George_Phillips 2008.
3. George Phillips Bond, NNDB, www.nndb.com/people/868/000164376.

Julius Schmidt (1825-1884)

Updated the map of the Moon

Johann Friedrich Julius Schmidt, a German astronomer, was born in Eutin, Germany on October 25, 1825. When he was 14 years old, he came into the possession of a copy of *Selenotopographische Fragmente* by Johann Schröter (1745-1816), and this influenced a lifelong interest in selenography (the study of the surface and physical features of the Moon).

While going to school in Hamburg, he visited Altona Observatory, where he became acquainted with the map of the Moon made by Wilhelm Beer (1797-1850) and Johann Mädler (1794-1874).

In 1845, Schmidt obtained a position as an assistant at an observatory in Düsseldorf, and a year later joined the Bonn Observatory under Friedrich Argelander (1799-1875). In 1853, he became director of Baron von Unkrechtsberg's private observatory at Olmütz. In 1858, he became director of Athens Observatory, where he spent the rest of his career.

He spent most of his career making drawings of the Moon. In 1866, he claimed that Linné crater had considerably changed its appearance, which began a controversy that continued for many decades.

Schmidt's 1874 map of the Moon was the first to surpass the celebrated map of Beer and Mädler. In 1878, he edited and published all 25 sections of a Moon map by Wilhelm Lohrmann (1796-1840). Lohrmann had completed his map in 1836, but had died in 1840.

In addition to his study of the Moon, Schmidt made many other observations, including the discovery of Nova Cygni 1876, also known as Q Cygni, in 1876.

Schmidt also did meteorological observations in many places in Greece, and regularly sent data to the Observatory of Paris. These results were presented in his work *Beitrage zur Physikalischen von Griechland* (1864).

With the help of volunteers, he recorded more than 3000 earthquakes, and published his *Studienn uber Erdbeben* (1975). He observed the Santorini volcano from the erection in 1866, and published the study of this and of three other volcanoes (Etna, Vesuvius, Stromboli) in 1874.

Schmidt died in Athens, Greece on February 7, 1884. The King and Queen of Greece attended the funeral oration at his observatory. **Schmidt crater** on the Moon is jointly named for him and two other people of the same last name.

1. 1858-1884: the years of Julius Schmidt in the Athens Observatory, Institute of Astronomy and Astrophysics, www.astro.noa.gr/History/h_1858-1884.
2. Johann Friedrich Julius Schmidt, Wikipedia, http://en.wikipedia.org/wiki/Johann_Friedrich_Julius_Schmidt.

Richard Carrington (1826-1875)

Made a systematic study of sunspots

Richard Christopher Carrington, a British amateur astronomer, was born in Chelsea, England on May 26, 1826. The second son of a wealthy brewer, he was originally expected to pursue a career in the Church, and in 1844 began studies in theology at Trinity College, Cambridge, graduating in 1848. By then, however, he had become interested in astronomy.

In 1849, he joined the Durham University Observatory, but resigned the position in 1852. Using his family fortune, he built his own house and observatory at Redhill, Surrey.

Although his work on asteroids and planets while at Durham University Observatory was enough to secure membership in the Royal Astronomical Society, Carrington's first major astronomical undertaking was the compilation between 1854 and 1857 of his *Catalogue of 3735 Circumpolar Stars*. Published in 1857, Carrington's catalog was highly praised and earned him the Gold Medal of the Royal Astronomical Society in 1859.

Impressed by the 1843 sunspot cycle discovery by Heinrich Schwabe (1789-1875), and aware of the lack of systematic sunspot observations, Carrington picked up the subject where Schwabe left off. Improving on Schwabe's projection/drawing method, he drew and recorded the positions of sunspots for eight years. He made several discoveries, including the Sun's differential rotation, the migration of spots toward the equator in the course of the cycle, the Sun's rotation axis, and the first and unexpected observation of a white-light flare. A white-light flare is a major flare in which small parts become visible in white light. Such flares are usually strong X-ray, radio, and particle emitters.

Carrington made the initial observations leading to the establishment of Spörer's law, which predicts the variation of sunspot latitudes during a solar cycle. The law was discovered and named by John Eddy (1931-).

Carrington's astronomical work came to an abrupt end when he had to take over his family's brewing business due to the death of his father in 1858.

In 1865, having fallen into ill health, he sold the family brewery and retired to an isolated spot at Churt, Surrey, where he established a new observatory, but never resumed serious astronomical work. He died there of a brain hemorrhage on November 27, 1875.

1. Carrington, Richard Christopher (1826-1875), The Internet Encyclopedia of Science, www.daviddarling.info/encyclopedia/C/Carrington.
2. Richard Christopher Carrington (1826-1875), High Altitude Observatory, www.hao.ucar.edu/Public/ education/bios/carrington.

Giovanni Donati (1826-1873)

The first to obtain and analyze the spectrum of a comet

Giovanni Battista Donati, an Italian astronomer, was born in Pisa, Italy on December 16, 1826. He graduated from the university in Pisa, and afterward joined the staff of the Observatory of Florence in 1852. He was appointed director of the Observatory in 1864. One of his major responsibilities as the director was to supervise the work at Arcetri, not far from Florence, where a new observatory was being set up.

Between 1854 and 1864 he discovered six new comets, including the spectacular 1858 comet C/1858 L1, now named comet **Donati** in his honor. When the comet was nearest the Earth, its triple tail had an apparent length of 50°, more than half the distance from the horizon to the zenith and corresponding to more than 45 million miles. At the time, it was still 228 million miles away, but turned out to be the brightest comet to visit close to the Earth during the 19th century. The orbital period of the comet was estimated at more than 2,000 years, and so it will not return until about the year 4000.

Donati also pioneered spectroscopy of comets to identify the chemical elements that make up the comet's body and tail. He discovered that the spectrum changed when a comet approached the Sun; that is, heating caused it to emit its own light, rather than reflect sunlight. He concluded that the composition of comets is, at least in part, gaseous.

His spectroscopic observations of the 1864 comet contained three prominent lines, which Donati named alpha, beta, and gamma. These same lines were seen in an 1866 comet by Angelo Sochi (1818-1878), and shown by William Huggins (1824-1910) in 1868 to be due to the presence of carbon. Since comets are thought to represent the primordial composition of the Solar System, this technique is used today to gain better understanding of how the Solar System formed.

In 1870, he observed an eclipse using a spectroscope of his own design that incorporated five prisms. Later he constructed an even more complex spectroscope having no fewer than 25 prisms.

While attending the International Meteorological Congress in Vienna, Donati fell ill of cholera and died a few hours after returning to Florence, on September 20 1873. **Donati crater** on the Moon is also named for him.

1. Donati, Giovanni Battista (1826-1873), The Internet Encyclopedia of Science, www.daviddarling.info/encyclopedia/D/Donati.
2. Donati's Comet, Microsoft Encarta Online Encyclopedia, http://au.encarta.msn.com/ encyclopedia_761568371/Donati's_Comet, 2008.
3. Giovanni Battista Donati, BookRags, www.bookrags.com/research/giovanni-battista-donati-scit-0512345.

Asaph Hall (1829-1907)

Discovered the two moons of Mars, Phobos and Deimos

Asaph Hall, an American astronomer, was born in Goshen, Connecticut on October 15, 1829. He attended the district schools until he was 13 years old. At 16 he was apprenticed to a carpenter, and he worked at that trade sporadically. He attended the Norfolk Academy to study mathematics one winter, spent a year and a half at Central College at McGrawville in New York, and received special instruction in astronomy during three months at the University of Michigan.

In 1857, after a period as a schoolmaster in Ohio and some months working as a carpenter, Hall secured a position at the Harvard Observatory. This gave him the opportunity to attend lectures and informally complete his education. He immediately proved to be an excellent observer, and in 1859 began to send papers, chiefly on the orbits of comets and asteroids, to scientific journals.

In 1862, he went to Washington as an aide in the Naval Observatory, and the following year he was appointed professor of mathematics there. In 1872, he was made chief of the Naval Observatory.

In 1875, Hall was given responsibility for the Naval Observatory's 26-inch telescope, the largest refractor in the world at the time. It was with this telescope that he discovered Mar's satellites, Phobos and Deimos. He also noticed a white spot on Saturn, which he used as a marker to ascertain the planet's rotational period.

Hall's nearly 500 published papers include investigations of the orbits of various satellites, the mass of Mars, the perturbations of the planets, and the advance of Mercury's perihelion (the point nearest the Sun). He also investigated the parallax of the Sun, the distances of Alpha Lyrae and 61 Cygni, the mass of Saturn's rings, and the orbits of double stars, along with the solution of many mathematical problems suggested by these investigations.

In addition, he traveled to Vladivostok, Russia in 1874 and San Antonio, Texas in 1882 to observe two transits of Venus.

Following his retirement from the Naval Observatory in 1891, Hall taught at Harvard, and continued to work in astronomy. He published his final paper in September 1906, and died at Annapolis, Maryland on November 22, 1907.

1. Asaph Hall, NNDB, www.nndb.com/people/591/000165096.
2. Asaph Hall, Virtual American Biographies, www.famousamericans.net/asaphhall.
3. Asaph Hall, Wikipedia, http://en.wikipedia.org/wiki/Asaph_Hall.
4. Encyclopedia of World Biography on Asaph Hall, BookRags, www.bookrags.com/biography/asaph-hall.

Simon Newcomb (1835-1909)

Studied the motion of the Moon and the planets and re-determined various astronomical values

Simon Newcomb, a Canadian-American mathematician and astronomer, was born in Wallace, Nova Scotia on March 12, 1835. His father was an itinerant New England schoolteacher. Apprenticed at the age of 16 to an herbalist doctor, Newcomb ran away two years later to the United States.

He taught at country schools in Maryland for several years, and in 1857 was appointed a "computer" in the Nautical Almanac Office, then located at Harvard University. While there, he attended the Lawrence Scientific School, receiving a B.S. in 1858.

In 1861, Newcomb was commissioned professor of mathematics in the U.S. Navy, and shortly thereafter was assigned to the Naval Observatory and Nautical Almanac Office. There, he began his mathematical investigations of such fundamental questions as the orbits of Neptune and Uranus, the motion of the Moon, and the right ascensions of the equatorial fundamental stars. His revision of the value of the solar parallax, published in 1867, remained standard until 1895 when it was superseded by his own revision.

In 1877, Newcomb was appointed superintendent of the American Ephemeris and Nautical Almanac Office. He immediately began to reform the entire basis of fundamental data involved in the computation of the ephemeris, which is a table of values that gives the positions of astronomical objects in the sky at a given time.

As early as 1867, he had suggested the desirability of accurately determining the velocity of light as a means of obtaining a reliable value for the radius of the Earth's orbit. In 1878, he began his experiments, for a while collaborating with Albert Michelson (1852-1931), whose later works far overshadowed Newcomb's efforts in this line.

Newcomb also published a number of mathematics textbooks and several astronomy books for a popular audience, including *Popular Astronomy* (1878), *The Stars* (1901), *Astronomy for Everybody* (1902), and his autobiographical *Reminiscences of an Astronomer* (1903). He also wrote a novel, *His Wisdom, the Defender* (1900), and three books on economics, a subject on which he was considered an authority. He died of bladder cancer in Washington, D.C., on July 11, 1909. Asteroid **855 Newcombia** and **Newcomb crater** on the Moon is named after him.

1. Encyclopedia of World Biography on Simon Newcomb, BookRags, www.bookrags.com/biography/simon-newcomb.
2. Simon Newcomb, NNDB, www.nndb.com/people/473/000103164.
3. Simon Newcomb, Wikipedia, http://en.wikipedia.org/wiki/Simon_Newcomb.

Giovanni Schiaparelli (1835-1910)

Linked meteor showers to comet remnants and mistakenly described water-filled canals on Mars

Giovanni Virginio Schiaparelli, an Italian astronomer was born in Savigliano, Piedmont, Italy on March 14, 1835. He graduated from the University of Turin and studied at the Royal Observatory in Berlin under Johann Encke (1791-1865).

After a brief spell at Pulkova Observatory in Russia, he joined the staff of Milan's Brera Observatory in 1860, and became its director two years later. He was the first to demonstrate that the Perseid and Leonid meteor showers were associated with comets and that comets follow similar paths through space.

In 1877, he produced the most detailed map of Mars ever published, filling in additional features over the next decade. It became a standard reference in planetary cartography. For naming major Martian features, he used Latin and Mediterranean place names taken from ancient history, mythology, and the Bible. This scheme he devised survives to this day.

Schiaparelli's original map contained a curious network of linear markings, which crisscrossed the Martian surface and joined one dark area to another. These lines, which Schiaparelli referred to as canali, he named after famous rivers, both fictional and real.

The romantic and evocative names he chose proved to have a powerful influence over some of his contemporaries, and strengthened the opinion of some astronomers, such as Percival Lowell (1855-1916), that intelligent life existed on Mars. Schiaparelli himself clearly favored a view of Mars in which the dark areas were seas and the brighter regions land.

In 1879, he refined his original map, making some changes, such as the apparent invasion of a bright area known as Libya, which is near Syrtis Major. This encouraged him in his belief that Syrtis Major was a shallow sea, which at times flooded the lands around it. He believed the Martian atmosphere to be rich in water vapor, and asserted that the canali comprised a true surface-water system.

Schiaparelli was a prolific astronomer whose research ranged widely, but whose name is forever associated with Mars and the controversy over the Martian canals. He died in Milan, Italy on July 4, 1910. Named after him are **Asteroid 4062 Schiaparelli**, **Schiaparelli crater** on the Moon, and **Schiaparelli crater** on Mars.

1. Giovanni Virginio Schiaparelli, NNDB, www.nndb.com/people/497/000095212.
2. Giovanni Virginio Schiaparelli, Wikipedia, http://en.wikipedia.org/wiki/Giovanni_Schiaparelli.
3. Schiaparelli, Giovanni Virginio (1835-1910), The Internet Encyclopedia of Science, www.daviddarling.info/encyclopedia/S/Schiaparelli.

Norman Lockyer (1836-1920)

Co-discovered the element helium in the Sun

Joseph Norman Lockyer, a British astronomer, was born in Rugby, England on May 17, 1836. He initially worked as a civil servant in the British War office, pursuing astronomy in his spare time. His rising reputation as a solar physicist led him to be appointed the director of the Solar Physics Observatory in South Kensington in 1885.

In 1868, he fitted a spectrograph to a telescope to allow him to study prominences and the outer solar atmosphere on a routine basis, as opposed to only during a total eclipse. He coined the name "Chromosphere" for the outer layers of the solar atmosphere.

In France, Jules Janssen (1824-1907) simultaneously and independently proceeded along very similar lines, and came to similar conclusions. The following year, working in collaboration with Janssen, Lockyer identified a chromospheric spectral line of an unknown chemical element, which he named Helium. Helium was finally isolated in the laboratory in 1895 by William Ramsay (1852-1916), following which Lockyer was knighted.

In 1869, Lockyer founded the journal *Nature*, to this day one of the leading general scientific Journals. He edited the Journal for 50 years.

In 1890, he became interested in possible astronomical alignments of ancient Greek and Egyptian monuments and temples. In 1901, he extended his studies to Stonehenge. Once approximate alignment of a given monument had been identified, he dated the monument by assuming exact alignment at time of construction, and interpreted the difference in terms of the precession of the Earth's orbital axis. He derived an age of 1848 BC for the construction of Stonehenge. Stonehenge is now believed to have been built in several stages between 3000 and 1500 BC.

After his retirement in 1911, Lockyer established an observatory near his home. Originally known as the Hill Observatory, the site was renamed the **Norman Lockyer Observatory** after his death.

Although many of his hypotheses and conclusions were not universally well received and often did not survive the test of time, he is to be credited with founding the field of Archeoastronomy. He died on August 16, 1920 in Salcombe Regis, Devonshire. Also named for him are **Lockyer crater** on the Moon and **Lockyer crater** on Mars.

1. J. Norman Lockyer (1836-1920), High Altitude Astronomy, www.hao.ucar.edu/Public/education/bios/lockyer.
2. Joseph Norman Lockyer, Wikipedia, http://en.wikipedia.org/wiki/Norman_Lockyer.
3. Lockyer, (Joseph) Norman (1836-1920), The Internet Encyclopedia of Science, www.daviddarling.info/ encyclopedia/L/Lockyer.

Henry Draper (1837-1882)

Made the first photograph of a stellar spectrum

Henry Draper, an American physician and astronomer, was born in Prince Edward County, Virginia on March 7, 1837. His father, John Draper (1811-1882), was a physician, chemist, botanist, and professor at New York University (NYU). John Draper was the first to photograph the Moon through a telescope. Henry's mother was the daughter of the personal physician to the Emperor of Brazil.

In 1857, at the age of 20, Draper graduated from NYU medical school. He then traveled in Europe for a year. In Ireland, he visited, and was greatly influenced by William Parson (1800-1867). When he returned from Europe, he began preparing his own telescope glass mirror, which he installed in his new observatory on his father's estate at Hastings on Hudson, New York.

Draper worked first as a physician at Bellevue Hospital, and later as both a professor and dean of medicine at New York University.

In 1861, he married Anna Mary Palmer, daughter of a multi-millionaire. She became his lab assistant and was an expert at coating and developing Draper's wet-plate photographs.

Draper became one of the pioneers of the use of astrophotography. He took the first photograph of a star's spectrum in 1872, and eight years later took the first photographs of a nebula.

In 1881, he took the first wide-angle photograph of the tail of a comet and the first spectrum photograph of a comet's head. He also took high-quality photographs of Jupiter, Mars, Venus, numerous bright stars, and the spectra of the Sun and Moon.

For his astrophotography activities, he received numerous awards, including honorary law degrees from NYU and the University of Wisconsin-Madison. He also received a Congressional medal for directing the United State's expedition to photograph the 1874 transit of Venus.

Draper's career was cut short when he became ill after a Rocky Mountain hunting trip, and died on November 20, 1882. His widow funded the **Henry Draper Medal** for outstanding contributions to astrophysics.

In 1897, the Harvard project completed the ***Henry Draper Catalog* of *Stellar Spectra***. It was the first comprehensive classification of stars according to their spectra. The small **Draper crater** on the Moon is also named in his honor.

1. Draper, Henry (1837-1882), The Internet Encyclopedia of Science, www.daviddarling.info/encyclopedia/D/Draper_Henry.
2. Henry Draper, Wikipedia, http://en.wikipedia.org/wiki/Henry_Draper.
3. Henry Draper, NNDB, www.nndb.com/people/739/000167238.

Robert Ball (1840-1913)

Discovered six new nebulae

Robert Stawell Ball, an Irish astronomer, was born on July 1, 1840. His father, in his position as Secretary of Dublin Zoo, often took immediate charge of new animals at their home before they were transported to the Zoo. Robert was educated at Dr. John Lardner Burke's School in Dublin, then later at Dr. Brindley's school in Tarvin, England.

His father died in 1857, and he returned to Ireland. Shortly afterward, he entered Trinity College, Dublin to study mathematics and science. It was only with great difficulty that his mother was able to pay his university fees.

William Parsons (1800-1867), the 3rd earl of Rosse, had constructed the largest reflecting telescope in the world at Birr Castle. The 72-inch telescope became operational in 1845, and was used mainly to observe spiral galaxies, which Rosse had classified as nebulae. In about 1864, Rosse sought a tutor for his three sons, and Ball was recommended to him for the post. Ball accepted on condition that he could use of the 72-inch telescope, which Rosse agreed to.

Ball took up the position with Parsons in 1865. There, he discovered six previously unknown nebulae, which were then listed in the *New General Catalogue.* He also observed the Leonid meteor shower on the night of November 13-14, 1866. After two years working for Rosse, Ball was offered the new chair of applied mathematics and mechanics in the Royal College of Science in Dublin, which he accepted.

In physics, the main mathematical topic on which Ball did research was dynamics, in particular the theory of screws. His first paper on the topic was written in 1869. He was invited to address the British Association meeting on the topic in 1870. Another major mathematical publication around this time was *Experimental Mechanics* (1871).

In 1874, Ball became Royal Astronomer of Ireland and Professor of Astronomy in Trinity College Dublin.

In 1876, he published *The Theory of Screws: A Study in the Dynamics of a Rigid Body*. He was knighted in 1886.

In 1892, he became the Lowndean Professor of Astronomy and Geometry at Cambridge, and in 1893 the director of the Cambridge Observatory. He died in Cambridge on November 25, 1913 and is buried in St. Giles cemetery near his predecessor, John Couch Adams (1819-1892).

1. Robert Stawell Ball, MacTutor, www-groups.dcs.st-and.ac.uk/~history/Biographies/Ball_Robert.
2. Sir Robert Ball, Victorian astronomer and Lecturer, www.freewebs.com/ziksby.

Hermann Vogel (1841-1907)

Pioneered the use of astronomical spectroscopy and photography

Hermann Carl Vogel, a German astronomer, was born in Leipzig, Kingdom of Saxony on April 3, 1841. He first studied at the polytechnical school in Dresden, and then returned to Leipzig to study natural sciences at the University, where he became an assistant at the Observatory.

In 1870, increasingly interested in astronomy, he became director of the Bothkamp Observatory near Kiel. He joined the staff of the Potsdam Astrophysical Observatory even before it opened in 1876, and served as its director from 1882 to 1907.

Vogel pioneered the use of the spectroscope in astronomy. He made spectroscopic analyses of stars, planets, comets, and the Sun. He was the first to demonstrate the Sun's rotation by measuring Doppler shifts of its receding and approaching limbs. He also chemically analyzed planetary atmospheres.

He is best known for a discovery he made in 1890—the spectra of certain stars shifted slightly over time, moving toward the red and then later toward the blue. His interpretation of this result was that the star was moving toward and then away from the Earth, and that the accompanying spectral shifts were the result of the Doppler effect. These stars appeared to be orbiting around a hidden center of mass, and thus were double-star systems. However, in each case the companion star could not be seen using a telescope, and so these double-star systems were designated spectroscopic binaries. He was also among the first astronomers to measure the radial velocities of bright stars.

By obtaining periodic Doppler shifts in the components of Algol, Vogel proved that the latter was a binary star. Thus, Algol was one of the first known spectroscopic binaries, and is also known to be an eclipsing binary. An eclipsing binary star is a binary star in which the orbit plane of the two stars lies so nearly in the line of sight of the observer that the components undergo mutual eclipses.

Vogel was awarded the Gold Medal of the Royal Astronomical Society (1893), Henry Draper medal (1893), Landskroener Medal of Achievement (1898), Richard C. White Purple Honors Medal (1899), and the Bruce Medal (1906). He died on August 13, 1907. **Vogel crater** on the Moon and a crater on Mars are named after him.

1. Hermann Carl Vogel, The Bruce Metalist, http://phys-astro.sonoma.edu/brucemedalists/vogel/Vogel.
2. Hermann Carl Vogel, Wikipedia, http://en.wikipedia.org/wiki/Hermann_Carl_Vogel.
3. Vogel, Hermann Carl (1841-1907), The Internet Encyclopedia of Science, www.daviddarling.info/encyclopedia/V/Vogel.

Alexander Bickerton (1842-1929)

Revived the catastrophic hypothesis of the origin of the Solar System

Alexander William Bickerton, a British chemist and astronomer, was born at Alton in the county of Hampshire, England on January 7, 1842. His father was a builder's clerk. After being employed, first in a railway workshop and later in an engineer's office, he set up a cabinet-making establishment, using machinery of his own invention. The enterprise failed, but Bickerton had begun studying science under a teacher working for the Science and Art Department of the school.

He organized and taught science classes for working men in Chelsea. He believed that a science class should be made "as entertaining as a music hall and as sensational as a circus." He was soon addressing crowded lecture rooms. In 1870, he became a science lecturer at the Hartley Institute, Southampton, but in 1874 he accepted the Chair of Chemistry at the new university college at Christchurch, New Zealand.

As the years passed, he became known as a teacher of exceptional ability, but at the same time his agitation for university reform made powerful enemies on the Board of Governors of Canterbury College.

In 1878, he began to formulate his theory of partial impact, ascribing the sudden appearance of bright new stars, followed by a rapid waning in brilliance, to collisions between dark bodies moving through space. Having struck each other a glancing blow, these bodies developed an intense degree of heat and created one or more new stars of exceptional, though temporary, brilliance. The theory had first been suggested by Georges Buffon (1707-1788).

In 1894, Bickerton's enemies on the Board of Governors appointed a committee to inquire into the management of his department, but, having powerful friends, he survived what in fact was an attempt to prepare the way for his dismissal.

A few years later, he founded a "federative home" at Wainoni, hoping to initiate a new form of society. The rumor spread that the principles of sexual morality were being ignored by the "federators." The Board finally secured Bickerton's dismissal in 1902.

He was the author of a number of books, including *The Romance of the Heavens*, *The Romance of the Earth*, and *The Perils of a Pioneer*. Some months before his death in London on January 21, 1929, as an act of restitution, Canterbury College appointed him Professor Emeritus.

1. Bickerton, Alexander William (1842-1929), The Internet Encyclopedia of Science, www.daviddarling.info/encyclopedia/B/Bickerton.
2. Bickerton, Alexander William, An Encyclopedia of New Zealand 1966, www.teara.govt.nz/1966/B/BickertonAlexanderWilliam/BickertonAlexanderWilliam/en.

Camille Flammarion (1842-1925)

Did more to encourage public interest in astronomy than anyone else of his day

Nicolas Camille Flammarion, a French astronomer and author, was born in Montigny-le-Roi, Haute-Marne, France on February 26, 1842. He was the brother of Ernest Flammarion (1846-1936), founder of the Groupe Flammarion publishing house.

Flammarion served for some years at the Paris Observatory and at the Bureau of Longitudes, but in 1883 he set up a private observatory at Juvisy (near Paris) and continued observations, especially of double and multiple stars and of the Moon and Mars.

His first book, *La Pluralité es Mondes Habité* (1862), secured his reputation as both a great popularizer and a leading advocate of extreme pluralism (belief that there exist numerous other worlds harboring life and, in particular intelligent life). By 1882, the book had gone through 33 editions, and continued to be translated and reprinted well into the 20th century.

Flammarion's passionate belief in life on other worlds was nurtured by his readings of previous pluralist authors, such as Bernard Fontanelle (1657-1757), Christiaan Huygens (1629-1695), Jérôme Lalande (1732-1807), and David Brewster (1781-1868).

Flammarion, and another French writer, J. H. Rosny (1856-1940), were the first to popularize the notion of beings that were genuinely alien and not merely minor variants on humans and other terrestrial forms.

His best-selling work, *Astronomie Populaire* (1880), is filled with speculation about extraterrestrial life. An entire chapter is taken up in arguing the case for lunar life, while Mars he considers to be an Earth almost similar to ours. In 1892, he speculated further on Mars in his *La Planè Mars et ses Conditions D'habitabilité.*

Concerning the Martian canals, he suggested they may be due to superficial fissures produced by geological forces or perhaps even to the rectification of old rivers by the inhabitants for the purpose of the general distribution of water. As to Martian life, he concluded that Mars may be inhabited by a race superior to our own.

Flammarion was the author of more than 70 books. He died on June 3, 1925 and is buried on Juvisy Observatory grounds. **Flammarion crater** on the Moon and **Flammarion crater** on Mars are named after him.

1. Camille Flammarion, Biographies, SGNY, www.sgny.org/main/Biographies/bio_CF.
2. Camille Flammarion, Wikipedia, http://en.wikipedia.org/wiki/Camille_Flammarion.
3. Flammarion, (Nicolas) Camille (1842-1925), The Internet Encyclopedia of Science, www.daviddarling.info/encyclopedia/F/Flammarion.

Thomas Chamberlin (1843-1928)

Co-formulated the planetismal hypothesis on the origin of the planets in the Solar System

Thomas Chamberlin, an American geologist and astronomer, was born in Mattoon, Illinois on September 25, 1843. He was the son of a Methodist minister and farmer. Thomas's discovery of fossils in a local limestone quarry aroused his interest in geology. In 1866, he received a B.A. from Beloit College, and after a year of graduate study at the University of Michigan became professor of natural sciences at the State Normal School at Whitewater, Wisconsin. He then went to Beloit College as professor of geology. He worked for the Wisconsin Geological Survey from 1873, serving as chief geologist for the period 1876 to 1882. From 1881 until 1904, he was in charge of the glacial division of the U.S. Geological Survey.

After a period as president of the University of Wisconsin (1887-1892) he became head of the department of geology and director of the Waller Museum at the University of Chicago. In 1893, he founded the *Journal of Geology* and was editor in chief until 1922.

Apart from his work on the geological surveys, Chamberlin's most significant work was in the field of glaciation. He showed that drift deposits are composed of at least three layers, and went on to establish four major ice ages, which were named the Nebraskan, Kansan, Illinoian, and Wisconsin after the states in which they were most easily studied.

Together with the astronomer, Forest Moulton (1872–1952), Chamberlin formulated, in 1906, the planetismal hypothesis on the origin of the planets in the Solar System. They proposed that a star had passed close to the Sun, causing matter to be pulled out of both. Within the gravitational field of the Sun, this gaseous matter would condense into small planetesimals, and eventually into planets. This was stated in *The Origin of the Earth* (1916). He also published the theory in *The Two Solar Families* (1928), but it has little support today as it cannot account for the distribution of angular momentum in the Solar System.

Chamberlin died in Chicago on November 15, 1928. There are buildings named for him on the Beloit College and University of Wisconsin-Madison campuses. The lunar crater **Chamberlin** and a crater on Mars are named in his honor.

1. Thomas C. Chamberlin, The University of Chicago Centennial Catalogues, www.lib.uchicago.edu/projects/centcat/centcats/fac/facch02_01.
2. Thomas Chrowder Chamberlin, Answers.com, www.answers.com/topic/thomas-chamberlin.
3. Thomas Chrowder Chamberlin, Wikipedia, http://en.wikipedia.org/wiki/Thomas_Chrowder_Chamberlin.

Edward Pickering (1846-1919)

A pioneer in the fields of stellar spectroscopy and photometry

Edward Pickering, an American astronomer, was born in Boston, Massachusetts on July 19, 1846. He was the older brother of astronomer William Pickering (1858-1938). He attended Lawrence Scientific School, graduating with a B.S. in 1865. He then taught mathematics at that institution for a year before moving to Massachusetts Institute of Technology, where he became the Thayer professor of physics in 1868.

In 1876, Pickering accepted the directorship of the Harvard Observatory, despite having no experience as an observational astronomer. The choice of a physicist, however, placed Harvard in the leadership of the new astronomy—one which used the methods of physics to seek a knowledge of stellar structure and its evolution.

He was able to hire several women in his observatory due to a number of donations from the widow of Henry Draper (1837-1882).

Pickering became an expert in stellar photometry, a field barely explored with large instruments at the time. When he began the work, even the magnitudes of the stars were not fixed on any generally accepted scale. Pickering established a widely accepted color scale and employed the meridian photometer—his own invention—to achieve unprecedented accuracy in determining the magnitudes of 80,000 stars.

In 1885, he began the compilation on some 300,000 glass plates, a complete photographic chart of the stellar universe down to the eleventh magnitude.

Pickering was also a leader in stellar spectroscopy, laying the foundation for the method of spectral classification now universally accepted, and obtaining material for the *Draper Catalogue*, which contained 200,000 stars. He and Hermann Vogel (1841-1907) independently discovered the first spectroscopic binary stars. He also discovered a new series of spectral lines, now known as the **Pickering series**, which turned out to be due to ionized helium.

By the time of his death on February 3, 1919, he was recognized as one of the two or three outstanding astronomical researchers in America. **Pickering crater** on the Moon is named both for him and William Pickering. There is also a **Pickering crate**r on Mars in his honor, as well as minor planet **784 Pickeringia**.

1. 3.14 Edward Pickering Biography, Astro Blogging, http://pvastro0718.blogspot.com/2008/03/314-edward-pickering-biography.
2. Encyclopedia of World Biography on Edward Charles Pickering, BookRags, www.bookrags.com/biography/edward-charles-pickering.
3. Pickering, Edward Charles (1846-1919), The Internet Encyclopedia of Science, www.daviddarling.info/encyclopedia/P/PickeringE.

Johann Palisa (1848-1925)

Discovered 120 asteroids and published catalogs with the position of nearly 5,000 stars

Johann Palisa, an Austrian astronomer, was born in Troppau, Silesia (now Czech Republic) on December 6, 1848. From 1866 to 1870, he studied mathematics and astronomy at the University of Vienna; however, he did not receive his degree until 1884. In 1870, he became an assistant at the University observatory in Vienna, and in the following year he took a position at the observatory in Geneva.

At 24 years of age, Palisa became director of the Austrian Naval Observatory in Pola (now Pula) in 1872. He discovered his first asteroid, 136 Austria, in 1874. He subsequently discovered another 27 minor planets and one comet at Pola—all with a 6-inch refractor telescope.

When the new Vienna observatory was opened in 1880, he was offered a subordinate position. He accepted it because he would be able to use the observatory's large 27-inch refractor, at that time the largest telescope in the world. There, Palisa discovered another 94 asteroids. In addition, he discovered eight objects that were included by John Dreyer (1852-1926) in the *New General Catalogue*, as well as four nebulae listed in the *Index Catalogue*. He observed the total solar eclipse on May 6, 1880.

At the end of the 19th century, Palisa and Max Wolf (1863-1932) in Heidelberg joined forces and worked on the *Palisa-Wolf-Sternkarten*, the first photographic star atlas. Two years later, Palisa published his *Sternenlexikon*, a star catalog covering the sky between declinations minus one degree and plus 19 degrees. In 1908, Palisa became Vice Director of the Vienna observatory. He retired in 1919.

With 122 minor planets to his name, Palisa is still the most successful Austrian discoverer of asteroids, as well as the most successful visual discoverer in the history of minor planet research. Some of his notable discoveries include 153 Hilda, 216 Kleopatra, 243 Ida, 253 Mathilde, 324 Bamberga, and the Amor asteroid 719 Albert.

Palisa died in Vienna on May 2, 1925. For his work, he was honored by having minor planet **914 Palisana** named for him, as well as the **Palisa crater** on the Moon. The **Johann Palisa Observatory and Planetarium** was founded in 1980 by the Technical University of Ostrava (known as Báňska) in the Czech Republic.

1. Johann Palisa, Wikipedia, http://en.wikipedia.org/wiki/Johann_Palisa.
2. Raab, Herbert, Johann Palisa, the most successful visual discoverer of asteroids, www.astrometrica.at/Papers/Palisa.pdf.
3. The Johann Palisa Observatory and Planetarium, Ostrava, www.ostrava.cz/jahia/Jahia.

Jacobus Kapteyn (1851-1922)

Discovered the first evidence of the rotation of our Galaxy

Jacobus Cornelius Kapteyn, a Dutch astronomer, was born in Barneveld, Netherlands on January 19, 1851. At age 18, he entered the University of Utrecht and in 1875 obtained his doctorate in physics. He worked for three years at the Leiden Observatory before becoming the first Professor of Astronomy and Theoretical Mechanics at the University of Groningen, where he remained until his retirement in 1921.

The University of Groningen had no observatory, and for years Kapteyn unsuccessfully attempted to secure funds to establish one. So, in 1896 he established at Groningen not an observatory but a laboratory where stellar photographs taken elsewhere could be analyzed.

In 1885, he offered to help David Gill (1843-1914) measure and reduce the photographs Gill had taken of the southern sky from his observatory at the Cape of Good Hope. The project took 14 years. The resulting star catalog contained almost a half million entries.

By 1889, Kapteyn had developed new methods for determining stellar parallaxes. This work soon evolved into studies on stellar proper motions (apparent angular rate of motion of a star), and by 1896 he had found indications that, contrary to accepted belief, stars do not move about at random in space. By 1905, he had proof that they do not. It was later realized that Kapteyn's data had been the first evidence of the rotation of our Galaxy.

In 1906, he proposed the "Kapteyn Plan of Selected Areas" for enlisting the help of astronomers throughout the world to determine the apparent magnitudes, parallaxes, spectral types, proper motions, and radial velocities of as many stars as possible in over 200 patches of sky.

On the basis of the results, he proposed a model of our galaxy, now known as the **Kapteyn universe**. He pictured the Solar System as being nearly centrally embedded in a dense, almost ellipsoidal, concentration of stars, which thinned out rapidly a few thousand light-years away from the center.

Kapteyn died in Amsterdam on June 18, 1922. Named in his honor are the **Astronomical Laboratory Kapteyn** at the University of Groningen, the **Kapteyn crater** on the Moon, **Asteroid 818 Kapteynia**, and **Kapteyn's Star.**

1. Encyclopedia of World Biography on Jacobus Cornelis Kapteyn, BookRags, www.bookrags.com/biography/jacobus-cornelis-kapteyn.
2. Jacobus Cornelius Kapteyn (1851 - 1922), A Short Biography, www.strw.leidenuniv.nl/~heijden/kapteynbio.
3. Kapteyn, Jacobus Cornelius (1851-1922), The Internet Encyclopedia of Science, www.daviddarling.info/encyclopedia/K/Kapteyn.

Walter Maunder (1851-1928)

Studied sunspots and the solar magnetic cycle

Edward Walter Maunder, a British astronomer, was born in London, England on April 12, 1851. His father was a minister of the Wesleyan Society. Edward attended King's College in London while working as a bank teller to support his studies, but never graduated.

In 1873, he took a position as a spectroscopic assistant at the Royal Observatory, headed by Astronomer Royal George Airy (1801-1892). After the death of his first wife in 1888, Maunder met and married Annie Russell (1868-1947), a Cambridge-trained mathematician with whom he collaborated for the remainder of his life. In 1916, she became one of the first women accepted by the Royal Astronomical Society.

Part of Maunder's job at the Observatory involved photographing and measuring sunspots. In doing so, he observed that the solar latitudes at which sunspots occur varies in a regular way over the course of an 11 year cycle. He published the results in the form of the "butterfly" diagram.

He examined old records from the observatory's archives to determine whether there were other such periods. These studies led him in 1893 to identify the period 1645 to 1715 when sunspots were rare. Gustav Spörer (1822-1895) had earlier identified the same period. The period has come to be known as the **Maunder Minimum.**

In 1882, Maunder observed what he called an "auroral beam." The phenomenon is yet to be explained. It was perhaps an early recorded observation of a noctilucent cloud (bright cloudlike atmospheric phenomena visible in a deep twilight).

He observed Mars and was a skeptic of the notion of Martian canals. His work led him to believe that seeing canals arose as an optical illusion. Also, he was convinced that there could not be life, as in our world, on Mars because the mean temperature was too low.

Maunder was the first editor of the *Journal of the British Astronomical Association*, a position later taken by his wife, Annie. His older brother, Thomas Maunder (1841-1935), was a co-founder, and secretary of the Association for 38 years.

Maunder died on March 21, 1928. In addition to being an astronomer he was also an esteemed biblical scholar. Craters on Mars and the Moon were named in his and Annie's honor.

1. Edward Walter Maunder (1851-1928), Université de Montréal, www.astro.umontreal.ca/ ~paulchar/ grps/histoire/newsite/bio/emaunder_e.
2. Edward Walter Maunder, Wikipedia, http://en.wikipedia.org/wiki/Edward_Walter_Maunder.
3. Maunder, Edward Walter (1851-1928), The Internet Encyclopedia of Science, www.daviddarling.info/encyclopedia/M/Maunder.

John Dreyer (1852-1926)

Published the *New General Catalogue of Nebulae and Clusters of Stars*

John Louis Emil Dreyer, a Danish-Irish astronomer, was born in Copenhagen on February 13, 1852. From his schooldays in Copenhagen, Dreyer showed unusual ability in history, mathematics, and physics.

In 1874, at the age of 22, he went to Ireland to work as the assistant of Lawrence Parsons, 4th Earl of Rosse (1840-1908), at Birr Castle. Parsons was the son and successor of William Parsons, 3rd Earl of Rosse (1800-1867). It was William Parsons who built the massive Leviathan telescope, the famous 72-inch telescope, which at the time was the world's largest telescope.

Dreyer began observing and surveying deep sky objects (star clusters, nebulae, and galaxies), and in 1878 published a supplement of about 1,000 new "nebulae" to the *General Catalogue* of John Herschel (1792-1871). That same year, Dreyer went to Dublin and worked as an assistant at Dunsink Observatory.

In 1882, he went to Armagh Observatory in Northern Ireland, where he served as director until 1916. Armagh Observatory was badly funded in those days and had old, smaller instruments. Though Dreyer obtained a new 10-inch refractor, the lack of funding for an assistant precluded him from a continuation of traditional positional astronomy, so he concentrated on compiling new catalogs from older observations.

Dreyer's major contribution was the monumental *New General Catalogue of Nebulae and Clusters of Stars*, which listed 7840 objects. He followed this with two supplementary *Index Catalogues* (1895, 1908), which contained a further 5386 objects. It is the order in which they appear in these catalogs that defines the name of many prominent galaxies, nebulae, and star clusters.

In 1890, he published a biography of his childhood hero *Tycho Brahe* (1546-1601), a noted astronomer from his native country. In his later years, Breyer edited Brahe's publications and unpublished correspondence, which eventually filled 15 volumes. He also wrote *History of the Planetary Systems from Thales to Kepler* in 1906.

Breyer won the Gold Medal of the Royal Astronomical Society in 1916. He was president of that society from 1923 to 1925. He died in Oxford, England on September 14, 1926. A crater on the Moon is named after him.

1. Dreyer, John Louis Emil (1852-1926), The Internet Encyclopedia of Science, www.daviddarling.info/encyclopedia/D/Dreyer.
2. John Louis Emil Dreyer, Armagh Observatory, www.arm.ac.uk/history/dreyer.
3. John Louis Emil Dreyer, Wikipedia, http://en.wikipedia.org/wiki/John_Louis_Emil_Dreyer.

Henri Poincaré (1854-1914)

Formulated a preliminary version of the special theory of relativity

Jules Henri Poincaré, a French mathematician and theoretical physicist, was born in Nancy, France on April 29, 1854. His father, Leon Poincaré (1828-1892), was a professor of medicine at the University of Nancy. Henri's brother, Raymond Poincaré (1860-1934), became president of the French Republic during World War I. During his childhood, Henri was seriously ill for a time with diphtheria.

Poincaré entered the École Polytechnique in 1873. There, he studied mathematics. On graduation, he went on to study at the École des Mines, where he continued to study mathematics in addition to mining engineering, and on graduation received the degree of ordinary engineer. Poincaré obtained his doctorate from the University of Paris in 1879.

He made many original fundamental contributions to pure and applied mathematics, mathematical physics, and celestial mechanics.

Beginning in 1881, he taught at the University of Paris, eventually holding the chairs of Physical and Experimental Mechanics, Mathematical Physics and Theory of Probability, and Celestial Mechanics and Astronomy. Poincaré is considered the originator of the theory of analytic functions of several complex variables.

He sketched a preliminary version of the special theory of relativity in which he stated that the velocity of light is a limit velocity and that mass depends on speed. His principle of relativity stated that no mechanical or electromagnetic experiment can discriminate between a state of uniform motion and a state of rest, and he derived the Lorentz transformation.

Poincaré revolutionized celestial mechanics, inaugurating a rigorous treatment and initiating studies of the stability of the Solar System.

He found that a three body system is often chaotic in the sense that a small perturbation in the initial state, such as a slight change in one body's initial position, might lead to a radically different later state than would be produced by the unperturbed system. If the slight change isn't detectable by our measuring instruments, then we wouldn't be able to predict which final state will occur. So, Poincaré's research proved that the problem of determinism and the problem of predictability are distinct problems.

Poincaré died in Paris, France on July 17, 1912. Named after him are **Poincaré crater** on the Moon, and **Asteroid 2021 Poincaré**.

1. Henri Poincaré, Biographies/Images of Physicists and Astronomers, www.mlahanas.de/Physics/Bios/HenriPoincare.
2. Henri Poincaré, Wikipedia, http://en.wikipedia.org/wiki/Henri_Poincar%C3%A9.
3. Jules Henri Poincaré (1854-1912), The Internet Encyclopedia of Philosophy, www.iep.utm.edu/p/poincare.

Percival Lowell (1855-1916)

Argued for the existence of a canal network on Mars

Percival Lawrence Lowell, an American astronomer, was born into a wealthy family in Boston, Massachusetts on March 13, 1855. His brother was president of Harvard University for 24 years. Lowell graduated from Harvard in 1876 with distinction in mathematics. Following graduation, he devoted the next 17 years to the family business and three extended stays in the Orient.

As early as 1890, he began to correspond with William Pickering (1858-1938) on the subject of Mars, and decided to become involved in the study of that planet.

In an address to the Boston Scientific Society in 1894, he stated that there is strong reason to believe that we are on the eve of pretty definite discovery in the matter of an investigation into the condition of life on other worlds, like or unlike man. As to the nature of the canali of Giovanni Schiaparelli (1835-1910), he had no doubt—they were the work of some sort of intelligent beings.

Together with Pickering and Andrew Douglass (1867-1962), Lowell set up two sizable telescopes in an observatory in Flagstaff, Arizona and began sketching "canals" and other intricate features on the Martian surface, including dark spots where the "canals" intersected. Pickering had first reported seeing these spots, which he referred to as lakes, in 1892. Now, Lowell saw them, too, and renamed them oases. In his mind, they were the tracts of vegetation, irrigated by meltwater brought by the canals from the poles. Lowell launched a publicity campaign to announce his theory about the grand hydrological schemes of an alien race, and published a number of books on the subject.

Lowell's greatest contribution to planetary studies came during the last decade of his life, which he devoted to the search for Planet X, a hypothetical planet beyond Neptune. Lowell erroneously believed that the planets Uranus and Neptune were displaced from their predicted positions by the gravity of the unseen Planet X. His work, however, formed the beginning of the effort that did eventually lead to the discovery of Pluto.

Lowell continued to argue vociferously for the existence of a canal network on Mars until his death on November 12, 1916. He is buried on Mars Hill near his observatory.

1. Lowell, Percival (1855-1916), The Internet Encyclopedia of Science, www.daviddarling.info/encyclopedia/L/LowellP.
2. Percival Lowell, Percival Lowell Observatory Archives, www.lowell.edu/Research/library/paper/archive_home.
3. Percival Lowell, Wikipedia, http://en.wikipedia.org/wiki/Percival_Lowell.

James Keeler (1857-1900)

Discovered the dark narrow gap in the outer part of the A ring of Saturn

James Edward Keeler, an American astronomer, was born in La Salle, Illinois on September 10, 1857. His father was a partner in the La Salle Iron Works. Keeler saw the solar eclipse that swept the United States in 1869; this had a significant influence on him. He ordered a two-inch achromatic lens and two smaller lenses from a Philadelphia optical house. From these, he constructed a telescope with which he viewed the Moon, Jupiter, Saturn, nebulae, and other celestial objects.

He received a B.A. in Physics and German, with Minors in mathematics, chemistry, and astronomy, from Johns Hopkins University in 1881. After graduation, he worked as the assistant of Samuel Langley (1834-1906) at the Allegheny Observatory near Pittsburgh.

Keeler was probably the first to observe the gap in Saturn's rings, now known as the Encke Division. He did so using the 36-inch refractor at Lick Observatory in 1888. The second major gap in the A Ring, discovered by Voyager, was named the **Keeler Gap** in his honor.

Keeler was one of the pioneers in utilizing Spectroscopy to study the composition of light from stars, nebulae, and other celestial objects. He was also one of the pioneers of the new branch of science known as Astrophysics, which is the application of physics to better understand celestial bodies. Along with George Hale (1868-1938), Keeler founded and edited the *Astrophysical Journal*, which remains a major journal of astronomy today.

He was appointed director of the University of Pittsburgh's Allegheny Observatory in 1891. In 1895, his spectroscopic study of the rings of Saturn revealed that different parts of the rings reflect light with different Doppler shifts, due to their different rates of orbit around Saturn. This was the first observational confirmation of the theory of James Maxwell (1831-1879) that the rings are made up of countless small objects, each orbiting Saturn at its own rate.

Keeler discovered two asteroids, one in 1899 and one in 1900, although the second was lost and only recovered about 100 years later.

He died in San Francisco, California on August 12, 1900. In addition to the Keeler Gap in Saturn's rings, craters on Mars and the Moon are named in his honor, as is the **Asteroid 2261 Keeler**. And a 14,240-foot peak near Mount Whitney is named the **Keeler Needle**.

1. James E. Keeler, NNDB, www.nndb.com/people/961/000167460.
2. James Edward Keeler, Wikipedia, http://en.wikipedia.org/wiki/James_Edward_Keeler.
3. Walsh, Glen A., Keeler, James Edward, http://johnbrashear.tripod.com/bio/KeelerJ.

Edward Barnard (1857-1923)

Discovered several satellites of Jupiter

Edward Emerson Barnard, an American astronomer and photographer, was born into poverty in Nashville, Tennessee on December 16, 1857. To help his family, he was employed at the age of nine in the studio of a Nashville photographer, where he remained for 16 years.

His interest in astronomy began in 1876 when he read a stray copy of a popular book on astronomy. He then constructed his first telescope with a one-inch lens from a broken spyglass.

Bernard's discovery of a number of unexpected comets led to a fellowship at Vanderbilt University, where he received a B.S. in 1887. He was then appointed junior astronomer at the recently established Lick Observatory (California), which had a new 36-inch telescope, then the largest in the world. There, he discovered the fifth satellite of Jupiter, Amalthea (the last satellite to be found without photographic aid). He followed this by locating the faint and distant sixth, seventh, eighth, and ninth satellites. He also began his photography of the Milky Way.

Barnard accepted a position at the Yerkes Observatory (Wisconsin) in 1895, and in 1897 he began observing with the 40-inch photographic telescope. He then began the micrometric triangulation of some of the globular clusters, which he continued for nearly 25 years, hoping to detect motions of the individual stars.

On May 29, 1897, Barnard narrowly escaped death when, just hours after he left the observatory's dome, the 37-ton elevating floor, used to lift observers to the level of the telescope's eyepiece, collapsed after a supporting cable broke.

The observatory's acquisition in 1904 of the 10-inch Bruce photographic telescope gave added impetus to Barnard's photography of comets and his mapping of the Milky Way.

In all, Barnard collected 1400 negatives of comets and nearly 4000 plates of the Milky Way and other star fields. His published papers number more than 900.

He played a prominent role at the turn of the twentieth century in denouncing the existence of Martian canals.

Bernard died on February 6, 1923, and was buried in Nashville after a funeral in the rotunda of the Yerkes Observatory.

1. Barnard, Edward Emerson (1857-1923), The Internet Encyclopedia of Science, www.daviddarling.info/encyclopedia/B/Barnard.
2. Edward Emerson Barnard 1857-1923, Tennessee Encyclopedia of History and Culture, http://tennesseeencyclopedia.net/imagegallery.php?EntryID=B007.
3. Encyclopedia of World Biography on Edward Emerson Barnard, BookRags, ww.bookrags.com/biography/edward-emerson-barnard.

Williamina Fleming (1857-1911)

The first to discover white dwarfs

Williamina (Mina) Stevens Fleming, a Scottish-American amateur astronomer, was born in Dundee, Scotland on May 15, 1857. She taught in Dundee from age 14 until her marriage to James Fleming in 1877. The couple immigrated to Boston when she was 21. A year later, she was abandoned by her husband while pregnant with their child. To support herself and the baby, Fleming worked as a maid in the home of Edward Pickering (1846-1919), the director of the Harvard Observatory.

Pickering was said to have been unhappy with the work performed by his male employees and declared that his maid could do a better job. He hired Fleming in 1881 to do clerical work and some mathematical calculations at the Observatory. Fleming soon proved that she was capable of doing more. She devised a system of classifying stars according to their spectra, a distinctive pattern produced by each star when its light is passed through a prism. She used this system, which was later named after her, to catalog successfully over 10,000 stars over the next nine years. This work was published in 1890 in a book titled *Draper Catalogue of Stellar Spectra*.

Her duties were expanded and she was put in charge of dozens of young women hired to do mathematical computations, the work nowadays done by computers. She also edited all publications issued by the observatory. The quality of her work was so good that in 1898 Harvard appointed her curator of astronomical photographs. This was the first such appointment given to a woman.

In 1906, Fleming was the first American woman elected to the Royal Astronomical Society. In 1907, she published a study of 222 variable stars she had discovered. Her achievement is especially noteworthy when one takes into account that she had no formal higher education.

In 1910, she published her discovery of white dwarfs, which are stars that are very hot and dense and appear bluish or white in color. They are believed to be stars in a final stage of their existence. She also discovered 10 novae (exploding stars) and more than 200 variable stars.

Fleming died of pneumonia in Boston, Massachusetts on May 21, 1911. **Fleming crater** on the Moon is named jointly after her and Alexander Fleming (1881-1955), the discoverer of penicillin.

1. Bois, Denuta, Williamina Paton Stevens Fleming, Distinguished Women of Past and Present, www.distinguishedwomen.com/biographies/flemingw, 1966.
2. Fleming, Williamina Paton Stevens (1857-1911), The Internet Encyclopedia of Science, www.daviddarling.info/encyclopedia/F/Fleming.html.
3. Williamina Paton Stevens Fleminghttp, Wikipedia, http://en.wikipedia.org/wiki/Williamina_Fleming.

Mary Blagg (1858-1944)

Played an important role in standardizing the names of features on the Moon

Mary Adela Blagg, a British amateur astronomer, was born in Cheadle, Staffordshire, England on May 17, 1858. She lived her entire life in that locale. She was the daughter of a solicitor. She trained herself in mathematics by reading her brother's textbooks.

In 1875, she was sent to a finishing school in Kensington, where she studied algebra and German. She later worked as a Sunday school teacher and was the branch secretary of the Girls' Friendly Society, a social and service organization for girls and young women affiliated with the Anglican Church.

In middle age, she became interested in astronomy after attending a university extension course. Her tutor suggested she work in the area of selenography, which is the study of the surface and physical features of the Moon. In particular, he suggested she address the problem of developing a uniform system of lunar nomenclature. There were several major lunar maps at the time, but they all had significant discrepancies in terms of naming the various features.

In 1905, Blagg was appointed by the newly-formed International Association of Academies to build a collated list of all of the lunar features. She worked with S. A. Saunder on the long, tedious task. The result was published in 1913. She also did considerable work on variable stars in collaboration with Herbert Turner (1861-1930). These works were published in a series of 10 articles in the *Monthly Notices*. Turner admitted that the large majority of the work had been performed by Blagg.

After the publication of several research papers for the Royal Astronomical Society, she was elected as a fellow in 1916—the first woman to be allowed entry into that society.

In 1920, Blagg joined the Lunar Commission of the newly-formed International Astronomical Union. They tasked her with continuing her work on standardizing the nomenclature. For this task, she collaborated with Karl Müller (1866-1942) of Vienna. Together, in 1935, they produced a two volume set, *Named Lunar Formations*, which became the standard reference on the subject.

During her life, she performed volunteer work, including caring for Belgian refugee children during World War I. She died on April 14, 1944. **Blagg crater** on the Moon is named for her.

1. Blagg, Mary Adela (1858-1944), The Internet Encyclopedia of Science, www.daviddarling.info/encyclopedia/B/Blagg.
2. Mary Adela Blagg, Wikipedia, http://en.wikipedia.org/wiki/ Mary_Adela_Blagg.

William Pickering (1858-1938)

Discovered Saturn's ninth moon, Phoebe

William Henry Pickering, an American astronomer, was born in Boston, Massachusetts on February 15, 1858. He was the younger brother of astronomer Edward Pickering (1846-1919). William studied at the Massachusetts Institute of Technology (MIT), graduating in 1879. He then accepted a faculty position at MIT.

In 1887, he moved to Harvard College Observatory, where Edward was director. There, in 1899, he discovered Saturn's ninth moon, Phoebe, from photographic plates. This was the first planetary satellite with retrograde motion to be discovered. Pickering also believed he had discovered a tenth moon in 1905, which he called Themis; however, Themis was never verified and its existence has been discredited.

In 1903, Pickering produced a photographic atlas of the Moon, *The Moon: A Summary of the Existing Knowledge of our Satellite*.

He speculated in 1907 that the Moon was once a part of the Earth that had broken away from where the Pacific Ocean now lies. He also proposed some sort of continental drift that broke up America, Asia, Africa, and Europe, which once formed a single continent, because of the Moon separating from Earth. And in 1908, he hypothesized about the idea of flying machines that could be used to drop dynamite on the enemy in time of war.

He studied craters on the Moon, and hypothesized that changes in the appearance of the crater, Eratosthenes, were due to lunar insects. He also claimed to have found vegetation on the Moon. In 1919, he predicted the existence and position of a Planet X, near the constellation of Gemini, based on anomalies in the positions of Uranus and Neptune, but a search of Mount Wilson Observatory photographs failed to find the predicted planet.

Pickering constructed and established several observatories, including the private Flagstaff Observatory of Percival Lowell (1855-1916), who considered the "canals" on Mars to be the work of some sort of intelligent beings. Pickering went further than Lowell, when in 1903 he claimed to observe signs of life on the Moon.

He spent much of the later part of his life at his private observatory in Jamaica. He died on January 17, 1938. **Pickering crater** on the Moon is jointly named after him and Edward Pickering.

1. Pickering, William Henry (1858-1938), The Internet Encyclopedia of Science, www.daviddarling.info/encyclopedia/P/PickeringWH.
2. William Henry Pickering Wikipedia, http://en.wikipedia.org/wiki/William_Henry_Pickering.
3. William Henry Pickering, American astronomer (1858–1938), Answers.com, www.answers.com/topic/william-henry-pickering.

Svante Arrhenius (1859-1927)

Argued that life arrived on Earth in the form of microscopic spores that had been propelled across interstellar space

Svante August Arrhenius, a Swedish physicist, was born at Vik, near Uppsala, Sweden on February 19, 1859. His father, a surveyor and administrator on a Swedish estate, moved the family to Uppsala in the early 1860s to take a position as a supervisor at the University of Uppsala.

Arrhenius entered the University of Uppsala at age 17. There, he studied mathematics, chemistry, and physics, and earned his B.S. in 1878. In 1881, he transferred to the Swedish Academy of Sciences, where he studied electrical theory under Erik Edlund (1819-1888). In 1884, he returned to Uppsala to present his dissertation and obtain his doctoral degree.

Scientist had known for a century that some substances conducted electricity when dissolved in water, but not in their dry state. These were called electrolytes. Substances that would not conduct electricity when dissolved were called non-electrolytes. Arrhenius performed hundreds of experiments, measuring the conductivity of various compounds in solution, studying how the solutions' properties varied with the amount of compound dissolved, and measuring the boiling points and freezing points of the solutions. He concluded that each molecule of sodium chloride, for example, produces two particles. To explain this behavior, he proposed that when a molecule of an electrolyte dissolves in a solution, it separates into charged particles called ions, and that the electrical charges that the ions carry enable the solution to conduct electricity. The theory at the time was not universally accepted. However, in 1903 Arrhenius was awarded the Nobel Prize in chemistry for his electrolytic theory of dissociation.

Some of his later ideas were questionable. For instance, he believed that life on Earth had sprung from "spores" that had been driven through outer space by radiation pressure from starlight, a concept known as panspermia.

Arrhenius correctly believed that carbon dioxide gas in the atmosphere traps heat by allowing sunlight to reach Earth, but blocks the radiation of heat away from the planet, now known as the greenhouse effect. He died in Stockholm, Sweden on October 2, 1927. He is considered one of the founders of physical chemistry.

1. Arrhenius, Svante August (1859-1927), The Internet Encyclopedia of Science, www.daviddarling.info/encyclopedia/A/Arrhenius.
2. Arrhenius, Svante August, Microsoft Encarta Online Encyclopedia, http://au.encarta.msn.com/encyclopedia_761560437/Arrhenius_Svante_August, 2008.
3. World of Scientific Discovery on Svante August Arrhenius, BookRags, www.bookrags.com/biography/svante-august-arrhenius-wsd.

Daniel Barringer (1860-1929)

The first person to prove the existence of a meteorite crater on Earth

Daniel Barringer, an American mining engineering, was born on May 25, 1860. He was the son of Daniel Moreau Barringer (1806-1873), a U.S. Congressman from North Carolina between 1843 and 1849. Daniel graduated from Princeton University in 1879, and in 1882 graduated from the University of Pennsylvania's Law School. He later studied geology and mineralogy at Harvard University and the University of Virginia, respectively. Involvement in gold and silver mining ultimately made him a wealthy man.

In 1902, Barringer learned of a large crater located 35 miles east of Flagstaff, Arizona. The crater had previously been studied in 1891 by Grove Gilbert (1843-1918), who concluded that the crater could not be the result of a meteor impact, and could only be the result of an explosion.

Barringer, however, became convinced that the crater was of meteoritic origin. With both scientific and monetary aims in mind, he created the Standard Iron Company to mine the crater for the iron from the meteor he believed must be buried under the surface. The Company conducted drilling operations between 1903 and 1905, and concluded that the crater had indeed been caused by a violent impact, although they were unable to find the meteorite.

In 1906, Barringer and his partner, Benjamin Tilghman, presented papers to the U.S. Geological Survey outlining the evidence in support of the impact theory.

The mining of the crater continued until 1929 without finding the ten-million ton meteorite that Barringer assumed must be hidden. At this time, the astronomer, Forest Ray Moulton (1872–1952), performed calculations on the energy expended by the meteorite on impact, and concluded that the meteorite had most likely vaporized when it landed. By this time, Barringer was nearly bankrupted from his exploration of the crater.

Barringer died of a heart attack on November 30, 1929, shortly after reading the very persuasive arguments that no iron was to be found. A small crater on the far side of the Moon was named after Barringer posthumously. The Arizona crater was named **Barringer crater** in his honor. The impact theory has been confirmed with new evidence, most notably by Eugene Shoemaker (1928-1997) in the 1960s.

1. Daniel Barringer, Wikipedia, http://en.wikipedia.org/wiki/Daniel_Barringer_(geologist).
2. Daniel Moreau Barringer, The Internet Academy of Science, www.daviddarling.info/encyclopedia/B/Barringer.
3. Rabinowitz, Carla Barringer Daniel Moreau Barringer and the Battle for the Impact Theory, www.barringercrater.com/adventure/main.

Dorothea Klumpke (1861-1942)

The first woman to make astronomical observations from a balloon

Dorothea Klumpke, an American astronomer, was born in San Francisco, California on August 9, 1861. Her father was a German immigrant who had come to California in 1850 with the Gold Rush. In 1877, Dorothea started out studying music at the University of Paris, but later turned to astronomy, receiving her bachelor's degree in 1886. She then took a job at the Paris Observatory.

From 1886, the Paris Observatory was involved with the Carte du Ciel project, which required photographing the entire sky and showing stars as faint as the 14th magnitude. Dorothea found a niche at the International Congress of Astronomers as a linguist translating all the papers into French for the official records.

After studying mathematics and mathematical astronomy, in 1893 the University of Paris awarded her the D.Sc. degree. Despite being a woman, and in the face of fierce competition from 50 men, she secured the post of Director of the Bureau of Measurements at the Paris Observatory.

In 1896, she sailed to Norway on the Norwegian vessel Norse King to observe the solar eclipse of August 9, 1896. The eclipse was not a success because of obscuring clouds, but she met Isaac Roberts (1829-1904), a 67-year old Welsh widower and entrepreneur who had become a pioneer in astrophotography.

In 1899, France, Germany, and Russia organized a balloon expedition to observe a great meteor shower, now known as the Leonids. The French chose Klumpke to ride in a balloon to observe the shower, which turned out a failure.

In 1891, Dorothea and Isaac were married and stayed at his Sussex home. She left her job at the Paris Observatory to be with Isaac, whom she assisted in a project to photograph all 52 of the Herschel areas of nebulosity. Their marriage ended with Isaac's death in 1904.

Dorothea returned to Paris Observatory and spent 25 years processing the plates and Isaac's notes, periodically publishing papers on the results.

After 1934, she and her sister, Anna, moved back to San Francisco, where Klumpke died on October 5, 1942. Minor planets **339 Dorothea** and **1040 Klumpkea** were named in her honor, as was the **Klumpke-Roberts Award** of the Astronomical Society of the Pacific.

1. Dorothea Klumpke, Wikipedia, http://en.wikipedia.org/wiki/Dorothea_Klumpke.
2. Klumpke-Roberts, Dorothea (1861-1942), The Internet Encyclopedia of Science, www.daviddarling.info/encyclopedia/K/Klumpke-Roberts.
3. Stone, Don, Dorothea Klumpke Roberts, Pioneer Woman Astronomer, Eastbay Astronomical Society, www.aanc-astronomy.org/articles/dorothea.

William Campbell (1862-1938)

Pioneered the use of spectroscopy to study the motions of stars

William Wallace Campbell, an American astronomer, was born in Hancock, Ohio on April 11, 1862. He graduated from the University of Michigan in 1886 with a degree in engineering. While he was a student at Michigan he worked as an assistant at the Observatory, where he became fascinated by astronomy. After graduation, he joined the faculty at the University of Colorado. Two years later he returned to Michigan.

In 1890, Campbell spent the summer as a volunteer at the University of California's Lick Observatory on Mount Hamilton. There, he studied astrophysics. In 1900, he became the director of Lick Observatory.

His *Stellar Motions* (1913) and catalog of nearly 1,000 stellar velocities became the classic works in the field. Through his studies of planetary atmospheres, Campbell was also one of the first to declare (1894) that Mars is unsuitable to sustain life as known on Earth.

Campbell planned an expedition to Australia to observe the solar eclipse of 1922. The purpose of the observations was to test Einstein's theory that light passing near a massive object, such as the Sun, is bent by the gravitational field of the object. The eclipse observations by the Lick team showed conclusively that light passing next to the edge of the Sun is bent by the amount Einstein predicted.

Campbell discovered that nearly one-third of the brighter stars are spectroscopic binaries; that is, systems in which two stars are so close together that they appear to be a single star, but which are shown spectroscopically to be two stars moving in orbits around their center of mass.

In 1922, Campbell became President of the University of California. He served as president of the National Academy of Sciences from 1931 to 1935.

He was awarded the Henry Draper Medal (1906), Royal Astronomical Society Gold Medal (1906), and Bruce Medal (1915).

Suffering advancing blindness and difficulty with speech following a stroke, Campbell leapt to his death from a fourth-story window in San Francisco on June 14, 1938. **Campbell Hall**, named in his honor, houses UC Berkeley's Astronomy Department. Also named for him are **Campbell crater** on the Moon, a crater on Mars, and **Asteroid 2751 Campbell**.

1. William W. Campbell, NNDB, www.nndb.com/people/771/000168267.
2. William Wallace Campbell (1862-1938), astronomer and administrator, ScienceMatters@Berkeley, http://sciencematters.berkeley.edu/archives/volume2/issue9/legacy.
3. William Wallace Campbell, Microsoft Encarta Online Encyclopedia, http://uk.encarta.msn.com/encyclopedia_761551753/campbell_william_wallace, 2008.

Annie Cannon (1863-1941)

Proposed a scheme to organize and classify stars based on their temperatures

Annie Jump Cannon, an American astronomer, was born in Dover, Delaware on December 11, 1863. Her father was a shipbuilder and state senator. Her mother had a childhood interest in star-gazing, and passed that interest along to Annie. In 1880, Annie was sent to Wellesley College in Massachusetts, where she was stricken with scarlet fever. As a result, she became almost completely deaf. Nevertheless, she graduated in 1884 with a degree in physics and returned home. There, she existed in a state of boredom until Sarah Whiting (1847-1927, her teacher at Wellesley, hired her as an assistant. This allowed Cannon to take graduate courses in physics and astronomy at the college. While at Wellesley, Whiting inspired her to learn about spectroscopy. Also during those years, Cannon developed her skills in the new art of photography.

To gain access to a better telescope, she enrolled at Radcliffe Women's College at Harvard, which had access to the Harvard College Observatory. In 1894, Cannon became a member of "Pickering's women," the women hired by Harvard Observatory director Edward Pickering (1846-1919) to complete the *Draper Catalogue* mapping and defining all the stars in the sky.

Not long after the work on the *Catalogue* began, a disagreement occurred as to how to classify the stars. Antonia Maury (1866-1952) insisted on a complex classification system, while Williamina Fleming (1857-1911), who was overseeing the project for Pickering, wanted a more simplified straightforward approach. Cannon suggested a solution. She started by examining the bright southern hemisphere stars. To these stars she applied a third system, a division of stars into the spectral classes O, B, F, G, K, and M. She gave her system a mnemonic of "Oh Be a Fine Girl and Kiss Me." Known as the Harvard Classification Scheme, it was the first serious attempt to organize and classify stars based on their temperatures. In 1922, her system was adopted by the International Astronomical Union as the official system for the classification of stellar spectra.

Cannon was given an honorary doctorate in 1925 from Oxford University, England. She died in Cambridge, Massachusetts on April 13, 1941 after receiving a regular Harvard appointment as the William C. Bond Astronomer. The **Cannon crater** on the Moon is named after her.

1. Annie Jump Cannon, ***San Diego Supercomputer Center,*** www.sdsc.edu/ScienceWomen/cannon.
2. Annie Jump Cannon, Wikipedia, http://en.wikipedia.org/wiki/Annie_Jump_Cannon.
3. Cannon, Annie Jump (1863-1941), The Internet Encyclopedia of Science, www.daviddarling.info/encyclopedia/C/Cannon.

Maximilian Wolf (1863-1932)

A pioneer of astrophotography

Maximilian Franz Joseph Cornelius Wolf, a German astronomer, was born in Heidelberg, Germany on June 21, 1863. He was awarded a Ph.D. by the University of Heidelberg in 1888; studied in Stockholm, Sweden for two years; and then returned to spend the rest of his life at Heidelberg, where he founded and directed the Königstuhl Observatory. He also served as professor of astrophysics.

Wolf was a pioneer in developing and using astrophotographic techniques, including the dry plate (a glass plate coated with a light-sensitive, gelatinous emulsion) and the blink comparator, which quickly alternates between views of two photographic plates taken at different times. With such tools he discovered hundreds of variable stars and asteroids and about 5,000 nebulae.

Between 1919 and 1931, Wolf searched for high proper motion stars, and produced a catalog of 1,566 such objects. Proper motion is the apparent angular rate of motion of a star or other object across the line of sight on the celestial sphere.

Wolf discovered more than 200 asteroids with the Bruce double-astrographic camera, which is a telescope designed for the sole purpose of astrophotography. Astrographs are usually used in wide field surveys of the night sky, as well as detection of objects such as asteroids, meteors, and comets. The first asteroid Wolf discovered, 323 Brucia, was named after Catherine Bruce, who had donated $10,000 for the construction of the telescope. Because of Wolf's pioneering work in the use of astrophotographic techniques, asteroid discovery rates increased sharply.

He used wide-field photography to study the Milky Way, and used statistical treatment of star counts to prove the existence of clouds of dark matter.

In 1910, Wolf proposed to the Zeiss optics firm the creation of a new instrument, now known as the planetarium. A planetarium is an optical device for projecting images of celestial bodies and other astronomical phenomena onto the inner surface of a hemispherical dome. World War I occurred before this could be developed, but the Zeiss company returned to it after the war. The first successful planetarium was completed in 1923. Wolf died in Heidelberg, Germany on October 3, 1932.

1. Max Wolf, Wikipedia, http://en.wikipedia.org/wiki/Max_Wolf.
2. Maximilian Franz Joseph Cornelius Wolf, The Bruce Medalists, www.phys-astro.sonoma.edu/brucemedalists/wolf/Wolf.
3. Wolf, Max(imilian) Franz Joseph Cornelius (1863-1932), The Internet Encyclopedia of Science, www.daviddarling.info/encyclopedia/W/Wolf_Max.

Hermann Minkowski (1864-1909)

Developed the concept of the space-time continuum

Hermann Minkowski, a Russian-born German mathematician, was born in Alexotas, Russia on June 22, 1864. His father was a rag merchant. The family returned to Germany in 1872 to the city of Königsberg. His brother, Oskar Minkowski (1858-1931) was the physiologist who discovered the link between diabetes and the pancreas.

Hermann studied mathematics at the University of Königsberg. In 1881, the Paris Académie Royale des Sciences offered a prize for a proof describing the number of representations of an integer as a sum of five squares of integers—a proof that the British mathematician H. J. Smith had outlined in 1867. Minkowski derived the proof independently. In 1883, both Smith and Minkowski were awarded the prize.

After receiving his doctorate from the University of Königsberg in 1885, Minkowski taught at the University of Bonn until 1894. Then, after teaching at the University of Königsberg for two years, he moved to the University of Zurich, where he stayed until 1902.

Minkowski played an important role in the development of modern mathematics. His work formed the basis for modern functional analysis. He did much to expand the knowledge of quadratic forms. He also developed the mathematical theory known as the geometry of numbers.

In 1902, he moved to the University of Göttingen and turned his attention to relativity theory. Albert Einstein (1879-1955), who published his initial work in relativity in 1905, had taken nine classes from Minkowski in Zurich.

In 2005, Minkowski participated in an electron theory seminar that discussed the current theories of electrodynamics. With this background, he studied the competing theories of subatomic particles proposed by Einstein and Hendrik Lorentz (1853-1928). He was the first to realize that both theories led to the necessity of visualizing space as a four-dimensional, non-Euclidean, space-time continuum. It became the frame of all later developments of the theory, and led Einstein to his conception of generalized relativity.

The work was published seven years after Minkowski's death from a ruptured appendix in Göttingen on January 12, 1909. He was 44 years old. Named for him are the lunar crater **Minkowski** (for him and his nephew, Rudolph Minkowski) and **Asteroid 12493 Minkowski**.

1. Hermann Minkowski, BookRags, www.bookrags.com/research/hermann-minkowski-scit-06123.
2. World of Mathematics on Hermann Minkowski, BookRags, www.bookrags.com/biography/hermann-minkowski-wom.

Robert Aitken (1864-1951)

Studied the statistics, motions, and orbits of binary stars

Robert Grant Aitken, an American astronomer, was born in Jackson, California on December 31, 1864. Aitken was partly deaf and used a hearing aid. He graduated from Williams College in Williamstown, Massachusetts in 1887. He taught mathematics and astronomy from 1888 to 1895, first at Livermore College in Livermore, California and then at the University of the Pacific in Stockton, California.

He was appointed assistant astronomer at Lick Observatory on Mount Hamilton in California in 1895, and he served as director of the Observatory from 1930 to 1935.

Aitken was the first to make a systematic survey of northern stars for binary stars. The greater part of this survey was undertaken with the 36-inch refracting telescope at Lick Observatory. During Aitken's survey, 4400 new pairs of stars were discovered.

His exhaustive study of binary stars—their statistics, motions, and orbits—led to the publication in 1932 of his *New General Catalogue of Double Stars Within 120° of the North Pole*. The catalog was a successor to the *Burnham Double Star Catalogue* and was based on observations compiled by Sherburne Burnham (1838-1921) from 1906 to 1912 and by Eric Doolittle (1870-1920) from 1912 to 1919. Aitken began work on the catalog shortly after Doolittle's death.

The catalog contains observations made up to 1927. Aitken's book *The Binary Stars* (1918), which he revised in 1935, was considered the definitive text on the subject at the time.

Aitken also measured positions of comets and planetary satellites and computed orbits. He was editor of the *Publications of the Astronomical Society of the Pacific* from 1898 to 1942.

He was awarded the French Academy of Sciences Lalande gold medal (1906), Bruce Medal (1926), and the Gold Medal of the Royal Astronomical Society (1932). He was president of the Astronomical Society of the Pacific (1898, 1915) and president of the American Astronomical Society (1937-1940).

Aitken retired from Lick Observatory in 1935. He died on October 29, 1951. Named after him are **Asteroid 3070 Aitken** and **Aitken crater** on the Moon, which is part of the very large **South Pole-Aitken basin**.

1. Aitken, Robert Grant, Microsoft Encarta Online Encyclopedia, http://encarta.msn.com/encyclopedia_762507607/aitken_robert_grant, 2008.
2. Robert Grant Aitken, The Bruce Medalists, www.phys-astro.sonoma.edu/BruceMedalists/Aitken/Aitken.
3. Robert Grant Aitken, Wikipedia, http://en.wikipedia.org/wiki/Robert_Grant_Aitken.

John Plaskett (1865-1941)

Co-published the first detailed analysis of the structure of the Milky Way Galaxy

John Stanley Plaskett, a Canadian astronomer, was born in Hickson, Ontario, Canada on October 17, 1941. His father died when he was 16, and he quit high school to help tend the family farm. Following high school, he worked as a mechanic and handy-man. After being hired as foreman in the department of physics at the University of Toronto he enrolled in undergraduate courses, and eventually obtained a B.S. in physics.

In 1903, at the new Dominion Observatory in Ottawa, he helped design and construct instruments, as well as conduct research on stellar radial velocities. In 1905, he was promoted to astronomer, beginning his career at the age of 40.

Plaskett vastly improved the observatory's 15-inch telescope by designing and building a spectroscope for it. He lobbied the Canadian parliament for a 72-inch telescope, and when it was approved he was appointed to supervise the building of the instrument for the Dominion Astrophysical Observatory in Victoria. When completed, it was the world's largest telescope at the time. It was named the **Plaskett Telescope**.

Plaskett discovered many new binary stars, including **Plaskett's star**, a massive binary previously thought to be a single star. His work on the radial velocities of galactic stars confirmed the model of Jan Oort (1900-1992) of the rotation of the Milky Way Galaxy, and located the most probable location of its gravitational center. He added evidence to support the theory of Arthur Eddington (1882-1994) that interstellar matter is widely distributed throughout the Galaxy, and showed that the stationary lines of ionized calcium in the spectra of hot stars are caused by clouds of interstellar gas.

Together with Joseph Pearce (1893-1988), he published the first detailed analysis of the structure of the Milky Way Galaxy (1935), demonstrating that the Sun is two-thirds out from the center of the galaxy and rotates once in 220 million years.

After retirement in 1935, Plaskett oversaw the construction of the 82-inch telescope for the MacDonald Observatory at the University of Texas. He died in Esquimalt, British Columbia, Canada on October 17, 1941. Minor planet **2905 Plaskett** was named after him and his son, Harry Plaskett (1893-1980), also an astronomer.

1. John S. Plaskett, NNDB, www.nndb.com/people/843/000172327.
2. John Stanley Plaskett, scence.ca, www.science.ca/scientists/scientistprofile.php?pID=279.
3. Plaskett, John Stanley (1865-1941), The Internet Encyclopedia of Science, www.daviddarling.info/encyclopedia/P/Plaskett.

Andrew Douglass (1867-1962)

Discovered a correlation between tree rings and the sunspot cycle

Andrew Ellicott Douglass, an American astronomer, was born in Windsor, Vermont on July 5, 1867. For a time, he worked at the Lowell Observatory in Flagstaff, Arizona, where he was an assistant to Percival Lowell (1855-1916) and William Pickering (1858-1938). He had a falling out with them when his experiments made him doubt the existence of artificially created canals on Mars and visible cusps on Venus. Dismissed from his post, he went on to found the Steward Observatory.

Douglass is best known for founding dendrochronology, which is the counting of rings in tree trunks to determine the tree's age. While working at the Lowell Observatory, Douglass realized there was a relationship between climate and plant growth. He started to count the rings of pine and fir trees. The rings are made of xylem, which is a transport tissue used mainly to transport water from the root to the tree. Once a year, usually in the spring or summer, a new layer of xylem is formed, adding another year to the existing rings. In moist climates, the rings are wide, and in more arid climates, the rings are narrow. He published three volumes on the subject, entitled *Climates Cycles and Tree Growth.*

In 1914, because of Douglass's work on the relationship between climate and tree growth, the curator of the American Museum of Natural History asked him to participate in the archaeological investigation of the Southwest.

In 1916, Douglass began obtaining and analyzing archaeological tree samples. By comparing samples between two sites in New Mexico, he concluded that the Pueblo Bonito site actually predated the Aztec Ruin site by 40 to 45 years. These findings led to the realization that relative dating could be used on many of the other ruins in the Southwest.

Douglass began to wonder if trees, by putting on fat rings in wet years and thin ones in dry years, might show signs of a periodicity linked to the solar cycle. He tracked this relationship into past centuries by studying beams from old buildings, as well as Sequoias and other long-lived trees. He, in fact, did find climate effects from solar variations, particularly in connection with the 17th-century dearth of sunspots known as Maunder Minimum.

Douglass died in Tucson, Arizona on March 20, 1962.

1. A. E. Douglass, Wikipedia, http://en.wikipedia.org/wiki/A._E._Douglass.
2. Andrew Douglass, History of Science 2008, http://historyofscience2008.blogspot.com/2008/03/andrew-douglass.
3. Weart, Spencer, Changing Sun, Changing Climate?, The Discovery of Global Warming, www.aip.org/history/climate/solar, August 2008.

Kristian Birkeland (1867-1917)

Determined that the aurora borealis is linked to solar magnetic activity

Kristian Olaf Birkeland, an inventor and solar investigator, was born in Christiania (Oslo today) on December 13, 1867. He studied mathematical physics, primarily in Paris and Geneva. For a short time, he studied under Heinrich Hertz (1857-1894) in Bonn, Germany.

The results of his Polar expeditions to Norway's high-latitude auroral regions, conducted from 1899 to 1900, contained the first global pattern of electric currents in that region made from ground magnetic field measurements.

The discovery of X-rays by Wilhelm Röntgen (1845-1923) inspired Birkeland to develop vacuum chambers to study the influence of magnets on cathode rays. He noticed that an electron beam directed toward the terrella (a small magnetized sphere used as laboratory model of the Earth) was guided toward the magnetic poles and produced rings of light around the poles. He concluded that the aurora borealis could be produced in a similar way. He developed a theory in which energetic electrons were ejected from sunspots on the solar surface, directed to the Earth, and guided to the Earth's Polar regions by the geomagnetic field. There, they produced the visible aurora. His theory was not widely accepted until confirmed by satellite evidence in the 1960s.

To fund his research, Birkeland co-founded a commercial venture to produce artificial fertilizer. The company, Norsk Hydro, is today Norway's largest company. His technique for nitrogen fixation brought him into contention for a Nobel Prize.

Birkeland apparently was the first to predict that plasma was ubiquitous in space, stating that the whole of space is filled with electrons and flying electric ions of all kinds. He was the first person to successfully predict that the solar wind behaves as do all charged particles in an electric field.

In 1913, Birkeland traveled to Egypt to study the zodiacal light, and found himself stranded there after the outbreak of World War I. On June 15, 1917, while on the way back to Norway, he was found dead in a Tokyo hotel room after taking 20 times the recommended dose of the barbiturate Veronal. He has been called the first space scientist because of his pioneering experiments in the early 1900s.

1. Birkeland, Kristian Olaf Bernhard (1867-1917), The Internet Encyclopedia of Science, www.daviddarling.info/encyclopedia/B/Birkeland.
2. Kristian Olaf Birkeland, Wikipedia, http://en.wikipedia.org/wiki/Kristian_Birkeland.
3. World of Invention on Kristian Olaf Bernhard Birkeland, The Internet Encyclopedia of Science, www.bookrags.com/biography/kristian-olaf-bernhard-birkeland-woi.

Frank Dyson (1868-1939)

Organized expeditions which confirmed Einstein's theory of the effect of gravity on light

Frank Watson Dyson, an English/Scottish astronomer, was born in Measham, near Ashby-de-la-Zouch, England on January 8, 1868. His father supplemented the family income by preaching in nearby Netherseal as well as in Measham. Unable to afford a horse and trap, he made the six-mile round trip each Sunday on foot. Frank went on to study mathematics at Trinity College, Cambridge.

In 1894, he accepted the post of chief assistant at the Royal Observatory at Greenwich. His first assignment was to take responsibility for the observatory's contribution to an international scheme for creating a star catalog of the entire heavens. He quickly realized that far more knowledge was needed of the positions and motions of the stars. He made this a priority when he moved to Edinburgh in 1906 to take up the position of Astronomer Royal for Scotland.

Four years later, he was back at Greenwich, having been appointed the ninth Astronomer Royal of England. He is the only person ever to have held both these posts successively. In 1915, he was knighted.

Between 1900 and 1912, Dyson went on expeditions to Portugal, Sumatra, Tunis, and Paris to observe total eclipses. The research contributed to our knowledge of the Sun's outermost layers.

During the First World War, he realized that the forthcoming eclipse, on May 29, 1919 would be ideal for testing out the general theory of relativity of Albert Einstein (1879-1955), which postulated that light would be deflected by the Sun's gravitational field. If true, then stars observed close to the Sun during an eclipse would appear shifted in position.

To test this, Dyson, with Arthur Eddington (1882-1944), went on two expeditions, one to Brazil and the other to West Africa.

After a careful study of the photographic plates from the expeditions, Dyson announced that Einstein's prediction was in fact true. Einstein's general theory of relativity quickly gained widespread acceptance.

While on a family visit to Australia, Dyson suffered a stroke. During the voyage home he died on May 25, 1939, just a few days from Cape Town, and was buried at sea.

1. Dyson, Frank Watson (1868-1939), The Internet Encyclopedia of Science, www.daviddarling.info/encyclopedia/D/Dyson_Frank.
2. Dyson, Sir Frank: Astronomer Royal Was No Pipsqueak, You & Yesterday, www.youandyesterday.co.uk/articles/Dyson,_Sir_Frank: Astronomer_Royal_was_no_pipsqueak.
3. Frank Dyson, Greenwich 2000, wwp.greenwich2000.com/info/heritage/people/astronomers/dyson.

George Hale (1868-1938)

Showed that sunspots were cooler than the surrounding photosphere and that they were strongly magnetic

George Ellery Hale, an American astronomer, was born in Chicago, Illinois on June 29, 1868. He was educated at the Massachusetts Institute of Technology (MIT), the Observatory of Harvard College, and in Berlin.

As an undergraduate at MIT, he invented the spectroheliograph, with which he made his discoveries of the solar vortices and the magnetic fields of sunspots. The spectroheliograph is an instrument that captures a photographic image of the Sun at a single wavelength of light. The light is focused onto a photographic medium and the slit is moved across the disk of the Sun to form a complete image. It was developed independently by Henri Deslandres (1853-1948) in 1890.

In 1890, Hale was appointed director of the Kenwood Astrophysical Observatory. He was on the Astrophysics faculty at Beloit College from 1891 to 1893 and the faculty of the University of Chicago from 1893 to 1905. He was coeditor of *Astronomy and Astrophysics* from 1892 to 1895, and after that, editor of the *Astrophysical Journal*.

He helped found a number of observatories, including Yerkes Observatory (Chicago), Mount Wilson Observatory (San Gabriel Mountains of Southern California), and the Hale Solar Laboratory (Pasadena, California). At Mount Wilson, he hired and encouraged Harlow Shapley (1885- 1972) and Edwin Hubble (1889-1953), and did a great deal of promotion of astronomical institutions, societies, and journals.

Hale also played a central role in developing the California Institute of Technology into a leading research university and in building the Palomar Observatory near San Diego, California. He conceived and helped design the first giant, reflecting telescope—a reflector with a 200-inch mirror—which was installed at Mount Palomar Observatory in 1948. It was named the **Hale Telescope** in his honor.

In 1908, Hale discovered that sunspots were cooler than the surrounding photosphere and that they were strongly magnetic.

Hale died on February 21, 1938. Also named after him are the 22-year solar **Hale cycle**; **Asteroid 1024 Hale**, **Hale crater** on the Moon; **Hale crater** on Mars; **Hale Middle School**, Woodland Hills; California; **Hale House**, Shoreland Hall, University of Chicago; and **Hale Building**, Pasadena, California.

1. George Ellery Hale, Microsoft Encarta Online Encyclopedia, http://encarta.msn.com/encyclopedia_761563562/hale_george_ellery, 2008.
2. George Ellery Hale, Mount Wilson Observatory Association, www.mwoa.org/hale.
3. George Ellery Hale, Wikipedia, http://en.wikipedia.org/wiki/George_Ellery_Hale.

Henrietta Leavitt (1868-1921)

Discovered the period-luminosity relation of Cepheid variables

Henrietta Leavitt, an American astronomer, was born in Lancaster, Massachusetts, on July 4, 1868. Her father was a Congregationalist minister who had a parish in Cambridge. She attended Oberlin College, switching to what is now Radcliffe College. During her senior year, she took an astronomy course, which sparked her interest in the subject.

After receiving an A.B. in 1892, she spent a number of years at home because of an illness that left her severely deaf. After some traveling, she volunteered as a research assistant at Harvard College Observatory in 1895.

She was appointed by Edward Pickering (1846-1919) to the permanent staff in 1902. There, she became chief of photographic photometry. She remained at Harvard the rest of her life.

In 1907, Pickering asked Leavitt to establish a north polar sequence of star brightness to serve as a standard for the entire sky. This standard was desirable because the photographic process in astronomy was unreliable. Leavitt used 299 plates from 13 telescopes, and compared stars ranging from the fourth to the twenty-first magnitude in brightness.

In 1913, Leavitt's system was adopted by the International Committee on Photographic Magnitudes. She went on to establish brightness sequences for 108 areas in the sky.

Leavitt discovered 2,400 variable stars, which was about half those known at the time. Most notably, she studied photographs of the Magellanic Clouds taken at Harvard's observatory in Arequipo, Peru. Of the 1,800 variable stars, Leavitt detected that some were Cepheids (regularly change their brightness). In 1908, she found that the longer the period of pulsation, the brighter the star. By 1912, she had proven that the apparent brightness increased linearly with the logarithm of their periods. Cosmic distances, which previously could be determined only out to about 100 light-years, could now be determined out to distances of 10 million light-years. This was the first method that gave scientists a conclusive grasp of the immense size of the universe.

Leavitt died of cancer in Cambridge, Massachusetts on December 21, 1921. **Asteroid 5383 Leavitt** and the **Leavitt crater** on the Moon are named in her honor.

1. Henrietta Swan Leavitt, Wikipedia, http://en.wikipedia.org/wiki/Henrietta_Swan_Leavitt.
2. World of Physics on Henrietta Leavitt, BookRags, www.bookrags.com/biography/henrietta-leavitt-wop.
3. World of Scientific Discovery on Henrietta Swan Leavitt, BookRags, www.bookrags.com/biography/henrietta-swan-leavitt-wsd/

José Solá (1868-1937)

Played a part in discrediting the Mars canal hypothesis

José Comas Solá, a Spanish astronomer, was born in Barcelona, Spain on December 19, 1868. He studied at the College of Physical and Mathematical Sciences of the University of Barcelona and graduated in 1889.

Solá began his astronomical observations in 1886 at the private observatory of Rafael Patxot in Sant Feliu de Guixols. He began observing Mars, and continued this through a number of oppositions of the planet. As early as 1894, he produced a relief map of Mars using a Grubb refractor with a 6-inch diameter, producing a magnification of 180 times. A relief map is a three dimensional map that depicts the topography of a planet's surface.

Extending his observations to other planets, Solá determined the rotational period of Saturn in 1902.

In 1904, he became the founder and first director of the Fabra Observatory that belonged to the Academy of Sciences and Arts in Barcelona. The observatory was equipped in 1904 with a double refractor with 15 inch diameter and with 6.0 meters and 3.8 meters focal length.

Many of Solá's works concerned the planets and comets. He discovered two comets, the first in Spain in 300 years. He also discovered 11 minor planets. He observed Mars and Saturn, measuring the period of rotation of the latter.

His observations of Mars played a part in discrediting the canal hypothesis. Shortly after the opposition of 1909, he wrote: "the marvelous legend of the canals of Mars has disappeared ... after this memorable opposition."

Titan, Saturn's largest moon, is the 11th in distance from the planet and the second largest satellite in the Solar System. It has been the subject of much fascination since Solá reported that the edges of Titan looked darker than its center. He produced sketches that hinted at the possibility of an atmosphere.

He was the first president of the Sociedad Astrónomica de España y América and editor of the *Urania*.

Solá died in Barcelona on December 2, 1937. He is honored by the minor planet **1655 Comas Solá**. Also, a crater on Mars bears his name.

1. Comas Solá, José (1868-1937), The Internet Encyclopedia of Science, www.daviddarling.info/encyclopedia/C/ComasSola.
2. Josep Comas Solà, Wikipedia, http://en.wikipedia.org/wiki/Josep_Comas_Sol%C3%A1.
3. Plicht, Chris Comas Solá, José [Josep] (1868 - 1937), www.plicht.de/chris/files/c.htm#ComasSola.

Eugène Antoniadi (1870-1944)

A leading critic of the canal hypothesis of Mars

Eugène Michel Antoniadi, a Turkish-born astronomer of Greek descent, was born in Constantinople (now Istanbul, Turkey) on March 1, 1870. He spent most of his life in France, becoming a French citizen in 1928.

In 1893, he was invited by Camille Flammarion (1842-1925) to work at his observatory at Juvisy. There, Antoniadi claimed to have seen Martian canals as put forth by Percival Lowell (1855-1916). As late as 1903, he maintained the incontestable reality of the canal phenomenon.

His reputation as an observer had become so great that by the 1909 opposition of Mars he was given access to the largest telescope in Europe—the 33-inch refractor at the Meudon Observatory located near Paris. With this instrument, he was able to resolve the canals into streaks or borders of darker regions, and concluded that the geometrical canal network was an optical illusion. Quickly, observers in the United States, including George Hale (1868-1938), confirmed Antoniadi's opinion. Worldwide belief in the canal hypothesis went into rapid decline.

He was also a regular observer of the inner planets, Mercury and Venus. In his two works *La Planète Mars* (1930) and *La Planète Mercure* (1934), he published the results of many years' observations, and presented the best maps of Mars and Mercury to appear until the space probes of recent times. Like Giovanni Virginio (1835-1910), however, he believed Mercury always kept the same face toward the Sun, an assumption now known to be false. His observations seemed to confirm the rotation period of 88 days of Giovanni Schiaparelli (1835-1910), identical with the planet's period of revolution around the Sun. The modern accepted value is approximately 59 days.

Antoniadi also observed the great Martian storms of 1909, 1911, and 1924, noting after the last one that the planet had become covered with yellow clouds and presented a color similar to Jupiter. He claimed to have seen local obscurations, which he thought were due to material suspended in a thin Mercurian atmosphere.

He is also known for creating the **Antoniadi scale**, which is a system used to categorize the weather conditions when viewing the stars at night.

Antoniadi died on February 10, 1944. A crater on Mars and the **Antoniadi crater** on the Moon were named in his honor, as well as **Antoniadi Ridge** on Mercury.

1. Antoniadi Scale, Wikipedia, http://en.wikipedia.org/wiki/Antoniadi_scale.
2. Antoniadi, Eugène Michael (1870-1944), The Internet Encyclopedia of Science, www.daviddarling.info/encyclopedia/A/Antoniadi.
3. Eugène Michael Antoniadi, Answers.com, www.answers.com/topic/eug-ne-michael-antoniadi.

Forest Moulton (1872-1952)

Co-proposed that the planets of the Solar System arose from an encounter between the Sun and another star

Forest Ray Moulton, an American Astronomer, was born on April 29, 1872 in Le Roy, Michigan. His birth took place in a log cabin built by his father, a civil war veteran, on the family's 160 acre homestead claim. Ray was educated at home by his mother, and later was taught in a one-room school. He graduated from Albion College in south-central Michigan in 1895, and enrolled as a graduate student at the University of Chicago. In 1899, he earned a Ph.D. in astronomy and mathematics. He immediately became a member of the astronomy faculty there, rising to full professor in 1912.

When the United States entered World War I, Moulton was commissioned a major in the Army and Placed in charge of the Ballistics Branch of the Army Ordinance Department in Maryland. Following the war, he returned to the University of Chicago, where he taught theoretical astronomy and mathematics.

Along with Thomas Chamberlin (1843-1928), Moulton proposed that planetismals had coalesced to form the Solar System. Their hypothesis called for the close passage of another star to trigger this condensation, a concept that has since fallen out of favor. Moulton proposed correctly that satellites, which were discovered to be in orbit around Jupiter, were gravitationally-captured planetismals.

In 1926, Moulton left the University of Chicago to become director of the Utilities Power and Light Company. In 1937, he took a position as permanent secretary of the American Association for the Advancement of Science, then the nation's largest scientific society.

Moulton authored several books and articles, and was widely considered an excellent teacher. Some of his works include *An Introduction to Celestial Mechanics* (1902), *The Nature of the World and Man* (1926), and *Consider the Heavens* (1935). He also became the host of a popular radio science program. His books, lectures, and radio performances influenced astronomers, college students, and scientists.

He died on December 7, 1952. The **Moulton crater** on the Moon, the **Adams-Moulton methods** for solving differential equations, and the **Moulton plane** in geometry are named after him.

1. Forest Ray Moulton, Wikipedia, http://en.wikipedia.org/wiki/Forest_Ray_Moulton.
2. Forest Ray Moulton: Astronomer, Mathematician, Son of Homesteaders, Homestead National Monument of America, www.nps.gov/home/historyculture/upload/MW,pdf,MoultonBio,b.pdf.
3. O'Connor, J.J. and E. F. Robertson, Forest Ray Moulton, MacTutor, www-history.mcs.st-andrews.ac.uk/history/Biographies/Moulton.

Willem de Sitter (1872-1934)

Argued that there might be large amounts of matter which do not emit light

Willem de Sitter, a Dutch mathematician, physicist and astronomer, was born in Sneek, Netherlands on May 6, 1872. The son of a judge, he studied mathematics and physics at Groningen, where his interest in astronomy was aroused by Jacobus Kapteyn (1851-1922). He then joined the Groningen astronomical laboratory. He worked at the Cape Observatory in South Africa from 1897 to 1899. In 1908, he was appointed to the chair of astronomy at Leiden and in 1919 became director of the Leiden Observatory.

In 1913, de Sitter, based on observations of double star systems, argued that the velocity of light was independent of the velocity of the source. Up until then, scientist had unsuccessfully sought emission theories of light which depended on the velocity of the source, but which did not conflict with experimental evidence.

In 1916, he developed the concept of the **de Sitter universe**, a solution for Einstein's general relativity equations in which there is no matter and a positive cosmological constant. Unlike Einstein, who felt the universe had matter and was fixed in size, de Sitter maintained that relativity actually implied that the universe was expanding. Einstein eventually agreed with de Sitter. The de Sitter universe formed the basis of the Big Bang theory.

De Sitter's work led directly to the 1919 expedition by Arthur Eddington (1882-1944) to measure the gravitational deflection of light rays passing near the Sun.

In 1932, Einstein and de Sitter published a joint paper in which they proposed what became known as the **Einstein-de Sitter model** of the universe. They argued that there might be large amounts of matter which does not emit light and has not been detected. This matter, now called dark matter, has since been shown to exist by observing its gravitational effects.

De Sitter is also known for his research on the satellites of Jupiter and the rotation of the Earth.

He died in Leiden, Netherlands on November 20, 1934. Also named after him are the **De Sitter crater** on the Moon and **Asteroid 1686 De Sitter**. His son, Aernout, was director of the Bosscha Observatory in Lembang, Indonesia (then the Dutch East Indies), when he was captured by the Japanese at the outset of World War II, and died in a Sumatra labor camp in September of 1944.

1. J. J. O'Connor and E F Robertson, Willem de Sitter, MacTutor, www-gap.dcs.st-and.ac.uk/~history/Printonly/Sitter.
2. Willem de Sitter, Wikipedia, http://en.wikipedia.org/wiki/Willem_de_Sitter.
3. William de Sitter, NNDB, www.nndb.com/people/209/000168702.

Heber Curtis (1872-1942)

Played an important role in establishing the size of the universe

Heber Doust Curtis, an American astronomer, was born in Muskegon, Michigan on June 27, 1872. At the University of Michigan he studied Latin and Greek, receiving his B.A. in 1892 and a year later his M.A. His first job was teaching Latin at his old high school in Detroit.

At the age of 22 he became Professor of Latin at Napa College, California, where he had access to a refracting telescope and small observatory. He went on in 1902 to earn his Ph.D. in astronomy from the University of Virginia.

In 1898, he joined the staff of Lick Observatory, located on Mount Hamilton in California. In 1920, he became director of the University of Pittsburgh's Allegheny Observatory, and in 1930 he was appointed director of the University of Michigan's observatory. His early work, under the direction of William Campbell (1862-1938), involved measuring the radial velocities of stars. In 1910, he became interested in spiral nebulae.

The precise composition of spiral nebulae was the subject of dispute. The two main schools of thought were that they were either giant star clusters far beyond our own galaxy or simply clouds of debris and much closer. The scale of the universe itself was central to both theories.

In 1917, Curtis argued that the observed brightness of novae, found by him and by George Ritchey (1864-1945) on photographs of spiral nebulae, indicated that the nebulae lay well beyond our Galaxy. He also maintained that extremely bright novae, later identified as supernovae, could not be included with the novae as distance indicators.

By photographing the spiral nebulae, Curtis began to comprehend the actual vastness of space, viewing nebulae as islands in the universe. He estimated the Andromeda Nebula (now known as the Andromeda Galaxy) to be 500,000 light-years away, a view opposed by many, including Harlow Shapley (1885-1972), who proposed that the spiral nebulae lay within the Milky Way.

In 1920, at a meeting of the National Academy of Sciences, Curtis engaged in a famous debate with Shapley over the size of the Galaxy and the distance of the spiral nebulae. The matter lay unresolved, however, until 1924 when Edwin Hubble (1889-1953) re-determined the distance of the Andromeda Nebula and demonstrated that it was a galaxy in its own right. Curtis died in Ann Arbor, Michigan on January 9, 1942.

1. Bitterman, Jay, Heber Curtis, *Lake County Astronomical Society* http://bpccs.com/lcas/Articles/curtis.
2. Curtis, Heber Doust (1872-1942), The Internet Encyclopedia of Science, www.daviddarling.info/encyclopedia/C/Curtis.

Ejnar Hertzsprung (1873-1967)

Developed the first color-magnitude diagram for stars

Ejnar Hertzsprung, a Danish chemist and amateur astronomer, was born in Frederiksberg, Denmark on October 8, 1873. His father had studied astronomy, but could not get a job in that field, so he became a director of an insurance company. Ejnar studied chemistry at Copenhagen Technical Institute. Beginning in 1902, he began visiting the University Observatory and the privately owned Urania Observatory in Copenhagen.

In 1905, he set up a standard that he called the "absolute magnitude" of a star, defining the quantity as what a star's brightness would be at a distance of ten parsecs from the Earth (one parsec is equivalent to 3.26 light-years). Once he had determined the intrinsic brightness of a number of stars, it became apparent to him that there was a relationship between a star's color and its luminosity. He found that red stars generally are not as bright as blue stars, while yellow stars fall somewhere between the two. Furthermore, since the color of a star is related to its temperature, Hertzsprung concluded that a star's temperature and brightness are correlated. In 1907, he published a paper in which he suggested there must be a connection between the spectrum and the luminosity of stars. He sent a preprint to Karl Schwarzchild (1873-1916), the director of the Göttinger Observatory, who offered him a job. In 1909, they both moved to Potsdam Observatory.

During travel to the United States in 1910, Schwarzchild met Henry Russell (1877-1957), who had come to the same conclusion as Hertzsprung. The **Hertzsprung-Russell Diagram** was published in 1911. It showed the relationship between absolute magnitude, luminosity, classification, and effective temperature of stars. The diagram has played an important role in providing clues to astronomers about stellar evolution.

In 1913, Hertzsprung determined the distance to the Small Magellan Cloud, the first distance to an object outside the Milky Way, by using the delta-Cephei type of variable stars.

Hertzsprung worked at Leiden University Observatory from 1919 to 1945, becoming director there in 1935. Later in his life he made discoveries in the evolution of open star-clusters and variable stars. He retired in 1945, and died in Denmark on October 21, 1967 at the age of 94. **Hertzsprung crater** on the Moon and **Asteroid 1693 Hertzsprung** are named for him.

1. Ejnar Hertzsprung, Rundetaarn.DK, www.rundetaarn.dk/engelsk/observatorium/hertz.
2. Ejnar Hertzsprung, Wikipedia, http://en.wikipedia.org/wiki/Ejnar_Hertzsprung.
3. World of Physics on Ejnar Hertzsprung BookRags, www.bookrags.com/biography/ejnar-hertzsprung-wop.

Karl Schwarzschild (1873-1916)

Produced the first exact solutions to the Einstein field equations for general relativity

Karl Schwarzschild, a German Jewish physicist and astronomer, was born in Frankfurt am Main, Germany on October 9, 1873. He published a paper on celestial mechanics when he was only 16 years old. He then studied at Strasbourg and Munich, obtaining his doctorate in 1896 for a work on the theories of Henri Poincaré (1854-1912).

From 1897, he worked as assistant at the Observatory in Vienna, where he developed a formula to calculate the optical density of photographic material. This formula was important because it allowed more accurate photographic measurements of the intensities of faint astronomical sources.

From 1901 to 1909, Schwarzschild was director of the Göttingen Observatory. He moved to a post at the Astrophysical Observatory in Potsdam in 1909. At the outbreak of World War I, he joined the German army, despite being over 40 years old. He served on both the western and eastern fronts, rising to the rank of lieutenant in the artillery. While serving on the front in Russia in 1915, he developed a rare and painful disease known as pemphigus, which is an autoimmune disease of the skin and mucus membranes.

He published three papers, two on relativity theory and one on quantum theory. His papers on relativity produced the first exact solutions to the Einstein field equations, and a minor modification of these results gave a solution that now bears his name—the **Schwarzschild metric**.

Schwarzschild's second paper, which gives what is now known as the **Inner Schwarzschild solution,** is valid for the Sun and stars when viewed as a quasi-isotropic heated gas. When the mass density of this central body exceeds a particular limit, it triggers a gravitational collapse, which, if it occurs with spherical symmetry, produces a **Schwarzschild black hole**. This occurs when the mass of a neutron star exceeds about three times the mass of the Sun.

Schwarzschild's struggle with pemphigus may have contributed to his death in Potsdam on May 11, 1916. **Asteroid 4463 Marschwarzschild** is named for him. He was the father of astrophysicist Martin Schwarzschild (1912-1997).

1. Karl Schwarzschild, Microsoft Encarta Online Encyclopedia, http://encarta.msn.com/encyclopedia_761580124/karl_schwarzschild.
2. Karl Schwarzschild, NNDB www.nndb.com/people/713/000168209.
3. O'Connor, J.J. and E. F. Robertson Karl Schwarzschild, MacTutor, www-groups.dcs.st-and.ac.uk/~history/Printonly/Schwarzschild.

Vesto Slipher (1875-1969)

Provided some of the earliest evidence that the universe is expanding

Vesto Melvin Slipher, an American astronomer, was born in Mulberry, Indiana on November 11, 1875. He studied mechanics and astronomy at the University of Indiana, receiving a Ph.D. in 1909. He spent his entire career at Lowell Observatory in Flagstaff, Arizona, where he became director in 1926.

At the time of his birth, it was commonly accepted that the universe was composed of a single galaxy. Everything that was visible, whether a star or a mysterious, wispy nebula was a part of the Milky Way.

Percival Lowell (1855-1916) felt certain that the spiral clouds he and others observed were planetary systems that were being formed. He hoped that understanding these spirals would increase the knowledge about the formation of our Solar System. In 1901, he hired Slipher to work on the problem at the Lowell Observatory.

The process involved studying the spectrum of the light from the nebulas, and measuring any Doppler shift present. If the light of the nebula was shifted toward the blue end of the spectrum, it meant the nebula was moving toward us; if shifted toward the red, it was moving away.

By late 1912, Slipher had photographed the Andromeda nebula four times. On examination, the spectrum appeared more like the light of stars than that of contracting gases. In addition, the entire nebula was shifted toward the blue end of the spectrum, indicating the entire system was approaching the Earth at a speed of 186 miles per second. By 1914, Slipher had analyzed the light from twelve other spirals. Only two showed a blue shift; the others were red-shifted. It was obvious that velocities this high implied vast distances. These spirals had to be star systems located outside the Milky Way. The discovery that same year that the Milky Way galaxy was rotating caused a revision in the calculations. Of all the nebulas measured, only two actually had blue shifts. In addition, the approach of the Andromeda nebula was revised downward to 31 miles per second.

Slipher's work provided the foundation on which other scientists, such as Edwin Hubble (1889-1953), built a new theory of the universe.

Slipher retired in 1952, and died in Flagstaff, Arizona on November 8, 1969.

1. Slipher, Vesto Melvin (1875-1969), The Internet Encyclopedia of Science, www.daviddarling.info/encyclopedia/S/Slipher_Vesto.
2. Vesto Melving Slipher, Lowell Observatory Archiveswww.lowell.edu/Research/library/paper/vm_slipher.
3. Vesto Slipher, Wikipedia, http://en.wikipedia.org/wiki/Vesto_Slipher.
4. World of Scientific Discovery on Vesto Melvin Slipher, BookRags, /www.bookrags.com/biography/vesto-melvin-slipher-wsd.

James Jeans (1877-1946)

Developed a model for a single star evolving into a double-star system

James Hopwood Jeans, a British physicist, astronomer, and mathematician, was born in Ormskirk, Lancashire, England on September 11, 1877. In 1897, he entered Trinity College, Cambridge, and in 1903 received his master's degree, despite the fact that tuberculosis of the joints forced him to go to several sanatoriums (1902 and 1903). In 1904, he was appointed university lecturer in mathematics at Cambridge. During his illness, from which he recovered, he wrote his first book, *The Dynamical Theory of Gases*.

From 1905 to 1909, Jeans taught applied mathematics at Princeton University. He returned to Cambridge as Stokes lecturer in 1910, but two years later relinquished the position to devote full time to research and writing.

His interests mainly involved the kinetic theory of gases and the theory of radiation, especially applied to the new quantum theory of Max Planck (1858-1947) and others. John Strutt (1842-1919) and Jeans, in 1905, separately derived what later came to be called the **Rayleigh-Jeans law**. This law describes the spectral radiance of electromagnetic radiation at all wavelengths from a black body at a given temperature.

During World War I, Jeans proved that a rotating incompressible mass with increasing rotational velocity will first become pear shaped and then cataclysmically fission into two parts. This was a model for a single star evolving into a double-star system.

His analysis of rotating bodies led him to conclude that the theory of Pierre Laplace (1749-1827) that the Solar System formed from a single cloud of gas was invalid. He proposed, incorrectly, that the planets condensed from material drawn out of the Sun by a hypothetical catastrophic near-collision with a passing star.

In 1929, Jeans turned to popular science, writing especially in astronomy, and soon became very successful. His books included *The Universe around Us* (1929), *The Mysterious Universe* (1930), and *The Growth of Physical Science* (1947).

Jeans was knighted in 1928. He died at his home in Dorking, Surrey on September 16, 1946. **Jeans crater** on the Moon is named after him, as is **Jeans crater** on Mars.

1. Encyclopedia of World Biography on James Hopwood Jeans, Sir, BookRags, www.bookrags.com/biography/james-hopwood-jeans-sir
2. James Hopwood Jeans, Wikipedia, http://en.wikipedia.org/wiki/James_Hopwood_Jeans.
3. Jeans, James Hopwood (1877-1946), The Internet Encyclopedia of Science, www.daviddarling.info/ encyclopedia/J/JeansJ.
4. Sir James Hopwood Jeans, Answers.com, www.answers.com/topic/james-hopwood-jeans.

Henry Russell (1877-1957)

Discovered the relationship between a star's color and its brightness

Henry Norris Russell, an American astronomer, was born in Oyster Bay, New York on October 25, 1877. The son of a Presbyterian minister, he was educated at Princeton University, where he received a Ph.D. in 1900.

He spent the period 1902 to 2005 as a research student and assistant at Cambridge University, England. Returning to Princeton, he served as professor of astronomy from 1911 to 1927 and director of the university observatory from 1912 to 1947.

During the course of his observations, Russell discovered a relationship between a star's color and its brightness. Blue stars were the hottest, red stars were the coolest, and yellow stars were in between. For a cool, red star to appear bright, Russell reasoned that it either had to be very close or very large. A star with a large surface area could radiate a significant amount of light at low temperatures, so those very bright stars had to be giants. Hence, he called these stars "red giant" stars.

Using observations of a large number of stars, Russell created a diagram on which the stars' spectral classification, roughly equivalent to temperature, was plotted along the abscissa, and their luminosity was plotted along the ordinate. On this diagram, most stars fell along a diagonal line running from upper left (brightest stars) to lower right (dimmest stars). The stars on this diagonal line were called "main sequence stars." These are stable stars like our Sun. Stars not on the main sequence are typically those in advanced stages of stellar evolution, such as the red giants or supergiants.

This same relationship had been discovered by Danish astronomer Ejnar Hertzsprung (1873-1967) earlier, but it had received no publicity. The chart was subsequently named the **Hertzsprung-Russell diagram** (H-R diagram). The H-R diagram was used to deduce the path of stellar evolution. The original diagram served as a useful tool for stellar astronomers for nearly a century.

Russell died on February 18, 1957. Also named after him are the **Henry Norris Russell Lectureship** of the American Astronomical Society, **Russell crater** on the Moon, a Crater on Mars, and **Asteroid 1762 Russell**.

1. Evans, J.C., Henry Norris Russell (1877-1957), Physics & Astronomy Department, George Mason University, http://physics.gmu.edu/~jevans/astr103/ CourseNotes/Text/Lec04/russell.
2. Henry Norris Russell, Answers.com, www.answers.com/topic/henry-norris-russell.
3. Russell, Henry Norris (1877-1957), The Internet Encyclopedia of Science, www.daviddarling.info/encyclopedia/R/Russell.
4. World of Scientific Discovery on Henry Norris Russell, BookRags, www.bookrags.com/ biography/henry-norris-russell-wsd.

Bernhard Schmidt (1879-1935)

Invented the refractor-reflector telescope

Bernhard Schmidt, an Estonian Swedish optician, was born on March 30, 1879. He grew up on the island of Naissaar, off the coast of Reval (Tallinn), Estonia, then part of the Russian Empire. His father was a writer, farmer, and fisherman. At the age of 15, Bernhard lost his right hand and forearm in an accident involving experimentation with gunpowder. Later that year he built his own camera, photographed the local people, and sold the pictures.

Between 1895 and 1901, he worked in Tallinn as a photographer and in the electromotor works of Volta. In 1901, he left for Goeteborg, Sweden to attend a technical school, but instead ended up at Mittweida, Germany, preferring the more practically oriented school there. He supported himself by making and selling mirrors.

In 1904, Schmidt opened an optical workshop and offered his skills to observatories to improve their optics, lenses and mirrors. After awhile his name spread throughout Germany. In 1927, he sold his shop and moved to Hamburg to work as a freelance optician at the nearby observatory in Bergedorf. The director there was Richard Schorr (1867-1951), who had become aware of Schmidt's abilities.

After returning from observing the solar eclipse of 1929 in the Pacific Ocean, Schmidt began work on his now famous telescope. He completed the first one in 1930. It had a main mirror of 17.3 inches and a corrector plate of 14 inches. The focal ratio was 1.75 and the field of view was 7.5 degrees. The telescope corrected for the optical errors of spherical aberration, coma, and astigmatism, making available for the first time for astronomical research the construction of very large, wide-angled reflective cameras of short exposure time.

His design of a wide field camera for stars and celestial objects is today known as a **Schmidt Camera**. The original Schmidt Camera is now in the museum of the Hamburg Observatory. The main mirror was destroyed during a two-year lease to the German navy during World War II, when it was used to make observations of the British harbors off the French coast. The Navy paid for a replacement, which was made by the Zeiss optical shop in Jena.

Schmidt never married. He died of pneumonia on December 1, 1935. Minor planet **1743 Schmidt** was named in his honor.

1. Bernhard Schmidt, Wikipedia, http://en.wikipedia.org/wiki/Bernhard_Schmidt.
2. Plicht Christof A., Schmidt, Bernhard (1879-1935), Red Hill Observatory, www.plicht.de/chris/names.
3. Schmidt Camera, Answers.com, www.answers.com/topic/schmidt-camera-1.

Arthur Eddington (1882-1944)

Studied the evolution and constitution of stars

Arthur Stanley Eddington, a British astronomer, was born at Kendal, Westmorland, England on December 28, 1882. His father was the headmaster and proprietor of a school. Arthur moved with his mother and sister to Somerset after his father died in the typhoid epidemic of 1884. Arthur received his bachelor's degree in 1902 from Owens College, Manchester, and immediately went to Trinity College, Cambridge, where he received his B.A. in 1905. From 1906 to 1913 he was chief assistant to the Astronomer Royal at Greenwich, after which he returned to Cambridge as Plumian Professor of Astronomy.

His early work on the motions of stars was followed, from 1916 onward, by his work on the interior of stars. He published this work in his first major book, *The Internal Constitution of the Stars* (1926). He showed that for equilibrium to be maintained in a star, the inwardly directed force of gravitation must be balanced by the outwardly directed forces of both gas pressure and radiation pressure. He also proposed that heat energy was transported from the center to the outer regions of a star, not by convection as previously thought, but by radiation.

In his book, Eddington gave a full account of his mass-luminosity relationship, which was discovered in 1924. He showed that the more massive a star the more luminous it is. This allows the mass of a star to be determined if its intrinsic brightness is known.

Eddington realized that there is a limit to the size of stars. Relatively few should have masses exceeding 10 times the mass of the Sun. Exceeding 50 solar masses would cause the star to be unstable owing to excessive radiation pressure.

He wrote a number of books for both scientists and nonprofessionals, including *The Expanding Universe* (1933). It was through Eddington that Einstein's general theory of relativity reached the English-speaking world. He was greatly impressed by the theory, and was able to provide experimental evidence for it.

Eddington observed the total solar eclipse of 1919 and reported that a very precise and unexpected prediction made by Einstein in his general theory of relativity had been successfully observed; that is, the very slight bending of light by the gravitational field of the Sun.

Eddington was knighted in 1930. He died on November 22, 1944.

1. Arthur Eddington, Microsoft Encarta Online Encyclopedia, http://encarta.msn.com/encyclopedia_761564811/eddington_sir_arthur_stanley, 2008.
2. Arthur Stanley Eddington, Wikipedia, http://en.wikipedia.org/wiki/Arthur_Stanley_Eddington.
3. Sir Arthur Stanley Eddington, Answers.com, www.answers.com/topic/arthur-stanley-eddington.

Harlow Shapley (1885-1972)

Argued that the Milky Way Galaxy is far larger than generally believed

Harlow Shapley, an American astronomer, was born on a farm in Nashville, Missouri on November 2, 1885. After dropping out of school with a fifth-grade education, he later went back and finished high school.

In 1907, at the age of 22, he decided to study journalism at the University of Missouri. Learning that the School of Journalism had been postponed for a year, he decided to study Astronomy. After graduation, he received a fellowship to Princeton University for graduate work, where he studied under Henry Russell (1877-1957). He served as astronomer at Mount Wilson Observatory in California from 1914 to 1921.

Shapley used the period-luminosity relation of Henrietta Leavitt (1868-1921) for Cepheid variable stars to determine distances to globular clusters. He was the first to arrive at the conclusion that the Milky Way Galaxy is much larger than previously thought, and that the Sun's place in the galaxy is far from the center. His wife, Martha Betz (1890-1981), assisted him in astronomical research, both at Mount Wilson and at Harvard Observatory. She produced numerous articles on eclipsing stars and other astronomical objects.

On April 26, 1920, Shapley participated in the great debate with Heber Curtis (1872-1942) on the nature of nebulae and galaxies and the size of the universe. He argued against the theory that the Sun was at the center of the galaxy, and promoted the idea that globular clusters and spiral nebulae are within the Milky Way. He was incorrect about the latter point, but correct about the former. After the debate, he was hired to replace the recently deceased Edward Pickering (1846-1919) as director of the Harvard College Observatory.

In the 1940's, Shapley helped found the National Science Foundation. From 1941, he was on the original standing committee of the Foundation for the Study of Cycles.

In addition to astronomy, Shapley held a lifelong interest in myrmecology (the study of ants). He died of heart failure in Boulder, Colorado on October 20, 1972. Named after him are **Shapley crater** on the Moon, **Asteroid 1123 Shapleya**, **Shapley Supercluster**, planetary nebula **Shapley 1**, and the **Harlow Shapley Visiting Lectureships in Astronomy** of the American Astronomical Society.

1. Harlow Shapley, Microsoft Encarta Online Encyclopedia, /www.encarta.nl/ encyclopedia_761561771/Harlow_Shapley, 2008.
2. Harlow Shapley, Wikipedia, http://en.wikipedia.org/wiki/Harlow_Shapley.
3. Shapley, Harlow (1885-1972), The Internet Encyclopedia of Science, www.daviddarling.info/ encyclopedia/S/Shapley.

Robert Trumpler (1886-1956)

Discovered the absorption of light by interstellar dust

Robert Julius Trumpler, a Swiss-American astronomer, was born in Zürich, Switzerland on October 2, 1886. His father was a strict businessman. Robert obtained his early education in Switzerland before receiving his Ph.D. in Germany in 1910. World War I interrupted his plans, and he was mobilized and stationed in the Alps by the Swiss army.

In 1915, during the war, he immigrated to the United States and took a position at Allegheny Observatory at the University of Pittsburg. In 1918, he went to Lick Observatory at the University of California. In 1921, he became a naturalized citizen of the United States.

In 1922, Trumpler took part in an expedition to observe the solar eclipse of September 21 in Wallal, Australia. The purpose was to test Einstein's theory of general relativity. He also carefully observed and studied Mars during its oppositions in 1924 and 1926. He concluded that some of the Martian "canals" could be volcanic faults.

He is most noted for observing that the brightness of the more distant globular clusters was lower than expected, and that the stars appeared redder. He concluded that it was due to the interstellar dust scattered through the galaxy, which resulted in the absorption of light. Since globular clusters had been used to estimate the size of the Milky Way, this led to a reduction in the estimated size of about 40 percent. The absorption of light by interstellar dust was discovered independently by Russian astronomer Boris Vorontsov-Velyaminov (1904 -1994).

Trumpler also studied and cataloged open clusters to determine the size of the Milky Way galaxy. At first, he thought his analysis placed an upper limit on the Milky Way's diameter of about 10,000 parsecs, with the Sun located somewhat near the center, although he later revised this.

While cataloging open clusters he also devised a system for their classification by giving them designations for their central concentration, the range of brightness of their individual stars, and the number of stars in the cluster. This system of classifying opened clusters is still in use today.

In 1951, Trumpler retired to Rio del Mar, California. His health began to fail rapidly after being diagnosed with leukemia. He died in Berkeley, California on September 10, 1956. Named after him are **Trumpler crater** on the Moon and a crater on Mars,

1. Robert Julius Trumpler, Wikipedia, http://en.wikipedia.org/wiki/Robert_Julius_Trumpler.
2. Weaver, Harold F., Robert Julius Trumpler, Biographical Memoirs, www.nap.edu/readingroom/books/biomems/rtrumpler.
3. Robert Julius Trumpler (October 2, 1886 - September 10, 1956), SEDS, http://seds.org/Messier/xtra/Bios/trumpler.

Aleksandr Friedmann (1888-1925)

Formulated an early Big Bang theory

Aleksandr (Alexander) Friedmann, a Russian mathematician and cosmologist, was born in Saint Petersburg, Russia on June 16, 1888. His father was a composer. Aleksandr attended Saint Petersburg State University, where he studied mathematics from 1906 to 1910. In 1914, he was awarded a master's degree in pure and applied mathematics. In 1914 and 1915, during World War I, he served with the Russian air force as a technical expert and as a pilot. During the later part of the war, he lectured pilots on aerodynamics.

In 1916, Friedmann became the head of the Central Aeronautical Station in Kyiv, and moved with the Central Aeronautical Station to Moscow in 1917. Near the end of that year, the Central Aeronautical Station disbanded after the Russian Revolution of 1917. From 1918 to 1920, Friedmann was a professor of theoretical mechanics at Perm State University.

Friedmann returned to Saint Petersburg in 1920 and took up several concurrent appointments. He taught mathematics and mechanics at Petrograd University, was a professor of physics and mathematics at Petrograd Polytechnic Institute, and did research at the Petrograd Institute of Railway Engineering, the Naval Academy, and the Optical Institute.

In 1915, Albert Einstein (1879-1955) published the general theory of relativity, but the upheaval of World War I and the civil war raging in Russia caused Friedmann not to learn of it until near the end of the decade. In 1922, he published a solution to Einstein's equations in the German journal *Zeitschrift für Physik.* Einstein and Dutch mathematician Willem de Sitter (1872-1934) had both published solutions earlier, which assumed that the universe did not grow or shrink. Friedmann's solutions, however, indicated that it was possible for the universe to expand over time, and that it was also possible for the universe to undergo periodic expansions and contractions, depending on the value of curvature (0,+,or -) used. These models have become known as **Friedmann universes**.

Einstein initially disputed Friedmann's results, but in 1923 he concluded that the solution was correct.

Friedmann's work on a model of an expanding universe helped shape modern cosmology and is considered by many to be an early Big Bang theory. Friedmann died in Saint Petersburg on September 16, 1925.

1. Alexander Friedmann, Microsoft Encarta Online Encyclopedia, http://encarta.msn.com/encyclopedia_761586013/alexander_friedmann, 2008.
2. Friedmann, Aleksandr Aleksandrovich (1888-1925), The Internet Encyclopedia of Science, www.daviddarling.info/encyclopedia/F/Friedmann.

Edwin Hubble (1889-1953)

Laid the observational basis for the cosmological theory of the expanding universe

Edwin Hubble, an American lawyer and astronomer, was born in Marshfield, Missouri on November 20, 1889. His father, a lawyer, was in the insurance business. Hubble received his bachelor's degree in law from the University of Chicago in 1910. He then went as a Rhodes Scholar to Oxford University, England.

From 1913 to 1914, he practiced law and taught high school in Kentucky and Indiana. Then he decided to devote his life to astronomy, and left for the University of Chicago's Yerkes Observatory in Williams Bay, Wisconsin.

In 1917, Hubble earned his Ph.D. in astronomy. About the same time, the United States declared war on Germany. Hubble volunteered to serve in the U.S. Army. In 1919, he left the Army with the rank of major.

Having earlier received an invitation from George Hale (1868-1938) to work at Mount Wilson Observatory near Pasadena, Californian, Hubble accepted the offer. There, the 100-inch Hooker telescope was located. It was the largest telescope in the world until 1948. Hubble worked at Mount Wilson for the rest of his career, and it was there that he carried out his most important work.

In 1924, Hubble measured the distance to the Andromeda nebula, a faint patch of light with about the same apparent diameter as the moon. He showed it was about a 100,000 times as far away as the nearest stars. He concluded that it had to be a separate galaxy, comparable in size with our own Milky Way, but much further away.

In 1929, he showed that galaxies are moving away from us with a speed proportional to their distance. The explanation was simple, but revolutionary—the universe was expanding.

Hubble died of a coronary thrombosis in San Marino, California on September 28, 1953. In 1990, NASA launched the **Hubble Space Telescope**. Also name after him are **Asteroid 2069 Hubble, Hubble crater** on the Moon, **Edwin P. Hubble Planetarium** (located in the Edward R. Murrow High School in Brooklyn, New York), **Edwin Hubble Highway** (stretch of Interstate 44 passing through his birthplace), and **Hubble Middle School** in Wheaton, Illinois.

1. Edwin Hubble, Microsoft Encarta Online Encyclopedia, http://encarta.msn.com/encyclopedia_761551516/edwin_hubble.
2. Encyclopedia of World Biography on Edwin Powell Hubble, BookRags, www.bookrags.com/biography/ edwin-powell-hubble
3. Wands, David, Edwin Powell Hubble, MacTutor, www-groups.dcs.st-and.ac.uk/~history/Biographies/Hubble.

Harold Jones (1890-1960)

Determined the distance to the Sun by triangulating the distance of the asteroid Eros

Harold Spencer Jones, a British astronomer, was born in Kensington, London on March 29, 1890. Although born Jones, he preferred his surname to be Spencer Jones. He was educated at Jesus College at the University of Cambridge.

In 1913, he became chief assistant at the Royal Greenwich Observatory. He led a decade-long, worldwide effort to determine the distance to the Sun by triangulating the distance of the asteroid Eros when it passed near Earth in 1930-1931. He photographed Eros more than 1,200 times and reduced the data from other observers in one of the most impressive computational feats of the pre-computer era. His work changed astronomers' estimation of the distance of the Earth from the Sun.

In 1933, he became Astronomer Royal at the Cape of Good Hope. There, he worked on proper motions and parallaxes. He showed that small residuals in the apparent motions of the planets are due to the irregular rotation of Earth. He computed a more accurate value for the mass of the Moon than had ever been attained. After World War II, he supervised the move of the Royal Observatory to Herstmonceux, where it was renamed the Royal Greenwich Observatory.

In the first edition of his influential book, *Life on Other Worlds*, published in 1940, he echoed the conclusions of James Jeans (1877-1946) and others that life is not widespread in the universe. However, by the second edition in 1952, with the rise of the updated version of the nebular hypothesis of Carl von Weizsäcker (1912-2007), he was optimistic on the likelihood of life evolving on the surface of a suitable planet. Among his other books were *Worlds without End* (1935) and *A Picture of the Universe* (1947).

He was awarded the Gold Medal of the Royal Astronomical Society (1943), Royal Medal (1943), and Bruce Medal (1949).

In 1956, Spencer Jones became secretary general of the International Council of Scientific Unions and served in this capacity until his death in London, England on November 3, 1960. Named after him are **Spencer Jones crater** on the Moon, **Jones crater** on Mars, and **Asteroid 3282 Spencer Jones**.

1. Harold Spencer Jones, NNDB, www.nndb.com/people/580/000170070.
2. Sir Harold Spencer Jones, Microsoft Encarta Online Encyclopedia, http://encarta.msn.com/encyclopedia_762508803/jones_sir_harold_spencer.
3. Spencer Jones, Harold (1890-1962), The Internet Encyclopedia of Science, www.daviddarling.info/encyclopedia/S/SpencerJones.

Harold Jeffreys (1891-1989)

Calculated the surface temperatures of the four outer planets to be more than minus 100 Centigrade

Harold Jeffreys, a British mathematician, statistician, geophysicist, and astronomer, was born in Fatfield, County Durham, England on April 22, 1891. He graduated from Durham University and then went on to study mathematics, astronomy, and geophysics at St. John's College, Cambridge, graduating in 1917.

He worked on war-related work in the Cavendish Laboratories from 1915 to 1917. He was then employed at the Meteorological Office from 1917 to 1922, where he worked on hydrodynamic problems. He returned to Cambridge, joining the faculty as lecturer in mathematics and reader in geophysics, and went on to become Plumian professor of astronomy and experimental philosophy, a post he held from 1946 to his retirement in 1958.

Much of Jeffreys' interest centered on the Solar System and on the theory of geophysics, which is the physics of the Earth and its environment. In 1923, he calculated the surface temperatures of the four outer planets—Jupiter, Saturn, Uranus, and Neptune—to be more than 100° below zero Centigrade. This was in sharp conflict with the then-prevailing view that these outer planets were red-hot. His findings were later verified by direct observation, and led to a complete revision of theories on the composition and structure of the outer planets.

Using observations on the Earth's bodily tides, Jeffreys in 1926 gave the first quantitative estimate of the rigidity of the Earth's core and established that most of the core is probably molten. He was the senior author of tables produced from 1930 to 1940, giving the travel times of earthquake waves through the interior of the Earth.

Jeffreys also contributed to theories of seismic wave propagation, mutations of the Earth's axis, mountain building, convection currents inside the Earth, tidal problems, and a theory on the internal structures of other terrestrial planets.

Additionally, he made significant contributions to the general theory of dynamics, aerodynamics, meteorology, relativity theory, and plant ecology. He was knighted in 1953 for his outstanding contributions to the scientific community. He died in Durham on March 18, 1989.

1. Encyclopedia of World Biography on Harold Jeffreys, Sir, BookRags, www.bookrags.com/biography/ harold-jeffreys-sir.
2. J.J. O'Connor and E. F. Robertson, Harold Jeffreys, MacTutor, www-history.mcs.st-andrews.ac.uk/Biographies/Jeffreys.
3. Sir Harold Jeffreys, Microsoft Encarta Online Encyclopedia, http://encarta.msn.com/encyclopedia_762508763/Jeffreys_Sir_Harold, 2008.

Seth Nicholson (1891-1963)

Discovered four moons of Jupiter and made the first infrared observation of celestial objects

Seth Barnes Nicholson, an American astronomer, was born in Springfield, Illinois on November 12, 1891. He was educated at Drake University in Des Moines, Iowa, where he became interested in astronomy. In 1914, at the University of California's Lick Observatory, while observing the recently-discovered eighth Jupiter moon, Pasiphaë, with the 36-inch Crossley reflector, he discovered a ninth moon, Sinope. He computed its orbit for his Ph.D. thesis in 1915.

He then went to Mount Wilson Observatory, located in Los Angeles County, California, where he spent his entire career. He discovered three more Jovian moons— Lysithea and Carme in 1938 and Ananke in 1951. Sinope, Lysithea, Carme, and Ananke originally were simply designated Jupiter IX, Jupiter X, Jupiter XI, and Jupiter XII. They were not given their present names until 1975. Nicholson himself declined to propose names.

His main assignment at Mount Wilson was observing the Sun with the 150-foot solar tower telescope and producing annual reports on sunspot activity and magnetism, which he did for decades. He also made a number of eclipse expeditions to measure the brightness and temperature of the Sun's corona; discovered a Trojan asteroid, 1647 Menelaus; and computed orbits of several comets, as well as that of Pluto.

In the early 1920s, Nicholson and Edison Pettit (1889-1962) made the first systematic infrared observations of celestial objects. They used a vacuum thermocouple to measure the infrared radiation from the Moon, and thus its temperature. This led to the theory that the Moon is covered with a thin layer of dust which acts as an insulator. They also used infrared to observe planets, sunspots, and stars. Their temperature measurements of nearby giant stars led to some of the first determinations of stellar diameters.

Nicholson was twice president of the Astronomical Society of the Pacific, and editor of its publications from 1943 to 1955. He died in Los Angeles, California on July 2, 1963. Named after him are **Asteroid 1831 Nicholson**, **Nicholson crater** on the Moon; **Nicholson crater** on Mars; and **Nicholson Regio** on Ganymede.

1. Nicholson, Seth Barnes (1891-1963), The Internet Encyclopedia of Science, www.daviddarling.info/encyclopedia/N/Nicholson.
2. Seth Barnes Nicholson, The Bruce Medalists, www.phys-astro.sonoma.edu/BruceMedalists/Nicholson/Nicholson.
3. Seth Barnes Nicholson, Wikipedia, http://en.wikipedia.org/wiki/Seth_Barnes_Nicholson.

Milton Humason (1891-1972)

Measured the redshifts of 620 galaxies

Milton Lasell Humason, an American astronomer, was born at Dodge Center, Minnesota on August 19, 1891. Sent to summer camp at Mount Wilson Observatory in Los Angeles County, California when he was 14 years old, he became so fascinated by the activities there that he somehow convinced his parents to let him drop out of school to return. He ended up working as a mule driver for the pack trains that traveled the trail between the Sierra Madre and Mount Wilson during construction work on the Observatory.

In 1911, he married the daughter of the Observatory's engineer, and became a foreman on a relative's ranch. When he learned of an opening for a janitor at Mount Wilson Observatory in 1917, he took the job.

Astronomers Edwin Hubble (1889-1953) and George Hale (1868-1938) noticed Humason's interest in astronomy and named him to the scientific staff in 1919. His first project was a search of the sky for planets beyond the orbit of Neptune. He barely missed discovering Pluto because the image fell on a flawed part of the photographic plate.

Humason's most important work was done jointly with Hubble in 1928, when he systematically began to observe the speed at which the distant galaxies are receding from us by observing the red shift of light. The farther the light is shifted to the red end of the spectrum, the faster the galaxy is receding. He took measurements for many hundreds of galaxies, developing new techniques along the way.

Working with others in 1956, he used his data to refine Hubble's law, which states that the speed at which a galaxy is receding from us is directly proportional to its distance from us. The new refinement allowed for greater galactic velocities in the distant past, which corresponded closely to the theory put forward by Georges Lemaître (1894-1966) and George Gamow (1904-1968), who proposed that the universe was created with a big bang.

During his studies on galaxies, Humason also discovered comet 1961e, notable for its large perihelion distance. He retired in 1957, and died in Mendocino, California on June 18, 1972. **Humason crater** on the Moon is named for him. The Boston band, Big Dipper, wrote a song about him called ***Humason*** on their album *Heavens*.

1. Humason, Milton Lasell (1891-1972), The Internet Encyclopedia of Science, www.daviddarling.info/encyclopedia/H/Humason.
2. Milton L. Humason, Wikipedia, http://en.wikipedia.org/wiki/Milton_L._Humason.
3. World of Scientific Discovery on Milton La Salle Humason, BookRags, WWW.bookrags.com/biography/milton-la-salle-humason-wsd.

Edward Appleton (1892-1965)

Discovered the layer of the ionosphere that reflects high-frequency radio waves

Edward Victor Appleton, a British physicist, was born in Bradford, Yorkshire, England on September 6, 1892. At age 16, he entered the University of London, and two years later was awarded a scholarship to study physics at Cambridge University. He graduated from Cambridge in 1913, and immediately began working in crystallography with Lawrence Bragg (1890–1971). With the advent of World War I he enlisted in the Royal Engineers and was assigned to signal duty as a commissioned officer. At the end of the war he returned to the Cavendish Laboratory at Cambridge, where, in collaboration with Balthazar van der Pol (1889-1959), he began an investigation of the operation of radio vacuum tubes. This led in 1932 to the publication *Thermionic Vacuum Tubes*, a monograph that served as an introduction to the physical principles underlying these electronic components.

In 1901, Guglielmo Marconi (1874-1937) succeeded in transmitting radio waves across the Atlantic Ocean for the first time. English physicist Oliver Heaviside (1850-1925) and American physicist Arthur Kennelly (1861-1939) postulated that this transmission was made possible by the presence of a layer of ionized gases in the upper atmosphere, which were believed to reflect radio waves back toward the Earth.

The experiments for which Appleton is best known were made using the British Broadcasting Corporation's transmitter at Bournemouth and recording the strength of its signal when received at Cambridge. By varying the frequency of the transmitted signal and noting the interference between the direct (ground) waves and the reflected (sky) waves, Appleton was able to prove that the Heaviside-Kennelly layer was located at a height of 60 miles above the surface of the Earth. In subsequent experiments he and coworkers discovered two other layers, now known as the **Appleton layers**, one of which was situated at a height twice that of the Heaviside-Kennelly layer and the other somewhat lower than it.

In 1924, Appleton was made Wheatstone Professor of Physics at King's College, University of London and remained there until 1936.

In 1941, he was knighted and in 1947 he was awarded the Nobel Prize in Physics. At the time of his death on April 21, 1965 he was Vice-Chancellor of the University of Edinburgh.

1. Edward V. Appleton, NobelPrize.org, http://nobelprize.org/nobel_prizes/physics/laureates/1947/appleton-bio.
2. Edward Victor Appleton, Wikipedia, http://en.wikipedia.org/wiki/Edward_Victor_Appleton.
3. Sir Edward Victor Appleton, Answers.com, www.answers.com/topic/edward-victor-appleton.

Walter Baade (1893-1960)

Discovered that there are two types of stars in a galaxy

Wilhelm Heinrich Walter Baade, a German astronomer, was born in Schröttinghausen, Westphalia Germany on March 24, 1893. His father was a schoolteacher. Walter's interest in astronomy began in high school. He went on to obtain a Ph.D. at Göttingen in 1919. He then spent 11 years on the staff of the University of Hamburg. He moved to the United States in 1931 to work at Mount Palomar and Mount Wilson observatories, where he remained until returning to Germany in 1958.

In the 1930s, Baade and his colleague, Fritz Zwicky (1898-1974), were among the first to suggest that neutron stars are the remnant cores of massive stars that ended their lives as supernovae.

Because Los Angeles, California was blacked out During World War I, the skies were exceptionally dark, allowing Baade to make a detailed study of the Andromeda galaxy. Previously, only the stars in the outer part of the galaxy's spiral arms, which were primarily blue-white in color, had been resolved in work by Edwin Hubble (1889-1953). Baade discovered that the stars near the core of the Andromeda galaxy were reddish. It became apparent to Baade that there were two distinct classes of stars in a galaxy. He believed that Population I stars contained heavier elements that were created over the eons in the interiors of other stars, and that these elements had been ejected into the galaxy as a result of supernovae. They were then incorporated into the Population I stars when they formed. The reddish stars in the galactic core, which Baade called Population II stars, are older, having formed long ago before the heavier elements built up.

Following World War II, Baade used the 200-inch Palomar telescope to identify over 300 Cepheid variable stars in the Andromeda galaxy. In 1952, he established a new period-luminosity curve, and determined that the distance to the Andromeda galaxy was actually around 2,000,000 light-years. Moreover, the universe was twenty times larger than originally estimated. The greater size of the universe meant that distant galaxies had to be much bigger and brighter than originally believed. The Milky Way became nothing more than an average galaxy among millions.

Baade died in Göttingen, Germany on June 25, 1960. Named after him are **Asteroid 1501 Baade**, **Baade crater** on the Moon, **Vallis Baade** on the Moon, one of the two Magellan telescopes, and **Baade's Window** (an area relatively free of dust near the Galactic Center in Sagittarius).

1. Walter Baade, Wikipedia, http://en.wikipedia.org/wiki/Walter_Baade.
2. Wilhelm Heinrich Walter Baade, Answers.com, www.answers.com/topic/walter-baade.
3. World of Scientific Discovery on Wilhelm Heinrich Walter Baade, BookRags, www.bookrags.com/biography/wilhelm-heinrich-walter-baade-wsd.

Ernst Öpik (1893-1985)

Believed comets originated in a cloud orbiting far beyond the orbit of Pluto

Ernst Julius Öpik, an Estonian astronomer/astrophysicist, was born in Port Kunda, Estonia on October 23, 1893. He went to University of Moscow to specialize in the study of minor bodies, such as asteroids, comets, and meteors. After graduation, he became director of the Astronomy Department at Tashkent. He completed his doctorate at the University of Tartu. From 1921 to 1944, he was an Associate Professor at Tartu University, and from 1930 to 1934 he was a visiting scientist at Harvard University.

In 1922, he correctly predicted the frequency of craters on Mars long before they were detected by space probes. In 1932, he postulated a theory concerning the origins of comets in our Solar System. He believed that they originated in a cloud orbiting far beyond the orbit of Pluto. This cloud is now known as the Oort cloud, named after Jan Oort (1900-1992), or alternatively the **Öpik-Oort Cloud**. The Oort cloud is an immense spherical cloud surrounding the planetary system and extending approximately three light years from the Sun. This vast distance is considered the edge of the Sun's orb of gravitational influence.

Öpik fled his native country in 1944 because the approaching Red Army raised fear among Estonians. Living as a refugee in Germany, he became rector of the Baltic University in Exile in the displaced persons camps.

Among Öpik's other discoveries were the first computation of the density of a degenerate body (white dwarf 40 Eri B) in 1915; the first accurate determination of the distance of an extragalactic object (Andromeda Nebula) in 1922; the first composite theoretical models of dwarf stars, showing how they evolve into giants in 1938; and a new theory of the origin of the Ice Ages in 1952. His statistical studies of Earth-crossing comets and asteroids are fundamental to the understanding of the motions of these objects and how they impact on Earth.

He spent 1948 to 1981 at the Armagh Observatory in Northern Ireland. He remained there despite offers of well-paying jobs in America. He won the Gold Medal of the Royal Astronomical Society in 1975 and the Bruce Medal in 1976.

Öpik died in Bangor, County Down, North Ireland on September 10, 1985. **Asteroid 2099 Öpik** was named in his honor.

1. A Short Biography of Ernst Öpik, Ormagh Observatory, www.arm.ac.uk/history/opik/biog.
2. Ernst Öpik, Wikipedia, http://en.wikipedia.org/wiki/Ernst_%C3%96pik.
3. Julius Öpik, The Bruce Medalists, www.phys-astro.sonoma.edu/brucemedalists/Opik/Opik.

Georges Lemaître (1894-1966)

Father of the Big Bang theory

Georges Henri Lemaitre, a Belgium mathematician, physicist, and astronomer, was born in Charleroi, Belgium on July 17, 1894. He studied humanities at a Jesuit school before entering the civil engineering school of the Catholic University of Leuven. When World War I broke out in Europe in 1914, he volunteered in the Belgian army and served as an artillery officer. After the war, he switched to the study of mathematics and physics, and also prepared for the priesthood. He earned a doctorate in 1920 from the Université Catholique de Louvain (UCL). His dissertation was entitled *L'Approximation des Fonctions de Plusieurs Variables Réelles* (Approximation of functions of several real variables). He was ordained as a priest in 1923.

Lemaître then studied at Cambridge University, England. In 1924-1925, he studied at Harvard University and Massachusetts Institute of Technology. During this time he became interested in the work of Vesto Slipher (1875-1969), who had shown that the spectral lines of stars in nebulae are shifted toward the red end of the spectrum, which implied that these nebulae are receding from the Earth.

In 1927, Lemaître accepted a full-time position at UCL and published his work *Un Univers Homogène de Masse Constante et de Rayon Croissant Rendant Compte de la Vitesse Radiale ses Nébuleuses Extragalactiques* (A Homogeneous Universe of Constant Mass and Growing Radius Accounting for the Radial Velocity of Extragalactic Nebulae). This paper explained the expanding universe within the framework of the general theory of relativity. Albert Einstein (1879-1955), himself, was skeptical. The theory's critics called it the Big Bang theory. Eventually, Einstein came to believe Lemaitre's theory, and the term Big Bang caught on.

Lemaître also studied cosmic rays and the three-body problem. His other published works include *Discussion sur L'évolution de L'univers* (Discussion on the Evolution of the Universe) in 1933 and *L'Hypothèse de L'atome Primitif* (Hypothesis of the Primeval Atom) in 1946.

He died in Louvain on the June 20, 1966. The **Lemaître crater** and the **Friedmann-Lemaître-Robertson-Walker metric** (an exact solution of Einstein's field equations of general relativity) are named for him.

1. Encyclopedia of World Biography on Georges Édouard Lemaître, Abbe, BookRags, www.bookrags.com/biography/georges-edouard-lemaitre-abbe.
2. Georges Lemaître, Wikipedia, http://en.wikipedia.org/wiki/Georges_Lema%C3%AEtre.
3. Nick Green, Georges-Henri Lemaitre Biography, About.com, http://space.about.com/cs/astronomerbios/a/lemaitrebio.

Bertil Lindblad (1895-1965)

Deduced that the rate of rotation of the stars in the outer part of the galaxy decreases with distance from the galactic core

Bertil Lindblad, a Swedish astronomer, was born in Örebro, Sweden on November 26, 1895. He obtained a B.S. in astronomy from the University of Uppsala in 1918 and a Ph.D. in Astronomy from the same university in 1920. He was director of Stockholm Observatory at the University of Stockholm from 1927 to 1965.

After studying the work of Jacobus Kapteyn (1851-1922) on moving stellar streams and the proposal of Howard Shapley (1885-1972) that the center of the Milky Way Galaxy lays tens of thousands of light-years away, Lindblad proposed the differential rotation theory. This theory states that the speed of rotation of stars about the galactic center depends on their distance from the center. In 1927, observations of stellar motion by Jan Oort (1900-1992) provided support for Lindblad's views. A class of resonances in rotating stellar or gaseous disks is named **Lindblad resonances** after Bertil Lindblad.

In the 1940s, Lindblad proposed the density wave theory to explain spiral structure of galaxies. This theory states that a wave produces a series of alternate condensations and rarefactions of the material through which it passes. According to this idea, the arms of a spiral galaxy represent regions of enhanced density (density waves) that rotate more slowly than the galaxy's stars and gas. As gas enters a density wave it becomes squeezed and makes new stars, some of which are short-lived blue stars that light the arms. He also studied star clusters, estimated the galactic mass of the Sun, and advanced the spectroscopic method for distinguishing between giant and main sequence stars.

As director of the Stockholm Observatory, he oversaw the four-year construction of the new Stockholm Observatory in suburban Saltsjöbaden, completed in 1931.

Lindblad was president of the International Astronomical Union from 1948 to 52, and he was briefly Chairman of the Nobel Foundation in 1965, but was overtaken by illness and served only a few months.

He died in Stockholm, Sweden on June 25, 1965. Also named after him are **Asteroid 7331 Balindblad** and lunar crater **Lindblad**. Inspired by his father, Per Olaf (1927-) specialized in the movements of stars.

1. Bertil Lindblad, NNDB, www.nndb.com/people/174/000170661.
2. Density Wave, The Internet Encyclopedia of Science, www.daviddarling.info/encyclopedia/D/densitywave.
3. Lindblad, Bertil (1895-1965), The Internet Encyclopedia of Science, www.daviddarling.info/encyclopedia/L/Lindblad.

Rudolph Minkowski (1895-1976)

Photographed two galaxies colliding

Rudolf Leo Bernhard Minkowski, a German-American astronomer, was born in Strasbourg, Germany on May 28, 1895. His father was a physician and physiologist. In 1913, Rudolf entered the University of Breslau to study physics. His studies were interrupted by World War I, during which he served in the German Army.

When the war was over, he studied in Berlin, and then returned to Breslau, where he finished his studies in 1921. He did his doctoral work under the supervision of Rudolf Ladenburg (1882-1952).

After working in Goettingen for a year, he moved to Hamburg in 1922 and worked at the University there. In 1933, Hitler assumed power in Germany. In 1935, because of his Jewish ancestry, Minkowski was no longer allowed to teach. Walter Baade (1893-1960), who earlier had left Germany and was at the Mount Wilson Observatory in Southern California, made available a post for Minkowski as a research assistant. Minkowski accepted the position and moved to California, where, from 1935 to 1960, he worked at Mount Wilson and Mount Palomar Observatories, both in Southern California.

Minkowski studied spectra, distributions, and motions of planetary nebulae. He more than doubled the number of these objects known. He investigated novae and supernovae and their remnants, especially the Crab Nebula. He headed the Palomar Observatory Sky Survey, which photographed the entire northern sky in the 1950s. With Baade, he identified the first optical counterpart to an extragalactic radio source, Cygnus A.

In 1952, his work established that galaxies can collide, and eight years later he used a 200-inch telescope to photograph the collision of two galaxies at a distance of an estimated 4.8-billion light years from Earth. At the time, this was the most distant event ever observed.

After his retirement from Mount Wilson and Mount Palomar observatories, he received an invitation from the Radio Astronomical Laboratory in Berkeley. He worked there from 1961 to 1965.

Minkowski died in Berkeley, California on January 4, 1976. Lunar crater **Minkowski** was named for him and his uncle, Hermann Minkowski (1864-1909).

1. Minkowski, Rudolph Leo Bernhard (1895-1976), The Internet Encyclopedia of Science, www.daviddarling.info/encyclopedia/M/Minkowski_Rudolph.
2. Plicht, Chris, Minkowski, Rudolf (1895 - 1976), www.plicht.de/chris/12minkow.
3. Rudolph Minkowski NNDB, www.nndb.com/people/872/000171359.
4. Rudolph Minkowski, Wikipedia, http://en.wikipedia.org/wiki/Rudolph_Minkowski.

Bernard Lyot (1897-1952)

Provided the first conclusive evidence that sandstorms blow across the surface of Mars

Bernard Ferdinand Lyot, a French astronomer and inventor, was born in Paris, France on February 27, 1897. After obtaining a degree in engineering at l'École Supérieure d'Électricité in Paris, he assisted Alfred Pérot (1863-1925) at l'École Polytechnique. During this time, he also studied mathematics, physics, and chemistry at the University of Paris.

From 1920 to 1952 he worked at the Meudon Observatory, which became part of the Paris Observatory in 1926. There, he was encouraged by its director, Henri Deslandres (1853-1948). His first astronomical research was aimed at measuring the polarization of sunlight reflected from planets and their moons to learn something about the composition of their surfaces. He constructed an extremely sensitive polariscope to investigate the polarization of light reflected from the planets.

Making many observations at the Pic du Midi Observatory, he found that the lunar surface behaves like volcanic dust and that Mars has sandstorms. He also investigated the atmospheres of the other planets.

While studying Mercury, he began to contemplate ways to eliminate the glare of the solar disk while carrying out observations very close to the Sun's limb. In 1929, he improved his polarimeter and developed an optical system that eliminated scattered light, creating the coronagraph, a device that allows observation of the solar corona without the need for an eclipse. By 1931, he was obtaining photographs of the corona and its spectrum.

For his solar studies, he developed monochromatic polarizing filters with transmission only one angstrom wide. He found new spectral lines in the corona, and he made the first motion pictures of solar prominences.

He also invented the **Lyot stop**, which is an optic stop that reduces the amount of flare caused by diffraction of other stops and baffles in optical systems.

Lyot received a number of awards, including the French Academy of Sciences Lalande Medal (1928) and the Royal Astronomical Society Gold medal (1939).

He suffered a heart attack while returning from an eclipse expedition in Sudan and died on April 2, 1952 at the age of 55. Named for him are the **Lyot crater** on the Moon, **Lyot crater** on Mars, and minor planet **2452 Lyot**.

1. Bernard Ferdinand Lyot, The Bruce Medalist, www.phys-astro.sonoma.edu/BruceMedalists/Lyot/Lyot.
2. Bernard Lyot, NNDB, www.nndb.com/people/234/000170721.
3. Bernhard Lyot, Wikipedia, http://en.wikipedia.org/wiki/Bernard_Lyot.

Otto Struve (1897-1963)

An early advocate of the search for extraterrestrial life

Otto Struve, a Russian-American astronomer, was born at Kharkov in Russia on August 12, 1897. He was the son of Gustav Struve (1858-1920), grandson of Wilhelm von Struve (1819-1905), great-grandson of Friedrich von Struve (1793-1864), and nephew of Karl Struve (1854-1920), all astronomers. During the Russian civil war, he fought on the side of the White Russian forces, and was wounded. When it was clear that they were losing the war, he retreated into exile. His father accompanied him as far as Sevastopol, where he died.

In the 1-1/2 years that Struve spent in exile in Gallipoli, Turkey, and later Constantinople, he learned that his brother Werner had died of tuberculosis and a younger sister had drowned. The widow of his uncle, Karl Struve, asked her late husband's successor at the Berlin-Babelsberg Observatory to write to the director of Yerkes Observatory in Chicago, Edwin Frost (1866-1935). He did, and Frost offered Struve a job. Struve then moved to the United States. He obtained his Ph.D. from the University of Chicago in 1923. He became a citizen in 1927, and eventually succeeded Frost as director of Yerkes Observatory. He would serve as director of four observatories in all.

He made detailed spectroscopic investigations of stars, especially close binaries stars and the interstellar medium, and discovered H II regions and gaseous nebulae. An H II region is a cloud of glowing gas and plasma in which star formation is taking place. He also contributed to the understanding of the broadening of spectral lines due to stellar rotation, electric fields, and turbulence. He worked to separate these effects from each other and from chemical abundances. He was a pioneer in the study of mass transfer in closely interacting binary stars.

Struve publicly expressed a belief that extraterrestrial intelligence was abundant, and so was an early advocate of the search for extraterrestrial life. In 1959, he became the first regular director of the National Radio Astronomy Observatory at Green Bank, West Virginia. He estimated in 1960 that there might be as many as 50 billion planets in our Galaxy alone.

He died on April 6, 1963. He had no children, and thus the Struve dynasty came to an end. Named after him are the **Struve crater** on the Moon (named for him and two other Struve astronomers), **Asteroid 2227 Otto Struve**, and the **Otto Struve Telescope** of McDonald Observatory.

1. Otto Struve, The Bruce Medalists, www.phys-astro.sonoma.edu/brucemedalists/struve/Struve.
2. Otto Struve, Wikipedia, http://en.wikipedia.org/wiki/Otto_Struve.
3. Struve, Otto (1897-1963), The Internet Encyclopedia of Science, www.daviddarling.info/encyclopedia/S/Struve.

Theodore Durham, Jr. (1897-1984)

Co-discovered that the atmosphere of Venus is composed primarily of carbon dioxide

Theodore Durham Jr., an American physician and astrophysicist, was born in New York City on December 17, 1897. By the age of 17 he had built an observatory on the grounds of his family's cottage in Northeast Harbor, where his father practiced medicine in the summers. Theodore received an A.B. in chemistry from Harvard University in 1921, an M.D. from Cornell University in 1925, and an A.M. and Ph.D. in physics in 1926 and 1927, respectively, from Princeton University.

He was a staff member of Mount Wilson Observatory from 1928 to 1947. His principal research activities included development of Coude spectrographs, introduction of the Schmidt camera in spectroscopy, studies of stellar atmospheres and interstellar material, studies of planetary atmospheres, development of photoelectric detectors for spectroscopy, and application of physical methods for research in medicine and surgery.

In 1932, Durham and Walter Adams (1876-1956) discovered that the atmosphere of Venus is principally composed of carbon dioxide. They demonstrated that if light were sent through a long pipe containing compressed carbon dioxide, the same spectrum could be reproduced on Earth, indicating that carbon dioxide, under higher pressure than the Earth's atmosphere, had been observed in the atmosphere of Venus. This was confirmed 35 years later in measurements transmitted from U.S. and Soviet spacecraft. Durham and Adams also showed that methane and ammonia are predominant in the outer atmosphere of Jupiter and Saturn.

After World War II, Durham spent several years applying physical methods to medical research, first at Harvard Medical School and then at the School of Medicine and Dentistry and the Institute of Optics at the University of Rochester.

In 1957, he joined the faculty of the Australian National University in Canberra, where he designed and installed a spectrograph for use in studying the composition of the stars of the Southern Hemisphere. From 1965 to 1970 he was a Senior Research Fellow at the University of Tasmania, Australia. After returning to the United States in 1970, he resumed his earlier association with the Harvard College Observatory.

Durham was Scientific Director of the Fund for Astrophysical Research from its founding in 1936 until his death on April 3, 1984

1. About Theodore Durham, Jr., Biography, Fund for Astrophysical Research, http://foundationcenter.org/grantmaker/fundastro/about.
2. Theodore Durham, The Internet Encyclopedia of Science, www.daviddarling.info/encyclopedia/D/Dunham.

Fritz Zwicky (1898-1974)

Coined the term "supernova"

Fritz Zwicky, a Swiss-American astronomer, was born in Varna, Bulgaria to Swiss parents on February 14, 1974. His father was the Bulgarian ambassador to Norway. Fritz received an advanced education in mathematics and experimental physics at the Swiss Federal Institute of Technology in Zürich. In 1925, he immigrated to the United States to work with Robert Millikan (1868-1953) at the California Institute of Technology (Caltech). He was appointed Professor of Astronomy at Caltech in 1942. He also worked for Aerojet Engineering Corporation from 1943 to 1961 and was a staff member of Mount Wilson and Mount Palomar Observatories for most of his career.

In 1934, Zwicky and Walter Baade (1893-1960) coined the term "supernova" and hypothesized that they were the transition of normal stars into neutron stars, as well as the origin of cosmic rays. In support of this hypothesis, Zwicky started hunting for supernovae, and found 120 of them. In 1937, he postulated that galaxy clusters could act as gravitational lenses by the previously discovered Einstein effect. A gravitational lens is formed when light from a very distant, bright source is bent around a massive object between the source object and the observer. It was not until 1979 that this effect was confirmed by observation of the twin quasar, Q0957+561.

Zwicky developed a generalized form of morphological analysis, which is a method for systematically structuring and investigating the total set of relationships contained in multi-dimensional, usually non-quantifiable, problem complexes· He wrote a book on the subject in 1969, and stated that he had made many of his discoveries using this method.

He devoted considerable time to the search for galaxies and the production of catalogs. From 1961 to 1968, he and his colleagues published a comprehensive six volume *Catalogue of Galaxies and of Clusters of Galaxies*. He also discovered the dark matter permeating the Coma Cluster of galaxies.

Zwicky had a difficult personality. He was fond of calling people "spherical bastards," because they were bastards every way he looked at them. He died in Pasadena on February 8, 1974. **Asteroid 1803 Zwicky**, the **Zwicky lunar crater**, and the **galaxy I Zwicky 18** were all named in his honor.

1. Fritz Swicky, Swedish Morphological Society, www.swemorph.com/zwicky.
2. Fritz Swicky, Wikipedia, http://en.wikipedia.org/wiki/Fritz_Zwicky.
3. Zwicky, Fritz (1898-1974), The Internet Encyclopedia of Science, www.daviddarling.info/encyclopedia/Z/Zwicky.

Ira Bowen (1898-1973)

Identified the mysterious "nebulum" spectral lines as lines of ionized oxygen and nitrogen

Ira Sprague Bowen, an American astronomer, was born in Seneca Falls New York on December 21, 1898. His father was a Methodist minister. Ira (Ike to his friends) studied at the Houghton Seminary and at Oberlin College, where he received a B.A. in 1919.

As a graduate student at the University of Chicago, he worked with Robert Milliken (1868-1953), whom he accompanied to the California Institute of Technology (Caltech). There, he received his Ph.D. His dissertation was on heat losses in the evaporation of water from lakes.

Bowen taught physics at Caltech from 1921 to 1945. His investigation of the ultraviolet spectra of highly ionized atoms led to his identification of the mysterious "nebulum" spectral lines of gaseous nebulae as lines of ionized oxygen and nitrogen. Nebulum was a hypothetical element proposed by William Huggins (1824-1910) as the cause of unfamiliar lines in the spectra of far-off nebulae. Bowen went on to explain most of the lines of gaseous nebulae.

During World War II, he was in charge of photographic work on the rocket project at the Jet Propulsion Laboratory in Pasadena, California.

As director of Mount Wilson Observatory (1946 to 1948) and Mount Palomar Observatory (1948 to 1964), he directed the completion of the 200-inch Hale telescope and 48-inch Schmidt telescope, and he designed many of their instruments, including a novel spectrograph.

Bowen invented the image slicer, which greatly increased the efficiency of gaseous nebulae observations; made substantial improvements to telescope technologies; and introduced the practice of baking photographic plates to improve their sensitivity. He also discovered what is now known as the **Bowen ratio**, the mathematical expression of the ratio of heat conduction to evaporative flux at the air-water interface, now a standard rule of meteorology and oceanography.

He was awarded the Henry Draper Medal (1942), Bruce Medal (1957), and Gold Medal of the Royal Astronomical Society (1966).

Bowen continued designing astronomical instruments after his official retirement. He died in Los Angeles, California on February 6, 1973. Also named after him are **Bowen crater** on the Moon and **Asteroid 3363 Bowen.**

1. Ira S. Bowen, NNDB, www.nndb.com/people/326/000167822.
2. Ira Sprague Bowen, Fellows Profile, www.ossc.org/bios/fellows-bowen.
3. Ira Sprague Bowen, The Bruce Medalist, /www.phys-astro.sonoma.edu/BruceMedalists/ Bowen/Bowen.

Willem Luyten (1899-1994)

Discovered many white dwarfs

Willem Jacob Luyten, a Dutch-American astronomer, was born in Semarang, Dutch Indies (now Indonesia) on March 7, 1899. His father was a teacher of French. At the age of 11, Willem observed Halley's Comet, which started his fascination with astronomy. In 1912, his family moved back to the Netherlands, where he studied astronomy at the University of Amsterdam and received his B.A. in 1918.

Luyten was the first student to earn his Ph.D. with Ejnar Hertzsprung (1873-1967) at Leiden University. He was only 22 years old at the time.

In 1921, he left for the United States, where he first worked at the University of California's Lick Observatory.

From 1923 to 1930, he worked at the Harvard College Observatory. He spent the years 1928 to 1930 in Bloemfontein, South Africa. After returning to the United States in 1931, he taught at the University of Minnesota from 1931 to 1967, and then served as Astronomer Emeritus from 1967 until his death.

Luyten studied the proper motions of stars, and discovered many white dwarfs. The proper motion of a star is the measurement of its change in position in the sky over time after improper motions are accounted for. This contrasts with radial velocity, which is the time rate of change in distance toward or away from the viewer. The improper motion of a star refers to the change of its coordinates on the sky not originating from the motion of the star itself.

He found proper motions of more than 120,000 stars. Later, he had the engineers at Control Data Corporation build him an automated, computerized plate scanner and measuring machine, and found 400,000 more. He also discovered some of the Sun's nearest neighbors, including a star that would be named after him, **Luyten's Star,** and the high–proper motion star system, **Luyten 726-8**, which was soon found to contain the flare star, UV Ceti.

Over his career, Luyten published some 500 research papers and wrote numerous popular articles for the *New York Times*, *Minneapolis Star and Tribune*, and other periodicals. He received the James Craig Watson Medal in 1964 and the Bruce Medal in 1968. He died in Minneapolis, Minnesota on November 21, 1994. Also named after him is **Asteroid 1964 Luyten.**

1. Upgren, Arthur, William Jacob Luyten, Biographical Memoirs, National Academy of Sciences, www.nap.edu/html/biomems/wluyten.
2. William Jacob Luyten, The Bruce Medalist, www.phys-astro.sonoma.edu/BruceMedalists/Luyten/Luyten.
3. William Jacob Luyten, Wikipedia, http://en.wikipedia.org/wiki/Willem_J._Luyten.

20th Century Astronomers

1900 to Present

Cecilia Payne-Gaposchkin (1900-1979)

Showed that hydrogen is the overwhelming constituent of the stars

Cecilia Helena Payne, a British-American astronomer, was born in Wendover, England on May 10, 1900. Her father, a London barrister, died when she was four years old. Her mother was a painter and musician. In 1919, Cecilia won a scholarship to Newnham College at Cambridge University, where she studied botany, chemistry, and physics. During her studies there she became fascinated with astronomy after attending a lecture on Einstein's theory of relativity.

On completion of her studies in 1923, she obtained a Pickering Fellowship from Harvard to study under Harlow Shapley (1885-1972). Boston, Massachusetts, became her home for the rest of her career.

In 1925, she was given an unofficial staff position at the Harvard Observatory. That same year, she was awarded the first Ph.D. in astronomy at Radcliffe. Her doctoral dissertation, entitled *Stellar Atmospheres*, was the first paper written on the subject. It was also the first research to apply the recent theory of ionization of Indian physicist Meghnad Saha (1893-1956) to show that hydrogen is the overwhelming constituent of the stars. Henry Russell (1877-1957) at Princeton, several years later, reached the same conclusions and published them, thereby receiving credit. In 1934, she married Russian astrophysicist Serge Gaposchkin, whom she had met in Europe.

It was not until 1938 that her work was recognized, and she was granted the title of astronomer. Finally, in 1956 she was promoted to professor, given an appropriate salary, and named Chairman of the Department of Astronomy—the first woman to hold such a position at Harvard.

Payne-Gaposchkin devoted a large part of her research to the study of stellar magnitudes and distances. She and Serge pioneered research into variable stars, including the structure of the Milky Way and the nearby Magellanic Clouds. Through their studies, they made over two million magnitude estimates of the variable stars in the Magellanic Clouds.

Over her career, she published several books, including *The Stars of High Luminosity* (1930 and *Variable Stars* (1938), written with her husband. In 1977, she received the Henry Norris Russell Prize from the American Astronomical Society. She died on December 7, 1979.

1. Cecilia Payne-Gaposchkin, Wikipedia, http://en.wikipedia.org/wiki/Cecilia_Payne-Gaposchkin.
2. Encyclopedia of World Biography on Cecilia Payne-Gaposchkin, BookRags, www.bookrags.com/biography/cecilia-payne-gaposchkin.
3. Payne-Gaposchkin, Cecilia Helena (1900-1979), The Internet Encyclopedia of Science, www.daviddarling.info/encyclopedia/P/Payne-Gaposchkin.

Jan Oort (1900-1992)

Proposed that a cloud of cometary material surrounds the Solar System at an enormous distance

Jan Oort, a Dutch astronomer, was born in the farming village of Franeker in Holland on April 28, 1900. He entered the University of Groningen in 1917, and earned his doctoral degree in 1926. He did research at the Leiden Observatory in 1924, and lived in the United States as a research associate at Yale University Observatory from 1924 to 1926.

In 1926, Oort joined the faculty of the University of Leiden, and became director of the observatory in 1945. He did his early studies under Jacobus Kapteyn (1851-1922), who placed the Sun at the center of a relatively small galaxy. In 1917, Harlow Shapley (1885-1972) proposed a far bigger galaxy. Oort provided observational evidence that confirmed the main features of Shapley's model.

Oort showed that our Solar System rotates around the distant center of our galaxy. It was the first direct evidence of the Milky Way's rotation. He also determined that the Sun is not close to the galaxy's center

After World War II, Oort and his associates at Leiden built a huge radio telescope to detect radio waves in hydrogen. They found evidence that stars are formed from hydrogen and dust clouds. The group provided the evidence of the spiral structure of our galaxy. And they investigated the processes occurring in the galactic core and the vast corona of hydrogen encircling the galaxy. They also investigated the origin of radio signal sources, including the Crab Nebula. They found that light from the Crab Nebula was polarized and produced by synchrotron emission.

Oort's observations showed that there is much more mass in the universe than can be detected visually, now known as “dark matter,” which is thought to make up more than 90 percent of the universe.

He theorized that in the distant past, a planet that occupied a position between Mars and Jupiter had exploded, sending a small percentage of the material to a region roughly 4,000 times as far away from our Sun as Pluto. Fragments of this material are occasionally pulled by the gravity of the outer planets or a passing star into an orbit around the Sun. The region that is the birthplace of comets became known as the **Oort Cloud**.

Oort died in Leiden on November 5, 1992. Other things named for him are **Asteroid 1691 Oort** and **Oort constants** of galactic structure.

1. Encyclopedia of World Biography on Jan Hendrik Oort, BookRags, www.bookrags.com/biography/jan-hendrik-oort.
2. Jan Oort, Wikipedia, http://en.wikipedia.org/wiki/Jan_Oort.
3. World of Scientific Discovery on Jan Hendrick Oort, BookRags, www.bookrags.com/biography/jan-hendrick-oort-wsd.

Donald Menzel (1901-1976)

Co-discovered that the Sun's corona contains oxygen

Donald Howard Menzel, an American astronomer, was born in Florence, Colorado on April 11, 1901. At 16 years of age, he enrolled at the University of Colorado to study chemistry. Observing a solar eclipse in 1918 led him to change to astronomy. He spent summer vacations at Harvard University as a research assistant to Harlow Shapley (1885-1972).

Menzel received his Ph.D. from Princeton in 1924. He then taught at the University of Iowa and Ohio State University before becoming an assistant astronomer at Lick Observatory in California in 1926.

While at Lick, he participated in many observing programs with the large telescopic equipment. His major work, however, was in the interpretation of the spectrum of the atmosphere of the Sun from photographs taken at various total solar eclipses. He participated in the observation of two such eclipses in the years 1930 and 1932.

Menzel discovered, with J. C. Boyce, in 1933 that the Sun's corona contains oxygen. In 1938, he became professor of astrophysics at Harvard, and in 1954 he was appointed director of the observatory.

In 1941, with W. W. Salisbury, he made the first of the calculations that led to radio contact with the Moon in 1946. He also did valuable work on planetary atmospheres and on the composition of the Sun. His work with Lawrence Aller (1913-2003) and James Baker (1914-2005) defined many of the fundamental principles of the study of planetary nebulae.

Menzel was a prominent skeptic of UFOs. He authored, or co-authored, three popular books in an attempt to debunk them: *Flying Saucers* (1953), *The World of Flying Saucers: A Scientific Examination of a Major Myth of the Space Age* (1963), and *The UFO Enigma: The Definitive Explanation of the UFO Phenomenon* (1977).

In 1968, he testified before the U.S. House Committee on Science and Astronautics that he considered all UFO sightings to have natural explanations.

Although he believed that the many reported UFO sightings could be explained by mundane objects, he saw no strong reason against the idea that planets inhabited by super-beings should not exist in great abundance in the universe.

Menzel retired from Harvard in 1971, and died on December 14, 1976.

1. Donald Howard Menzel, Wikipedia, http://en.wikipedia.org/wiki/Donald_Menzel.
2. Dr. Donald H. Menzel, National Capital Area Skeptics http://ncas.org/ufosymposium/menzel.html#biog.
3. Menzel, Donald Howard (1901-1976), The Internet Encyclopedia of Science, www.daviddarling.info/encyclopedia/M/Menzel.

Peter van de Kamp (1901-1995)

Initiated a search for unseen companions of 54 stars known to lie within 16 light-years of the Sun

Peter (Piet) van de Kamp, a Dutch-American astronomer, was born at Kampen in the Netherlands on December 26, 1901. He studied at the University of Utrecht, and then started his professional career at the Astronomical Laboratory of Groningen. In 1923, he went to the Leander McCormick Observatory at the University of Virginia for a year's residence. There, he assisted Samuel Mitchell (1874-1960) with his extensive stellar parallax program and Harold Alden (1890-1964) with the lengthy boss star project.

The following year, he went to the Lick Observatory in California as a Kellogg fellow. He received his Ph.D. from the University of California in astronomy in 1925. He then returned to the McCormick Observatory. His work there consisted of assisting with the parallax program and continuing the proper motion work that he and Alden had begun. Van de Kamp and Alexander Vyssotsky (1888-1973) spent eight years measuring 18,000 proper motions. He also did an investigation of general and selective absorption of light within the galaxy.

In the spring of 1937, van de Kamp left McCormick Observatory to take over as director of Swarthmore College's Sproul Observatory. He initiated a search for unseen companions of 54 stars known to lie within 16 light-years of the Sun. Over the next two decades, the Sproul group reported evidence for planetary bodies around several of the target stars, including 61 Cygni, Ross 614 and Lalande 21185.

In 1963, after making observations of Barnard's Star, he reported a periodic wobble in its motion, apparently due to a planetary companion. It was not until several decades later that a consensus formed that this had been a spurious detection.

Van de Kamp argued that there was no reason to doubt the existence of large numbers of extrasolar bodies, which are greater in size than Jupiter and smaller than stars. Recent confirmations of the existence of giant planets and brown dwarfs beyond the Solar System have verified that view.

Van de Kamp was also a very talented musician. He helped to organize an orchestra in Charlottesville, which he conducted. He died in on May 18, 1995.

1. Peter van de Kamp, Answers.com, www.answers.com/topic/peter-van-de-kamp.
2. Peter van de Kamp, Wikipedia, http://en.wikipedia.org/wiki/Peter_van_de_Kamp.
3. Stout, David, Peter van de Kamp, Astronomer And Musician at Swarthmore, 93, The New York Times, May 23, 1995.

George Gamow (1904-1968)

Found that stars tend to become hotter when their hydrogen is depleted

George Antonovich Gamow, a Russian physicist and astronomer, was born in the town of Odessa, Russian Empire (now in Ukraine) on March 4, 1904. In 1922-1923, he studied at Novorossia University. During the years 1923 to 1929, he studied optics and cosmology at the University of Leningrad. In 1926, he attended summer school at the University of Göttingen, where he developed his quantum theory of radioactivity, the first successful explanation of the behavior of radioactive elements. He then studied one year at the Institute of Theoretical Physics with Niels Bohr (1885-1962). Gamow's proposal that atomic nuclei can be treated as little droplets of "nuclear fluid" led to today's theory of fusion and fission.

While at Cambridge University (1929-1930), he worked on a formula for calculating the rate of thermonuclear reaction in the interior of stars. He predicted that stars become hotter when their hydrogen is depleted, which was opposed to the thinking at the time.

After three years back in the Soviet Union, in 1934 Gamow returned to the United States, where he eventually became Chairman of the Physics Department at George Washington University. During the early years there, he collaborated with Edward Teller (1908-2003) on the theory of beta-decay, and formulated the **Gamow-Teller Selection Rule for Beta Emission**. He also developed the theory of the internal structure of red giant stars. With Mario Schoenberg (1914-1990), he developed the theory of Urca process and with Ralph Alpher (1921-2007) the theory of the origin of chemical elements by the process of successive neutron capture. The Urca process is reaction which emits a neutrino, and which is assumed to take part in cooling processes in neutron stars.

During World War II, Gamow was involved with the Manhattan project developing an atomic bomb. He also contributed to the development of the hydrogen bomb, and he predicted that the Big Bang would have produced background microwave radiation. His prediction was later verified.

In 1954, he published papers on the information storage and transfer in a living cell. In these papers, he proposed the so-called genetic code, an idea later confirmed by experimental studies in laboratories.

In 1956, Gamow moved from George Washington University to the University of Colorado. He died on August 19, 1968.

1. Biography of George Gamow, http://personal.ifae.es/redondo/physics/biog/Gamow.pdf.
2. George Gamow, EPS, www.norskfysikk.no/nfs/old/epsbiografer/GAMOW.PDF.
3. The George Gamow Memorial Lecture Series, The University of Colorado, www.colorado.edu/physics/Web/Gamow/life.

Bruno Rossi (1905-1993)

Made the first discoveries regarding the nature of cosmic rays

Bruno Benedetto Rossi, an Italian-American experimental physicist, was born in Venice, Italy on April 13, 1905. His father was an electrical engineer. After receiving his doctorate from the University of Bologna in 1928, Bruno became an assistant at the Physics Institute of the University of Florence. In 1932, he moved to the University of Padua as professor of experimental physics. In the fall of 1938, he was expelled from his position because of the racial decrees of the fascist state against Jews. In 1939, he went to the University of Chicago, where he was given a temporary position as a research associate. He began a series of experiments that yielded the first proof of the decay of a fundamental particle, the mesotron (muon), and a precise measurement of its mean life at rest. In 1942, he was appointed associate professor at Cornell University.

During World War II, Rossi worked first as a consultant on radar development at the Radiation Laboratory of the Massachusetts Institute of Technology (MIT) and then at Los Alamos as co-director of the Detector Group responsible for development of instrumentation for experiments that supported the development of the atomic bombs.

In 1946, he was appointed professor of physics at MIT, where he established the Cosmic Ray Group to investigate the nature and origins of cosmic rays and the properties of the sub-nuclear particles produced in the interaction of cosmic rays with matter. In the late 1950s, when particle accelerator experiments had come to dominate experimental particle physics, Rossi turned his attention to exploratory research made possible by the availability of space vehicles. He initiated rocket experiments that discovered the first extra-solar source of X-rays—neutron star Scorpius X-1. Rossi was made Institute Professor at MIT in 1965.

Among his contributions to the electronic techniques of experimental physics are the inventions of the coincidence circuit, the time-to-amplitude converter, and the fast ionization chamber.

Rossi retired from MIT in 1970. From 1974 to 1980, he taught at the University of Palermo. He died of heart failure in Cambridge, Massachusetts on November 21, 1993. Named for him are the **Rossi X-ray Timing Explorer**, a NASA satellite X-ray observatory; **Bruno Rossi Prize** of the High Energy Astrophysics division of the American Astronomical Society, and the **Bruno Rossi Chair** at MIT.

1. Bruno Benedetti Rossi, Answers.com, www.answers.com/topic/bruno-rossi.
2. Bruno Rossi, NNDB, www.nndb.com/people/308/000172789.
3. Bruno Rossi, Wikipedia, http://en.wikipedia.org/wiki/Bruno_Rossi.

Karl Jansky (1905-1950)

First to identify radio waves from beyond the Solar System

Karl Guthe Jansky, an American physicist and radio engineer, was born in Norman, Oklahoma on October 22, 1905. His father was Dean of the College of Engineering at the University of Oklahoma. Karl received his B.S. in physics in 1927.

In 1928, he joined Bell Telephone Laboratories in Holmdel, New Jersey. Bell Labs wanted to investigate atmospheric and ionospheric properties using short wavelengths of about 10 to 20 meters for use in transatlantic radio-telephone service. Jansky was assigned the task of investigating sources of static that might interfere with radio voice transmissions. He built an antenna designed to receive radio waves at a frequency of 20.5 MHz (wavelength about 14.6 meters). It was mounted on a turntable that allowed it to be rotated in any direction so that the direction of a received signal could be pinpointed. It earned the name "Jansky's merry-go-round." A small shed to the side of the antenna housed an analog pen-and-paper recording system.

After recording signals from all directions for several months, Jansky categorized them into three types of static—nearby thunderstorms, distant thunderstorms, and a faint steady hiss of unknown origin. He spent over a year investigating the source of the third type of static. By comparing his observations with optical astronomical maps, he concluded that the radiation was coming from the Milky Way and was strongest in the direction of the center of the galaxy, in the constellation of Sagittarius.

His discovery was widely publicized, appearing in the *New York Times* in 1933. He published his paper *Electrical Disturbances Apparently of Extraterrestrial Origin* that same year. However, Bell Lab was not interested in funding further studies on the subject.

In 1937, Grote Reber (1911-2002) single handedly built a radio telescope in his Illinois backyard, and John Kraus (1910- 2004) started a radio observatory at Ohio State University after World War II.

Jansky died at age 44 in a Red Bank, New Jersey hospital due to a heart condition on February 14, 1950. Named for him are the unit used by radio astronomers for flux density of radio sources (the **Jansky**), **Jansky crater** on the Moon, and the NRAO postdoctoral fellowship program. **Jansky noise** refers to high frequency static disturbances of cosmic origin.

1. Karl Guthe Jansky, Wikipedia, http://en.wikipedia.org/wiki/Karl_Guthe_Jansky.
2. Karl Jansky and the Discovery of Cosmic Radio Waves, National Radio Astronomy Observatory, www.nrao.edu/whatisra/hist_jansky.
3. World of Scientific Discovery on Karl Jansky, BookRags, www.bookrags.com/biography/karl-jansky-wsd.

Gerard Kuiper (1905-1973)

Proposed that a disk of comet nuclei extends from the Solar System's planetary zone out to as much as 1,000 times the Earth-to-Sun distance

Gerard Peter Kuiper, a Dutch-American astronomer, was born in the village of Tuitjenhorn in North Holland on December 7, 1905. His father was a tailor. Gerard attended Leiden University, where he was taught by Ejnar Hertzsprung (1873-1967), Antonie Pannekoek (1873-1960), Willem de Sitter (1872-1934), Jan Woltjer (1891-1946), Jan Oort (1900-1992), and Paul Ehrenfest (1880-1933).

Kuiper finished his doctoral dissertation on binary stars with Hertzsprung in 1933. He then became a fellow under Robert Grant (1864-1951) at Lick Observatory in California.

In 1935, he went to Harvard College Observatory, and then took a job at Yerkes Observatory of the University of Chicago. He became an American citizen in 1937. In 1951, he proposed that a disk of comet nuclei extends from the Solar System's planetary zone out to as much as 1,000 times the Earth-to-Sun distance. This is now known as the **Kuiper Belt**.

Kuiper discovered Uranus's fifth moon, Miranda, and Neptune's second moon, Nereid. In addition, he discovered carbon dioxide in the atmosphere of Mars and the existence of a methane-laced atmosphere above Saturn's moon, Titan. Kuiper also pioneered airborne infrared observation in the 1960s, using a Convair 990 aircraft. In addition, he discovered several binary stars, which received **Kuiper numbers** to identify them.

In 1959, he won the Henry Norris Russell Lectureship of the American Astronomical Society. In 1960, he moved to Tucson, Arizona to found the Lunar and Planetary Laboratory at the University of Arizona. In the 1960s, he helped identify landing sites on the Moon for the Apollo program.

Other things named for Kuiper are **Asteroid 1776 Kuiper**, the **Kuiper crater** on the Moon, craters on Mars and Mercury, and the **Kuiper Airborne Observatory**. The **Kuiper Prize** is the most distinguished award given by the American Astronomical Society Division of Planetary Sciences. The prize recognizes outstanding contributors to planetary science. It is awarded annually to scientists whose achievements have most advanced our understanding of planetary systems.

Kuiper died in Mexico City while on vacation on December 23, 1973. He is usually regarded as the father of modern planetary astronomy.

1. Gerard Kuiper, Wikipedia, http://en.wikipedia.org/wiki/Gerard_Kuiper.
2. Kuiper, Gerard Peter (1905-1973), The Internet Encyclopedia of Science, www.daviddarling.info/encyclopedia/K/KuiperGP.
3. Kuiper, Gerard Peter, BookRags, www.bookrags.com/research/kuiper-gerard-peter-spsc-02.

William Morgan (1906-1994)

Developed the MK system for the classification of stars through their spectra and discovered the Milky Way's spiral structure

William Wilson Morgan, an American astronomer, was born in Bethesda, Tennessee on January 3, 1906. Both parents were home missionaries in the Southern Methodist Church, and went from town to town spreading the gospel. William received his A.B. in 1927 from Washington and Lee University in Lexington, Virginia and his Ph.D. from the University of Chicago in 1931. He then joined the Chicago faculty as Instructor the following year.

He was named the Bernard E. and Ellen C. Sunny Distinguished Service Professor in 1966. He served as Director of the Yerkes and McDonald Observatories from 1960 to 1963.

Morgan was widely recognized for his discovery of the spiral arms of the Milky Way and for his system of classifying stellar brightness and spectra.

In addition to his discovery of the Milky Way's spiral structure, he is known for his development of three classification systems: the MK system, a two-dimensional classification system for stellar spectra and luminosity; the Yerkes system of classification of the optical forms of galaxies; and a precisely defined photoelectric method called the UBV system for determining the brightness of stars.

Morgan developed the MK system with Philip Keenan (1908-2000). It allowed astronomers for the first time to determine the luminosity of stars directly from observations of their spectra (the detailed colors and bright and dark bands in the light emitted by all stars). Morgan then used the MK system to determine the distances of bright stars within the Milky Way, and in so doing discovered the galaxy's spiral structure.

In 1943, with Keenan and photographic assistant Edith Kellman (1911-2007), Morgan wrote the *Atlas of Stellar Spectra*, a fundamental work that is still widely used by astronomers today.

Morgan had many interests in the more conventional American pastimes, such as detective stories and sports. He was awarded the Bruce Medal (1958), Henry Draper Medal (1980), and Herschel Medal (1983). He died of a heart attack in Williams Bay, Wisconsin on June 21, 1994. **Asteroid 3180 Morgan** is named for him.

1. Morgan, William Wilson (1906-1994), The Internet Encyclopedia of Science, www.daviddarling.info/encyclopedia/M/Morgan.
2. Obituary: William Morgan, Astronomy & Astrophysics, The University of Chicago Chronicle, http://chronicle.uchicago.edu/940714/morgan, July 14, 1994.
3. Osterbrock, Donald E., William Wilson Morgan, Biographical Memoirs, National Academy of Sciences, www.nap.edu/readingroom/books/biomems/wmorgan.

Clyde Tombaugh (1906-1997)

Discovered Pluto

Clyde William Tombaugh, an American astronomer, was born in Streator, LaSalle County, Illinois on February 4, 1906. The family later moved to Burdett, Kansas. His father was a farmer. Tombaugh bought his first telescope from Sears. In 1928, he built his own more powerful telescope, and sent drawings of his observations of Jupiter and Mars to the Lowell Observatory. These resulted in a job offer in 1929. He was given the job of performing a systematic search for a trans-Neptunian planet (also called Planet X), which had been predicted by Percival Lowell (1855-1916) and William Pickering (1858 -1938).

On February 18, 1930, using images taken in January of the same year, Tombaugh noticed a moving object in his search, and subsequent observations showed it to be the object we now call Pluto.

Following his discovery of Pluto, Tombaugh earned astronomy degrees from the University of Kansas and Northern Arizona University.

He worked at the White Sands Missile Range in the early 1950s, and taught astronomy at New Mexico State University from 1955. He discovered 14 asteroids, beginning with 2839 Annette in 1929, mostly as a by-product of his search for Pluto and his further searches for other celestial objects.

In the 1990s, following the discovery of the Kuiper belt, Pluto began to be seen not as a planet orbiting alone, but as one of a group of icy bodies in that region of space, and not even the largest one.

In all, Tombaugh cataloged 29,548 galaxies, 3,969 asteroids, two comets, and one nova. His prediction that Mars would have impact craters was proved correct by the Mariner 4 space probe in the 1960s

On August 20, 1949, Tombaugh reported seeing several UFOs near Las Cruces, New Mexico. He was probably the most eminent astronomer to have reported seeing UFOs and to support the extraterrestrial hypothesis. He retired in 1973, and died in Las Cruces, New Mexico on January 17, 1997. Following the launch of the New Horizons probe to Pluto in 2006, it was announced that some of Tombaugh's ashes were being carried aboard the spacecraft. The launch was watched by Tombaugh's 93-year-old widow.

In 2006, the International Astronomical Union reclassified Pluto as a dwarf planet, rather than grouping it with the eight classical planets.

1. Clyde Tombaugh, Wikipedia, http://en.wikipedia.org/wiki/Clyde_Tombaugh.
2. Clyde W. Tombaugh 1906-1997, ICSTARS, www.icstars.com/HTML/icstars/graphics/clyde.
3. Tombaugh, Clyde W. (1906-1997), BookRags, www.bookrags.com/research/tombaugh-clyde-w-1906-1997-woes-02.

Bart Bok (1906-1983)

Proposed that small dark clouds that can be seen against the light of stars are made up of gas and dust that will eventually condense to form new stars

Bartholomeus (Bart) Jan Bok, a Dutch-American astronomer, was born in Hoorn, Netherlands on April 28, 1906. He received a B.S. from the University of Leiden in 1927. In 1929, he married fellow astronomer, Priscilla Fairfield (1896-1975), whom he had met at a meeting of the International Astronomical Union. Afterward, they collaborated closely on their astronomical work.

From 1929 to 1947, Bok taught at Harvard College Observatory. He received a Ph.D. from Groningen University in 1932. He was professor of astronomy at Harvard from 1947 to 1957. He then worked as director of Mount Stromlo Observatory in Australia for nine years before returning to the United States as director of Steward Observatory at the University of Arizona. He became a U.S. citizen in 1938.

Bok stated that there is no scientific or statistical truth to astrology. In 1975, he coauthored the statement *Objections to Astrology*, which was endorsed by 186 professional astronomers, astrophysicists, and other scientists, including 19 winners of the Nobel Prize. This led to the formation of the Committee for the Scientific Investigation of Claims of the Paranormal, of which Bok was a founding fellow.

He participated in several groups to view solar eclipses, the last near Irkutsk in Siberia in the summer of 1980. He mapped the spiral arm of the Milky Way and studied star clusters. He proposed that small dark clouds that can be seen against the light of stars are made up of gas and dust that will eventually condense to form new stars. The condensing gas and dust has come to be known as **Bok Globules**.

Bok was awarded the Bruce Medal (1977), Henry Norris Russell Lectureship (1982), and Klumpke-Roberts Award (1982).

He died of a heart attack at his home in Tucson, Arizona on August 5, 1983. Also named after him are the **Bok crater** on the Moon (with his wife), **Asteroid 1983 Bok** (with his wife), and the **Bok Telescope** at Steward Observatory.

The **Bart J. Bok Postdoctoral Fellowship** is given out by the Astronomy Department of the University of Arizona and Steward Observatory.

1. Bart Bok, Wikipedia, http://en.wikipedia.org/wiki/Bart_Bok.
2. Bart Jan Bok, NNDB, www.nndb.com/people/234/000167730.
3. Bok, Bartholomeus ("Bart") Jan (1906-1983), The Internet Encyclopedia of Science, www.daviddarling.info/encyclopedia/B/Bok.

Hans Bethe (1906-2005)

Worked out the details of the proton-proton chain, the main energy-producing reaction in stars less massive than the Sun

Hans Bethe, a German-American theoretical physicist, was born in Strasbourg, Alsace-Lorraine (now part of France) on July 2, 1906. His father was a physiologist. In 1928, at the age of 22, Hans earned his doctorate at the University of Munich. He taught physics at various universities in Germany from 1928 to 1933. After Hitler came to power, Bethe fled, first to England and then to Cornell University in Ithaca, New York.

In 1939, Bethe published *Energy Production in Stars* in which he advanced a theory of stellar fuel. He discovered that, by a series of transformations, carbon, acting as a catalyst, changes four atoms of hydrogen into an atom of helium. During these transformations, a very small loss of mass is converted into the enormous amount of energy which stokes the stars. For this achievement, and other work, Bethe was awarded the Nobel Prize in physics in 1967.

Bethe published the first theory of electron-positron pair creation and an improved theory of how charged particles interact. The latter is a key to the determination of the amount of radiation shielding required by nuclear reactors and by astronauts in space. It is also critical to the understanding of cosmic-ray phenomena, the scattering of mesons, and the energy levels of the hydrogen atom.

In 1941, Bethe became a naturalized citizen of the United States. Between 1943 and 1946 he worked as head of the Division of Theoretical Physics at the Los Alamos Scientific Laboratories, where the first nuclear bomb was being manufactured. He then returned to his teaching position at Cornell University. During the early 1950s, he played an important role in the development of the hydrogen bomb.

In 1996, the American Physical Society announced it would begin awarding the **Bethe Prize** for contributions to the field of physics. In 1997, Bethe called for a complete ban on nuclear testing He retired from Cornell in 1975, and died in Ithaca on March 6, 2005. Named after him are **Asteroid 30828 Bethe**, **Hans Bethe House** at Cornell University, and **Bethe ansatz** (a method for finding the exact solutions of certain one-dimensional quantum many-body models).

1. Encyclopedia of World Biography on Hans Albrecht Bethe, BookRags, www.bookrags.com/biography/hans-albrecht-bethe.
2. Hans Bethe, Wikipedia, http://en.wikipedia.org/wiki/Hans_Bethe.
3. World of Scientific Discovery on Hans Albrecht Bethe, BookRags, www.bookrags.com/biography/hans-albrecht-bethe-wsd

Fred Whipple (1906-2004)

Proposed the icy conglomerate hypothesis of comet composition

Fred Lawrence Whipple, an American astronomer, was born in Red Oak, Iowa on November 5, 1906. He attended Occidental College and the University of California, Los Angeles before receiving his doctorate in astronomy from the University of California, Berkeley.

In 1931, Whipple joined Harvard College Observatory in Cambridge, Massachusetts, and from 1950 to 1977 he was professor of astronomy there. In addition, he served as director of the Smithsonian Astrophysical Observatory, also in Cambridge, from 1955 until 1973. He was an adviser on space science to the National Aeronautics and Space Administration.

Many believed that comets were orbiting clouds of dust and small rock particles whose changes in shape were largely determined by gravitational forces. In 1950, Whipple proposed that comets comprise a nucleus consisting of frozen water and gases, such as methane, carbon dioxide, and ammonia, together with a variety of rocky debris. He believed that as a comet approaches the Sun, the Sun vaporizes the ice in the comet's nucleus, resulting in jets of dust and gas that either slow or accelerate the comet, causing slight variations in cometary orbits. He confirmed the spectacular coma (tail) was a product of solar wind. This icy conglomerate (dirty snowball) theory was initially highly controversial, but Whipple was eventually proven correct in 1986 when the European Space Agency spacecraft, Giotto, photographed Halley's Comet.

Among Whipple's other achievements was being the first to compute an accurate orbit for Pluto. And in the 1930s, his studies of the trajectories of meteors confirmed that meteoroids originated within the Solar System, rather than from interstellar space. He discovered six comets, all of which were named after him. His anticipation of space flight and artificial satellites led him to develop a satellite tracking system. In 1957, when the Soviet Union launched Sputnik 1, Whipple's network of amateur astronomers followed its progress.

After retirement, Whipple became professor emeritus at Harvard until he was 90. He died in Cambridge, Massachusetts on August 30, 2004. In 1982, the Mount Hopkins Observatory, located south of Tucson, Arizona, was renamed the **Whipple Observatory**.

1. Comet pioneer Fred Whipple dies, BBC News, http://news.bbc.co.uk/2/hi/science/nature/3614064.
2. Fred Lawrence Whipple, Microsoft Encarta Online Encyclopedia, http://ca.encarta.msn.com/encyclopedia_701879813/Whipple_Fred_Lawrence, 2008.
3. Whipple, Fred Lawrence (1906-2004), The Internet Encyclopedia of Science, www.daviddarling.info/encyclopedia/W/Whipple.

Bengt Strömgren (1908-1987)

Found that the chemical composition of stars was very much different than previously assumed

Bengt Georg Daniel Strömgren, a Danish astronomer/astrophysicist, was born in Gothenburg, Sweden on January 21, 1908. His father was professor of astronomy at the University of Copenhagen and director of the University Observatory. Bengt published his first paper when he was 14 years old. He graduated from Copenhagen University in 1925, after only two years, with a degree in astronomy and atomic physics. Two years later, he completed a doctoral degree.

He was appointed lecturer at the University of Copenhagen in 1932, and worked with Otto Struve (1897-1963) at the University of Chicago from 1936 to 1940. He then succeeded his father's professorship in Copenhagen.

In 1951, Strömgren went back to the United States and served as director of the Yerkes and McDonald Observatories until 1957. He then was appointed the first professor of theoretical astrophysics at the Institute for Advanced Study at Princeton. In 1967, he returned to Denmark.

In the late 1930s, Strömgren found the relative abundance of hydrogen in stars to be nearly 70 percent and helium to be about 27 percent. This was very much different than previously thought.

In the 1930s and 1940s, he engaged in pioneering work on emission nebulae, which are huge clouds of interstellar gas and dust shining by their own light. He showed that they consist largely of ionized hydrogen. At a certain distance from the star, known as the **Strömgren radius**, the emitted photons of radiation no longer possess sufficient energy to ionize the hydrogen, leading to a sharp boundary between ionized and non-ionized regions. These are now known as **Strömgren Spheres**. He found relations between the gas density, the luminosity of the star, and the size of the Strömgren sphere. And in the 1950s and 1960s, he pioneered photoelectric photometry with a four-color system, now called the **Strömgren photometric system**.

Strömgren was awarded the Bruce Medal (1959), Gold Medal of the Royal Astronomical Society (1962), and Henry Norris Russell Lectureship (1965). He died after a short illness on July 4, 1987. Also named after him are **Asteroid 1846 Bengt** and **Strömgren age**. **Asteroid** 1493 Sigrid is named after his wife.

1. Bengt Georg Daniel Strömgren, Answers.com, www.answers.com/topic/bengt-str-mgren.
2. Bengt Strömgren, Wikipedia, http://en.wikipedia.org/wiki/Bengt_Str%C3%B6mgren.
3. Strömgren, Bengt Georg Daniel (1908-1987), The Internet Encyclopedia of Science, www.daviddarling.info/encyclopedia/S/Stromgren.

Hannes Alfvén (1908-1995)

Showed that a plasma has an electric current that produces a magnetic field

Hannes Olof Gösta Alfvén, an electric power engineer and plasma physicist, was born in Norrköping, Sweden on May 30, 1908. His parents were both physicians. Originally trained as an electrical power engineer, Alfvén later moved to research and teaching in the fields of plasma physics, which is the study of gas-like mixtures consisting of electrically charged particles. He received his Ph.D. from the University of Uppsala in 1934. His dissertation was entitled *Investigations of the Ultra-short Electromagnetic Waves.*

Alfvén made many contributions to plasma physics, including theories describing the behavior of aurorae, the Van Allen radiation belts, the effect of magnetic storms on the Earth's magnetic field, the terrestrial magnetosphere, and the dynamics of plasmas in the Milky Way galaxy.

In 1934, he taught physics at both the University of Uppsala and the Nobel Institute for Physics in Stockholm, Sweden. In 1937, he argued that if plasma pervaded the universe, it could then carry electric currents capable of generating a galactic magnetic field. His theoretical work on field-aligned electric currents in the aurora was confirmed by satellite observations in 1974, resulting in the discovery of Birkeland currents.

In 1939, Alfvén postulated that magnetic storms occur when streams of plasma particles from the Sun enter the Earth's upper atmosphere and collide with neutral gas molecules, releasing energy that is seen as the light in the aurora.

In 1940, he became professor of electromagnetic theory and electrical measurements at the Royal Institute of Technology in Stockholm. In 1945, he acquired the position of Chair of Electronics. His title was changed to Chair of Plasma Physics in 1963. In 1967, after leaving Sweden and spending time in the Soviet Union, Alfvén moved to the United States. He worked in the departments of electrical engineering at the University of California, San Diego and the University of Southern California.

In addition to showing that plasma has an electric current that produces a magnetic field, Alfvén showed that, under certain conditions, the plasma binds, or freezes, the magnetic field, meaning that the plasma and the magnetic field move together. Physicists call this the frozen-in-flux theorem. Alfvén died in Djursholm, Sweden on April 2, 1995.

1. Hannes Alfvén Wikipedia, http://en.wikipedia.org/wiki/Hannes_Alfv%C3%A9n.
2. Hannes Alfvén, Biography, NobelPrize.org, http://nobelprize.org/nobel_prizes/physics/laureates/1970/alfven-bio.
3. Hannes Olof Gösta Alfvén, Microsoft Encarta Encyclopedia, 2003.

Viktor Ambartsumian (1908-1966)

The first to suggest that T Tauri stars are very young and to propose that nearby stellar associations are expanding

Victor Amazaspovich Ambartsumian, a Soviet Armenian astrophysicist, was born in Tbilisi, Georgia on September 18, 1908. His father was a philologist and writer. In 1924, Victor studied at the Physico-mathematical Department of Leningrad State Pedagogical Institute and then at Leningrad State University. He continued his postgraduate studies at Pulkovo Observatory.

After three years of affiliation with Leningrad University, in 1934 Ambartsumian founded and headed the first astrophysics chair there. From 1939 to 1941 he was the director of Leningrad University Observatory. He worked at the Pulkovo Observatory; the Byurakan Astrophysical Observatory, which he founded and directed; and Erevan University, While at Pulkovo he also taught at the University of Leningrad. He wrote the first Russian textbook on theoretical astrophysics.

Most of his research was devoted to invariance principles applied to the theory of radiative transfer, inverse problems of astrophysics, and the empirical approach to the problems of the origin and evolution of stars and galaxies. He was first to suggest that T Tauri stars (variable stars in the constellation Taurus) are very young and to propose that nearby stellar associations are expanding. He also showed that evolutionary processes, such as loss of mass, are occurring in galaxies. He worked on interstellar matter, radio galaxies, and active galactic nuclei.

In 1964, he became the founding editor-in-chief of *Astrofizika,* which was the leading astronomy publication in the Soviet Union. He kept that post until 1987. He served as president of the International Astronomical Union, and was the principal organizer of two major conferences on SETI (Search for Extra-Terrestrial Intelligence) at the Byurakan Observatory in 1964 and 1972.

Ambartsumian served as president of the Armenian Academy of Sciences from 1947 to 1993. He was awarded the French Academy of Sciences Janssen Medal (1956), the Bruce Prize (1960), and the Royal Astronomical Society Gold medal (1960).

He died in Byurakan on August 12, 1996. Named after him are minor planet **1905 Ambartsumian** and the Byurakan Astrophysical Observatory.

1. Ambartsumian, Viktor Amazaspovich (1908-1996), The Internet Encyclopedia of Science, www.daviddarling.info/encyclopedia/A/Ambartsumian.
2. Victor Amazaspovich Ambartsumian, The Bruce Medalists, www.phys-astro.sonoma.edu/ BruceMedalists/Ambartsumian/Ambartsumian.
3. Victor Amazaspovich Ambartsumian, Wikipedia, http://en.wikipedia.org/ wiki/ Viktor_Amazaspovich_Ambartsumian.

Subrahmanyan Chandrasekhar (1910-1995)

Developed a theory of white dwarf formation

Subrahmanyan Chandrasekhar, an Indian-born American astrophysicist and mathematician, was born in Lahore, India (now part of Pakistan) on October 19, 1910. His uncle was the Nobel Prize-winning Indian physicist, Chandrasekhara Raman (1888-1970). After high school, Chandrasekhar attended Presidency College in Madras, India, majoring in mathematics and physics. While in college, he published *The Compton Scattering and the New Statistics*. He completed a master's degree in 1930, and then attended Cambridge University, England, completing his Ph.D. in 1933. In 1936, he joined the faculty of the University of Chicago and the Yerkes Observatory.

Chandrasekhar's major fields of research were stellar evolution, stellar structure, and the processes of energy transfer within stars. It was known that stars could end their lives either dramatically and explosively as a supernova or as an extremely small dense star of low luminosity known as a white dwarf. What decided the particular path a star took was answered by Chandrasekhar in his *Introduction to the Study of Stellar Structure* (1939).

As stars evolve, he believed they release energy generated by their conversion of hydrogen into helium and even heavier elements. As they reach the end of their lives, stars have less hydrogen to convert, so they release less energy in the form of radiation. They eventually reach a stage where they are no longer able to generate the pressure needed to maintain their size against their own gravitational pull, and they begin to shrink, eventually collapsing into themselves and becoming white dwarfs, which are small objects of enormous density.

There is a point at which the mass of a star is too great for it to evolve into a white dwarf. Chandrasekhar calculated this mass to be 1.4 times the mass of the Sun. This has since become known as the **Chandrasekhar limit**.

Chandrasekhar became a United States citizen in 1953. He shared the 1983 Nobel Prize in physics with William Fowler (1911-1995) for his work on the structure and evolution of stars. He retired from the University of Chicago in 1980, and died from a heart attack in Chicago on August 21, 1995.

1. Subrahmanyan Chandrasekhar Biography, Notable Biographies, www.notablebiographies.com/Ch-Co/Chandrasekhar-Subrahmanyan.
2. Subrahmanyan Chandrasekhar, Answers.com, www.answers.com/topic/subrahmanyan-chandrasekhar.
3. World of Scientific Discovery on Subramanyan Chandrasekhar, BookRag, www.bookrags.com/biography/subramanyan-chandrasekhar-wsd.

Carl Seyfert (1911-1960)

Observed and described a group of galaxies now known as Seyfert's Sextet

Carl Keenan Seyfert, an American Astronomer, was born in Cleveland, Ohio on February 11, 1911. His father was a pharmacist. From 1929 to 1936, Carl studied at Harvard. Although initially studying medicine, inspired by Bartholomeus Bok (1906-1983) he turned to astronomy. He obtained an M.S. in 1933 and Ph.D. in 1936, both in astronomy. His Ph.D. dissertation, supervised by Harlow Shapley (1885-1972), was *Studies of the External Galaxies*. It focused on colors and magnitudes of galaxies.

Seyfert served as a staff member at McDonald Observatory in the Davis Mountains of West Texas from 1936 to 1940. There, he investigated the properties of B stars and large PM stars, and did work on variable stars. He also studied the distribution of colors, emission nebulae, and clusters in galaxies.

From 1940 to 1942, he was at Mount Wilson Observatory in the San Gabriel Mountains of Southern California as a National Research Council Fellow. He did pioneering research on nuclear emission in spiral galaxies. In 1943, he published a paper on galaxies with bright nuclei that emit radio waves at a rate between the rate of normal galaxies and that of radio galaxies and exhibit characteristically broadened emission lines. These galaxies are now called **Seyfert Galaxies**.

In 1946, Seyfert joined the faculty of Vanderbilt University in Nashville, Tennessee. In 1951, he observed and described a group of galaxies around galaxy NGC 6027. The group is now known as **Seyfert's Sextet**.

He was involved in instrumental innovations, including the use of photomultiplier tubes and television techniques and electronically controlled telescope drives. He obtained scientific results on variable stars, emission B stars in stellar associations, and the Milky Way structure.

In 1953, under Seyfert's direction, the new Arthur J. Dyer Observatory, equipped with a 24-inch reflector telescope, was completed at Vanderbilt. Seyfert held the position of director of the facility for the rest of his life.

Seyfert died at the age of 49 in Nashville in an automobile accident on June 13, 1960. Also named for him are a crater on the Moon (**Seyfert, 29.1N, 114.6E)** and the 24-inch telescope at Dyer Observatory.

1. Carl Keenan Seyfert (February 11, 1911 - June 13, 1960), SEDS, http://seds.org/MESSIER/xtra/Bios/seyfert.
2. Seyfert Galaxies, Microsoft Encarta Online Encyclopedia, http://encarta.msn.com/encyclopedia_761566139/seyfert_galaxy, 2008.

Luis Alvarez (1911-1988)

Proposed the asteroid impact theory to explain the extinction of the dinosaurs

Luis Walter Alvarez, an American physicist, was born in San Francisco, California on June 13, 1911. He was the son of a physician. He attended the University of Chicago, where he received his bachelor's degree in 1932, his master's in 1934, and his Ph.D. in 1936.

During World War II, Alvarez was a key participant in the Manhattan Project, including the Project Alberta on the dropping of the bomb. He flew as a scientific observer of the atomic bombing of Hiroshima. He and one of his students designed the exploding-bridgewire detonators for the spherical implosives used on the Trinity and Nagasaki bombs.

Alvarez also did important work relating to radar and navigation technologies. In particular, he developed the Ground Controlled Approach system (GCA), which pilots use to land a plane in low visibility conditions.

After the war, he went on to invent the synchrotron, a cyclic particle accelerator in which the magnetic field (turns the particles so they move in a circle) and the electric field (accelerates the particles) are carefully synchronized with the travelling particle beam. He also developed the Berkeley 40-foot proton linear accelerator, which was completed in 1947.

Alvarez was awarded the 1968 Nobel Prize in physics for the discovery of a large number of resonance states made possible through his development of the hydrogen bubble chamber. This was significant because it allowed scientists to record and study the short-lived particles created in particle accelerators.

In 1980, Alvarez and his son, Walter Alvarez (1940-), presented the asteroid-impact theory as an explanation for the presence of an unusual abundance of iridium associated with the geological event referred to as the K-T extinction boundary.

Ten years after this initial proposal, evidence of a huge impact crater called Chicxulub, off the coast of Mexico, strongly supported their theory. Since that time, the concept of impact by a large meteorite has become the most widely accepted explanation for the extinction of the dinosaurs.

In 1978, Alvarez was inducted into the National Inventors Hall of Fame.

He died on September 1, 1988.

1. Alvarez, Luis Walter, Microsoft Encarta Encyclopedia, 2003.
2. Luis Alvarez, Biography, NobelPrize.org, http://nobelprize.org/nobel_prizes/physics/laureates/1968/alvarez-bio.
3. Walter Alvarez, Wikipedia, http://en.wikipedia.org/wiki/Luis_Walter_Alvarez.

John Wheeler (1911-2008)

Coined the terms black hole and wormhole

John Archibald Wheeler, an American theoretical physicist, was born in Jacksonville, Florida on July 9, 1911. He received his doctorate from Johns Hopkins University in 1933. His dissertation, under the supervision of Karl Herzfeld (1892-1978), was on the theory of the dispersion and absorption of helium.

In 1937, he introduced the S-matrix, which became an indispensable tool in particle physics. He was a professor of physics at Princeton University from 1938 until 1976. He was a pioneer in the theory of nuclear fission, along with Niels Bohr (1885-1962) and Enrico Fermi (1901-1954). In 1939, he collaborated with Bohr on the liquid drop model of nuclear fission.

During World War II, Wheeler participated in the development of the atomic bomb under the Manhattan Project at the Hanford site. There, reactors were constructed to produce plutonium.

In 1957, while working on extensions to general relativity, he introduced the word "wormhole" to describe hypothetical tunnels in space-time. A trip through a wormhole could take much less time than a journey between the same starting and ending points in normal space.

In the 1960s, Wheeler made attempts at a unified field theory, which he called geometrodynamics, but it imperfectly explained the current state of physics, and he abandoned it.

The work of Wheeler and his students contributed greatly to the golden age of general relativity. He coined the term "black hole" in 1967 during a talk at the NASA Goddard Institute of Space Studies.

He was also a pioneer in the field of quantum gravity. He and Bryce DeWitt (1923-2004) developed the **Wheeler-DeWitt equation.** A wave function of the universe should satisfy this equation in a theory of quantum gravity. He received the Atomic Energy Commission's Enrico Fermi Award from President Lyndon B. Johnson in 1968.

Wheeler was director of the Center for Theoretical Physics at the University of Texas at Austin from 1976 to 1986. He produced one of the best quotes of physics, "Time is what prevents everything from happening at once." He died of pneumonia in Hightstown, New Jersey on April 13, 2008.

1. John A. Wheeler, Physicist Who Coined the Term 'Black Hole,' Is Dead at 96, The New York Times, www.nytimes.com/2008/04/14/science/14wheeler.html?_r=1&oref=slogin, April 14, 2008.
2. John Archibald Wheeler, NNDB, www.nndb.com/people/115/000099815.
3. John Archibald Wheeler, Wikipedia, http://en.wikipedia.org/wiki/John_Archibald_Wheeler.

William Fowler (1911-1995)

Described the processes of nuclear synthesis of chemical elements within stars

William (Willie) Alfred Fowler, an American astrophysicist, was born in Pittsburgh, Pennsylvania on August 9, 1911. He was raised in Lima, Ohio from the age of two when his father, an accountant, was transferred. Willie earned a B.S. from Ohio State University in 1933 and went on to receive a Ph.D. in nuclear physics at the Kellogg Radiation Laboratory at California Institute of Technology (Caltech) in 1936. His dissertation was entitled *Radioactive Elements of Low Atomic Number*. He joined the faculty at Caltech in 1939.

In 1954-1955, he spent a sabbatical year in Cambridge, England as a Fulbright Scholar with Fred Hoyle (1915- 2001). There, husband and wife team Geoffrey Burbidge (1925-) and Margaret Burbidge (1919-) joined them. In 1956, the Burbidges and Hoyle came to Kellogg.

The work for which Fowler is best known was developed in collaboration with Hoyle and the Burbidges. It was entitled *Synthesis of the Elements in Stars* and was published in *Reviews of Modern Physics* in 1957. It described the processes of nuclear synthesis of chemical elements within stars, and showed that all of the elements from carbon to uranium could be produced by nuclear processes in stars, starting with the hydrogen and helium produced in the Big Bang. The work became a cornerstone of modern astrophysics.

Fowler shared the 1983 Nobel Prize for physics with the astrophysicist Subrahmanyan Chandrasekhar (1910-1995) for his theoretical and experimental studies of the nuclear reactions of importance in the formation of the chemical elements in the universe.

Fowler won the Henry Norris Russell Lectureship of the American Astronomical Society in 1963, the Eddington Medal in 1978, and the Bruce Medal in 1979.

He received honorary degrees from the University of Chicago (1976), Ohio State University (1978), University of Liege (1981), Observatory of Paris (1981), and Denison University (1982). He was President of the American Physical Society in 1976.

Fowler died in Pasadena, California on March 14, 1995. He was a loyal fan of the Pittsburgh Pirates in the National Baseball League and of the Pittsburgh Steelers in the National Football League.

1. William A. Fowler, Autobiography, NobelPrize.org, http://nobelprize.org/nobel_prizes/physics/laureates/1983/fowler-autobio.
2. William A. Fowler, NNDB, www.nndb.com/people/965/000099668.
3. William Alfred Fowler, Wikipedia, http://en.wikipedia.org/wiki/William_Alfred_Fowler.

Grote Reber (1911-2002)

Conducted the first sky survey in the radio frequencies

Grote Reber, an American radio astronomer, was born in Wheaton, Illinois on December 22, 1911. He graduated from Armour Institute of Technology (now Illinois Institute of Technology) in 1933 with a degree in electrical engineering. He worked for various radio manufacturers in Chicago from 1933 to 1947. In 1933, when he learned of the work of Karl Jansky (1905-1950), he decided to build his own radio telescope in his backyard. His design was considerably more advanced than Jansky's, consisting of a parabolic sheet-metal mirror nine meters in diameter, focusing to a radio receiver eight meters above the mirror. The entire assembly was mounted on a tilting stand, allowing it to be pointed in various directions, although not turned. The telescope was completed in 1937.

In 1938, his third attempt at 160 MHz was successful, confirming Jansky's discovery. He then turned his attention to making a radio-frequency sky map, which he completed in 1941 and extended in 1943. He published a considerable body of work during this era, and was the initiator of the explosion of radio astronomy after 1945

His work uncovered a mystery. The standard theory of radio emissions from space was that they were due to blackbody radiation. If true, one would expect that there would be considerably more high-radiation than low-energy radiation due to the presence of stars and other hot bodies. Reber found the reverse—a considerable amount of low-energy radio signal. It was not until the 1950s that synchrotron radiation was offered as an explanation for these measurements.

In the 1950s, Reber turned to a field that was being largely ignored—very low frequency radio signals in the 0.5 to 3 MHz range. However, signals with frequencies below 30 MHz are reflected by the ionosphere. To overcome this, Reber moved to Tasmania, an island off the southern coast of Australia. There, on very cold, long, winter nights, the ionosphere de-ionizes, allowing the long radio waves to reach an antenna array.

Reber never married. He died in Tasmania on December 20, 2002. Some of his ashes were distributed among 24 major radio observatories around the world. Named after him are **Asteroid 6886 Grote**, the **Grote Reber Medal** for lifetime innovative contributions to radio astronomy, and the museum at the Launceston Planetarium.

1. Grote Reber, National Radio Astronomy Observatory, www.nrao.edu/whatisra/hist_reber.
2. Grote Reber, NNDB, www.nndb.com/people/090/000172571.
3. Grote Reber, Wikipedia, http://en.wikipedia.org/wiki/Grote_Reber.

Martin Schwarzschild (1912-1997)

First to take astronomical photographs from a hot-air balloon

Martin Schwarzschild, a German-American astronomer, was born in Potsdam, Germany on May 31, 1912. His father was astrophysicist Karl Schwarzschild (1873-1916), who died from a rare skin disease when Martin was four years old. His uncle was physicist Robert Emden (1862-1940). Martin earned a Ph.D. in Astronomy at the University of Göttingen in 1935. He taught astronomy at the University of Oslo in 1936-37.

Fleeing Nazi Germany, Schwarzschild came to the United States, and did a three-year postdoctoral fellowship at Harvard University from 1937 to 1940. He then taught at Columbia University from 1940 to 1947 and Princeton University from 1947 to 1979.

He became a naturalized U.S. citizen in 1942. He was twice decorated for his service in Army Intelligence during World War II. In the late 1940's, he was one of the first to use electronic digital computers for astronomical research.

Schwarzschild helped develop the theories of stellar structure and evolution and of galactic structure. He is best known for calculating the helium content of the Sun, showing how stars become red giants, and studying pulsating stars and differential solar rotation.

He headed the Stratoscope Project, which took scientific instruments to the limits of the Earth's atmosphere in hot-air balloons. In 1957, he and his team launched a balloon in Minneapolis. It traveled 81,000 feet above the Earth and took photographs that were described as the sharpest and most detailed ever taken of the Sun and the great gas storms that swirl on its surface. Photographs showed solar granules, sunspots, and convection in the solar atmosphere. It also allowed charting the infrared spectra of planets, stars, and the nuclei of galaxies.

Schwarzschild's 1958 book, *Structure and Evolution of the Stars*, became a standard text for a generation of astrophysicists. He was president of the American Astronomical Society from 1970 to 1972.

He retired from Princeton in 1979, and died of heart failure in Langehorn, Pennsylvania on April 10, 1997. At the time of his death, he was the Eugene Higgins Professor Emeritus of Astronomy at Princeton University. **Asteroid 4463 Marschwarzschild** is named after him.

1. Herszenhorn, David M., Martin Schwarzschild, 84, Innovative Astronomer, New York Times, http://query.nytimes.com/gst/fullpage.html?res=9C0DEED6113CF931A25757C0A961958260, April 12, 1997.
2. Martin Schwarzchild, The Bruce Medalists, www.phys-astro.sonoma.edu/BruceMedalists/Schwarzschild/Schwarzschild.
3. Martin Schwarzschild, NNDB, www.nndb.com/people/733/000168229.

Carl Weizsäcker (1912-2007)

Explained how stars can continue to radiate enormous amounts of energy for billions of years

Carl Friedrich von Weizsäcker, a German physicist, was born in Kiel, Schleswig-Holstein, Germany on June 28, 1912. His father was a diplomat. Carl studied physics, mathematics, and astronomy at the universities of Berlin, Göttingen, and Leipzig, obtaining his Ph.D. from Leipzig in 1933. Between 1933 and 1945, he taught successively at the universities of Leipzig, Berlin, and Strasbourg. In 1946, he returned to Göttingen as director of physics at the Max Planck Institute. In 1957, he was appointed professor of philosophy at Hamburg. In 1970, he moved to Starnberg as director of the Max Planck Institute on the Preconditions of Human Life in the Modern World.

In 1938, Weizsäcker addressed the problem of how stars can continue to radiate enormous amounts of energy for billions of years. Independently of Hans Bethe (1906-2005), he proposed a chain of nuclear-fusion reactions that proceed at the high temperatures occurring in the dense central cores of stars. In this sequence, called the carbon cycle, one carbon atom (the catalyst) and four hydrogen atoms undergo a transformation to one carbon atom and one helium atom. In the process, an immense amount of energy is released, which is eventually radiated from the star's surface as heat, light, and ultraviolet radiation. Because the stars are rich in hydrogen, they are able to continue radiating until their core hydrogen is consumed.

In 1944, Weizsäcker proposed a variation of the nebular hypothesis of Pierre de Laplace (1749-1827) to account for the origin of the planets. Beginning with the Sun surrounded by a disk of rotating gas, he argued that such a mass would experience turbulence and break into a number of smaller vortices and eddies. Where eddies meet, the conditions were said to be suitable for planets to form from the continuous aggregation of progressively larger bodies. However, the proposal did not explain the crucial point of how the planets acquired so much angular momentum. Hannes Alfvén (1908-1995) and Fred Hoyle (1915-2001) later proposed that forces generated by the Sun's magnetic field transmitted the momentum. Weizsäcker retired in 1980, and died on April 28, 2007. His brother, Richard, was president of Germany from 1984 to 1994.

1. Baron Carl Friedrich von Weizsäcker, Answers.com, www.answers.com/topic/carl-friedrich-von-weizs-cker.
2. Carl Friedrich von Weizsäcker, wikipedia, http://en.wikipedia.org/wiki/Carl_Friedrich_von_Weizs%C3%A4cker.
3. Weizsäcker, Carl Friedrich von (1912-2007, The Internet Encyclopedia of Science, www.daviddarling.info/encyclopedia/W/Weizsacker.

Horace Babcock (1912-2003)

The first to measure the distribution of magnetic fields over the solar surface

Horace Welcome Babcock, an American astronomer, was born in Pasadena, California on September 13, 1912. His father, Horace D. Babcock, was an astronomer at Mount Wilson Observatory in Pasadena, California and an expert in the then young science of laboratory spectroscopy.

In 1928, at the age of 16, Horace became an apprentice in the optical shop of Mount Wilson, where he learned the practical side of optical engineering and production. He also began observing, at first with his father and then alone. His first published paper was on solar spectroscopy in 1932. It involved the Paschen lines of hydrogen.

In 1930, Horace entered the California Institute of Technology (Caltech), and graduated in 1934 with a B.S. in physics. He then enrolled in astronomy at the University of California, Berkeley and received his Ph.D. in 1939. His thesis was on the rotation of the Andromeda Nebula (M31), whose radial velocities he measured spectroscopically.

Babcock joined the staff of Mount Wilson and Mount Palomar observatories in 1946, and worked closely with his father. Together, they constructed the first solar magnetograph, which allowed them to measure the magnetic field of the Sun.

Using another device of his own invention, Babcock was also able to discover magnetic fields in more distant stars. He also developed the **Babcock Model** of sunspots and their magnetism.

In 1953, he proposed the technique of adaptive optics—a system to deform telescope mirrors in real time to compensate for the blurring of images caused by distortions in the atmosphere. Adaptive optics has become a standard feature of modern telescopes.

From 1964 to 1978, Babcock directed the Mount Wilson and Palomar observatories. He pushed for the construction of an observatory in the Andes. The resulting Carnegie telescopes at Las Campanas in Chile's Atacama Desert are among the most powerful tools available for studying the center of the Milky Way galaxy.

Babcock died in Santa Barbara, California on August 29, 2003. **Asteroid 3167 Babcock** is named for him and his father.

1. Babcock, Harold Delos (1882-1968) and Horace Welcome (1912-), The Internet Encyclopedia of Science, www.daviddarling.info/encyclopedia/B/Babcock.
2. Horace Welcome Babcock, The Bruce Medalists, www.phys-astro.sonoma.edu/BruceMedalists/BabcockHW/BabcockHW.
3. McFarling, Usha Lee, Horace W. Babcock, 90; Invented Tools for Astronomical Measuring, LA Times, www.mailarchive.ca/lists/alt.astronomy/2003-09/0214.

Guillermo Haro (1913-1988)

Co-discovered nonstellar condensations in high density clouds near regions of recent star formation

Guillermo Haro, a Mexican astronomer, was born in Mexico City on March 21, 1913. He grew up during the turbulent time of the Mexican Revolution. He studied Philosophy at the National Autonomous University of Mexico (UNAM). He was hired in 1943 as an assistant at the newly founded Observatorio Astrofísico de Tonantzintla (OAT) in Pueblo. From 1943 to 1944, he worked at Harvard College Observatory.

In 1945, he returned to OAT, where he was responsible for the commissioning of the new 24-31-inch Schmidt Camera. With it, he studied extremely red and extremely blue stars. In 1947, he began working for the Observatorio de Tacubaya of UNAM. He detected a large number of planetary nebulae in the direction of the galactic center and discovered nonstellar condensations in high-density clouds near regions of recent star formation. The condensations also were discovered independently by George Herbig (1920-). They are now known as **Herbig-Haro objects**.

Haro and co-workers discovered flare stars in the Orion nebula region, and later in stellar aggregates of different ages. A **flare star** is a variable star that undergoes unpredictable and dramatic increases in brightness for a few minutes. It is believed that the flares are analogous to solar flares in that they are due to magnetic reconnection in the atmospheres of the stars. Haro continued searching for flare stars the remainder of his life.

In 1961, he published jointly with Willem Luyten (1899-1994) a list of 8746 blue stars in the direction of the north galactic pole. The work was done with the 48-inch Palomar Schmidt telescope using the three-color image technique developed at the Tonantzintla Observatory. At least 50 of the objects turned out to be quasars, which had not been discovered at the time.

Haro also discovered one supernova, more than 10 novae, one comet, and a number of T Tauri stars. T Tauri stars are a class of very young stars with masses less than about twice the Sun's, characterized by unpredictable changes in brightness.

During his career, Haro was director of the Observatory Astronómico de Tacubaya, the Instituto de Astoronomia, and the Observatory Astronómico Nacional. He died on April 26, 1988. The **Guillermo Haro Observatory** and **Guillermo Haro International Astrophysics Program** are named for him.

1. Flare Stars, Wikipedia, http://en.wikipedia.org/wiki/Flare_star.
2. Guillermo Haro, Wikipedia, http://en.wikipedia.org/wiki/Guillermo_Haro.
3. Vidrio, Lorena, Guillermo Haro, SJSU Virtual Museum, www.sjsu.edu/depts/Museum/haro.

Bernard Lovell (1913-)

Driving force behind the world's first giant radio telescope

Alfred Charles Bernard Lovell, a British astronomer, was born at Oldland Common, Gloucestershire, England on August 31, 1913. He received his Ph.D. from the University of Bristol in 1936. He was then appointed a lecturer in physics at the University of Manchester.

Lovell was occupied with radar research during World War II, and showed that radar echoes could be obtained from daytime meteor showers, which are invisible using optical astronomical techniques.

In 1951, he was appointed to the Chair of Radio Astronomy and the directorship of Jodrell Bank (now the Nuffield Radio Astronomy Laboratories) in Cheshire, a post he held until his retirement in 1981. In 1945, from two trailers of radar equipment that had been used in wartime defense work, Lovell constructed a giant radio telescope in a field at Jodrell Bank. He then began radio investigation of cosmic rays, meteors, and comets. He soon produced results on meteor velocities. The telescope came to public notice when it was used to track Sputnik in 1957. Lovell decided that a more permanent and ambitious telescope should be built.

Lovell began a ten-year struggle to finance a 250-foot steerable radio telescope with a parabolic dish with which to receive radio waves as short as 30 centimeters. The Department of Scientific and Industrial Research agreed to fund half if Lovell could raise the rest, which he was able to do by 1960.

Cambridge radio astronomers, under Antony Hewish (1924-), discovered pulsars, but they were limited to observing them for only the few minutes each day that the pulsars were on the Cambridge meridian. The steerable Jodrell Bank telescope could observe objects as long as they were above the horizon. Of the 50 pulsars discovered in the northern hemisphere before 1972, 27 of them were detected at Jodrell Bank. Jodrell Bank also showed that some quasars had angular diameters of one second of arc or less, which was surprisingly small for such large sources of energy.

Lovell was knighted in 1961. He wrote a number of books recounting the political, financial, and scientific story of Jodrell Bank. They include *The Story of Jodrell Bank* (1968), *The Jodrell Bank Telescopes* (1985), and his autobiography *Astronomer by Chance* (1996). The telescope at Jodrell Bank was named the **Lovell Telescope** in his honor.

1. Benard Lovell, Wikipedia, http://en.wikipedia.org/wiki/Bernard_Lovell.
2. Lovell, Sir Alfred Charles Bernard, Microsoft Encarta Online Encyclopedia, http://au.encarta.msn.com/encyclopedia_781534822/Lovell_Sir_Alfred_Charles_Bernard, 2008.
3. Sir Alfred Charles Bernard Lovell, Answers.com, www.answers.com/topic/bernard-lovell.

Yakov Zel'dovich (1914-1987)

Co-proposed a method for determining the Hubble constant

Yakov Borisovich Zel'dovich, a Russian mathematician and physicist, was born in Minsk (now Belarus) on March 8, 1914. Four months later, his family moved to Saint Petersburg. In 1931, at age 17, Yakov became a laboratory assistant at the Institute of Chemical Physics of the USSR Academy of Sciences. In 1939, he received a D.Sc. degree in Physics and Mathematics. In his doctoral work, he discovered the mechanism of the oxidation of nitrogen, which is now known as the **Zel'dovich Mechanism**. In 1941, he was evacuated to Kazan to avoid the Axis Invasion of the Soviet Union. In 1943, he moved to Moscow.

From 1943 to 1963, he participated in the Soviet Atomic Project. In 1952, he began work in the field of elementary particles and their transformations. He predicted the beta decay of a p-meson. Together with S. S. Gershtein, he noticed the analogy between the weak and electromagnetic interactions, and in 1960 he predicted the muon catalysis phenomenon. From 1965 until 1983, he was a head of the division at the Institute of the Applied Mathematics of the USSR Academy of Sciences. In 1965, he also became a professor at the Department of Physics of Moscow State University and head of the division of Relativistic Astrophysics at the Sternberg Astronomical Institute.

In 1965, Zel'dovich began working on the theory of the evolution of the hot universe, the properties of the microwave background radiation, the large-scale structure of the universe, and the theory of black holes. He predicted, with Rashid Sunyaev (1943-), that the cosmic microwave background should undergo inverse Compton scattering, which is now called the **Sunyaev-Zel'dovich effect**. This effect, often abbreviated as the SZ effect, is the result of high-energy electrons distorting the cosmic microwave background radiation (CMB). Some of the energy of the electrons is transferred to the low energy CMB photons. Observed distortions of the CMB spectrum are used to detect the density perturbations of the universe. Using the SZ effect, dense clusters of galaxies have been observed. The SZ effect also has been used to calculate the Hubble constant, which gives the rate of recession of distant astronomical objects per unit distance away. Zel'dovich died on December 2, 1987. **Asteroid 11438 Zel'dovich** was named in his honor.

1. Yakov Borisovich Zel'dovich, The Bruce Medalists, www.phys-astro.sonoma.edu/BruceMedalists/Zeldovich/Zeldovich.
2. Yakov Borisovich Zel'dovich, Wikipedia, http://en.wikipedia.org/wiki/Yakov_Zel'dovich.
3. Yakov Zel'dovich, The Nuclear Weapon Archive, http://nuclearweaponarchive.org/Russia/Zeldovch.

Adriaan Blaauw (1914-)

Showed that star formation is still taking place in the Milky Way galaxy

Adriaan Blaauw, a Dutch astronomer, was born in Amsterdam on April 12, 1914. He attended the University of Leiden, where he worked with Jan Oort (1900-1992) and Ejnar Hertzsprung (1873-1967). Afterward, he taught at Groningen University from 1938 to 1945. He obtained his Ph.D. from Groningen University in 1946. He then became a scholar at Yerkes Observatory at the University of Chicago during 1947-1948 and again from 1952 to 1957.

He was director of the Kapteyn Astronomical Institute at Groningen University from 1957 to 1970. He rebuilt the facility into a major research center concentrating on galactic structure. He was director of the European Southern Observatory in Chili from 1970 to 1974 and professor of astronomy at the University of Leiden from 1975 to 1981.

Blaauw's scientific contributions concerned the motions of star clusters and associations, runaway stars, star formation, and the determination of the cosmic distance scale. Runaway stars are heavy stars that travel through interstellar space with an anomalously high velocity. They have been known for several decades, but it has always been a problem to explain their high velocities.

He discovered that the zeta-Persei star cluster is relatively young (about 1.3 million years old). This was an important discovery because it showed that star formation is still taking place in the Milky Way galaxy and presumably in other similar galaxies. This, in turn, means that galaxies continue to evolve with time, rather than remaining static from the time of their formation until the last stars die billions of years later.

Blaauw became the first chairman of the board of directors of *Astronomy and Astrophysics* when the journal was formed by the merger of most European astronomy journals in 1968.

After retiring in 1981, he returned to the Kapteyn Institute at Groningen and wrote several historical articles and books, mostly on international organizations and collaborations in astronomy.

The **Blaauw Mechanism,** named for him, is a mathematical explanation of the disruption of a binary star system as one star throws off a shell of gas and the resulting loss of gravitational attraction changes the orbit of, or ejects altogether, the companion star.

1. Adriaan Blaauw, BookRags, www.bookrags.com/research/adriaan-blaauw-scit-0712345.
2. Adriaan Blaauw, The Bruce Medalists, www.phys-astro.sonoma.edu/BruceMedalists/Blaauw/Blaauw.
3. The Blaauw Mechanism, The Internet Encyclopedia of Science, www.daviddarling.info/encyclopedia/B/Blaauw_mechanism.

Lyman Spitzer, Jr. (1914-1997)

First to propose placing a large telescope in space

Lyman Strong Spitzer, Jr., an American astrophysicist, was born in Toledo, Ohio on June 26, 1914. He studied physics at Yale University, earning a B.A. in 1935. He then attended Cambridge University as a graduate fellow. He moved to Princeton in 1936 to study astrophysics with Henry Russell (1877-1957). After earning his Ph.D. in 1938, he spent 1938-1939 as a postdoctoral fellow at Harvard University.

Following a short period on the Yale faculty, he did underwater sound research in World War II, after which he returned briefly to Yale. At the age of 33, he succeeded Russell as chair of Princeton's astrophysical sciences department.

Spitzer's interest in interstellar matter dated back to the late 1930s when he noticed that elliptical galaxies contained old stars but no large amounts of interstellar gas, whereas spiral galaxies contained young stars and significant amounts of gas.

He worked on the theory of the heating and cooling of interstellar gases, stressing the presence and importance of interstellar magnetic fields, the likelihood of pressure equilibrium among various components, and the significance of interstellar dust grains.

Recognizing the importance of determining the transport coefficients in a fully ionized gas, Spitzer made the initial calculations of these quantities and also the first computations of toroidal containment, ohmic heating, and the diffusion losses of a confined plasma.

In 1951, he invented the stellarator, a device used to confine a hot plasma with magnetic fields to sustain a controlled nuclear fusion reaction.

In stellar dynamics, Spitzer showed how relaxation causes a stellar system to approach a singular state. Relaxation and the associated phenomenon of core collapse are accelerated by the existence of a spectrum of stellar masses, but retarded by the presence of binary stars.

In 1946, Spitzer proposed the development of a large space telescope. It came to realization as the Hubble Space Telescope, which was launched in 1990.

In the evening of March 31, 1997, after a full day at work, he collapsed and died at home in Princeton, New Jersey. NASA named the fourth Great Observatory after him, the **Spitzer Space Telescope.**

1. Lyman Spitzer, Jr. The National Aeronautics and Space Administration, http://hubble.nasa.gov/overview/spitzer_bio.php.
2. Lyman Spitzer, Jr., Physics Today, Volume 50, No. 10, pp. 123-124, October 1997.
3. Lyman Spitzer, Jr., The Bruce Medalists, www.phys-astro.sonoma.edu/brucemedalists/Spitzer/Spitzer.

James Van Allen (1914-2006)

Discovered the radiation belts that encircle the Earth

James Van Allen, an American physicist and astronomer, was born in Mount Pleasant, Iowa on September 7, 1914. He studied physics at both Iowa Wesleyan College and the University of Iowa, where he earned his Ph.D. in 1939. After three years as a research fellow at the Carnegie Institution of Science in Washington D.C., he entered the Navy and served four years as an ordinance and gunnery officer. After World War II, he became supervisor of the High Altitude Research Group and the Proximity Fuse Unit at Johns Hopkins University.

In 1951, he was appointed professor of physics at the University of Iowa, a post which later came to include astronomy. Between 1953 and 1954, he was engaged in Project Matterhorn, a study of thermonuclear reactions. From 1957 to 1958, he was involved in the coordination of events for International Geophysical Year.

Approximately 100 captured German V-2 rockets were shipped to White Sands Proving Ground to be tested. Van Allen used these missiles to carry scientific instruments to measure the levels of radiation encountered in their flight. In 1949, he pioneered what came to be known as the rockoon technique, involving the use of a balloon to lift a small rocket into the stratosphere prior to ignition, thereby eliminating the drag experienced by a rocket fired from the ground. When the V-2 program ended, he was put in charge of the Aerobee rockets, which were capable of carrying 150 pounds of scientific payload to an altitude of 300,000 feet. Projects undertaken included investigating and measuring solar radiation, sky brightness, atmospheric composition, and auroras.

In 1958, he fitted Explorer 1, and later Explorer 3, with Geiger counters to measure radioactivity at 800 kilometers and above. He then mapped the size and distribution of these radiation zones, and discovered that they took the form of two torroidal belts around the Earth at the level of the equator. The particles making up the radiation belts were trapped in the Earth's magnetic field, spiraling about the magnetic lines of force from pole to pole with an intensity that varied in relation to solar activity. These zones of intense radiation were named **Van Allen belts**.

Van Allen died on August 9, 2006. He was named *TIME* magazine Man of the Year in 1960.

1. Pioneering Astrophysicist James Van Allen Dies, NASA, /www.nasa.gov/vision/universe/features/james_van_allen, August 10, 2006.
2. Tatarewic, Joseph N., James A. Van Allen (1914-2006), AGU, www.agu.org/inside/awards/vanallen.
3. World of Scientific Discovery on James Van Allen, BookRags, www.bookrags.com/biography/james-van-allen-wsd.

Raymond Davis, Jr. (1914-2006)

Determined solar neutrino flux

Raymond Davis, Jr. was born in Washington, D.C. on October 14, 1914. His father worked in the photographic division of the National Bureau of Standards. Raymond received his bachelor and master's degrees from the University of Maryland in 1937 and 1939, respectively. He then went on to Yale University, where he received his Ph.D. in 1942.

After World War II, in which he served in the U.S. Army Air Corps, Davis worked as a research chemist for Monsanto Chemical Company. In 1948, he moved to the Brookhaven National Laboratory, where he worked as a research chemist. In 1984, he became a research professor in the astronomy department at the University of Pennsylvania.

During the 1950s, astrophysicists theorized that neutrinos, particles of extremely low mass, formed during fusion reactions within the Sun's internal furnace. As a result, the Sun threw off neutrinos by the trillion. The particles were named by Enrico Fermi (1901-1954), who believed they could never be detected. Davis became fascinated by the new theory.

In 1955, he built his first neutrino detector, located 20 feet underground at the Brookhaven National Laboratory to prevent cosmic rays from interfering with the results. It was filled with 1,000 gallons of chlorine compound. It was not successful.

In 1958, other researchers discovered there were more energetic neutrinos than the ones Davis had been seeking. These energetic neutrinos were thought to result from a different reaction in the Sun. Davis built another, more elaborate neutrino detector, which he completed in 1967. Occasionally, a neutrino would strike a chlorine atom, splitting a neutron into a proton and an electron and changing the chlorine into radioactive argon. Every couple of months, Davis drew the argon off and counted its atoms by measuring the rate of radioactive decay. The result indicated how many neutrinos had passed through the tank and reacted with the chlorine. The tank detected one neutrino every two days.

In addition to his work with neutrinos, Davis took part in the analysis of lunar samples gathered by the Apollo Moon missions of the 1970s, and measured the radioactivity of material from the Moon's surface.

He was awarded the 2002 Nobel Prize in Physics for his work. He died at his home in Blue Point, New York on May 31, 2006.

1. Raymond Davis, Jr., Autobiography, NobelPrize.org, http://nobelprize.org/nobel_prizes/physics/laureates/2002/davis-autobio.
2. Raymond Davis, Jr., Brookhaven National Laboratory, www.bnl.gov/bnlweb/raydavis.
3. World of Chemistry on Raymond Davis, Jr., BookRags, www.bookrags.com/biography/raymond-davis-jr-woc.

Fred Hoyle (1915-2001)

Championed the steady-state theory of the universe

Fred Hoyle, a British mathematician and astronomer, was born at Bingley, Yorkshire, England on June 24, 1915. His father worked in the wool trade. Fred studied mathematics and physics at Cambridge University, receiving his B.A. in 1936 and his M.A. in 1939. He spent most of his working life at the Institute of Astronomy at Cambridge, and served as its director for a number of years.

Hoyle worked with William Fowler (1911-1995) on the theory of nuclear reactions in stars, elaborating on gravitational, electrical, and nuclear fields, and on how various heavy elements are created within stars.

In 1948, he became involved with the steady state theory of cosmology, which proposed that as galaxies recede from one another over the eons new matter is created in the empty space in between, and new galaxies evolve. He believed in this theory because he felt its assertion that the universe would continue to expand indefinitely gave it a simplicity and symmetry lacking in the Big Bang theory.

Hoyle published several popular astronomy books, which included an explanation of the new steady state theory. The theory was opposed by George Gamow (1904-1968), who supported the Big Bang theory of creation. Gamow's objection to the steady state theory centered about the issue of the continuous creation of matter out of nothing throughout time. Hoyle, on the other hand, felt it was easier to accept continuous creation over the theory that all the matter in the universe was created instantly from nothing by a mysterious big bang.

The debate between the two sides remained unresolved until 1963 when Maarten Schmidt (1929-) discovered quasars. These unique objects did not fit into steady state cosmology, which required that the universe contain similar objects everywhere. The discovery of background microwave radiation by Arno Penzias (1933-) and Robert Wilson (1927-2002) in 1964 dealt a fatal blow to steady state theory.

In addition to his work as an astronomer, Hoyle was a writer of science fiction, including a number of books co-authored with his son Geoffrey. He was knighted in 1972, and died in Bournemouth, England, after a series of strokes on August 20, 2001. **Asteroid 8077 Hoyle** is named for him.

1. Fred Hoyle, Wikipedia, http://en.wikipedia.org/wiki/Fred_Hoyle.
2. Lovell, Bernard, Professor Sir Fred Hoyle, The Guardian, www.guardian.co.uk/news/2001/aug/23/guardianobituaries.spaceexploration, August 23 2001.
3. World of Scientific Discovery on Fred Hoyle, BookRags, www.bookrags.com/biography/fred-hoyle-wsd.

Philip Morrison (1915-2005)

The first to call upon the professional community to carry out a coordinated search for intelligent extraterrestrial signals

Philip Morrison, an American theoretical physicist, was born in Somerville, New Jersey on November 7, 1915. He received his B.S. from the Carnegie Institute for Technology in 1936. In 1940, he received his Ph.D. in theoretical physics from the University of California, Berkeley under the supervision of Robert Oppenheimer (1904-1967).

For the next two years, he taught physics at San Francisco State College and at the University of Illinois before joining the Manhattan Project toward the end of 1942. He was assigned to the Metallurgical Laboratory of the Manhattan Project at the University of Chicago, where he was a physicist and group leader, and later served in the same capacity at Los Alamos until 1946.

He brought the bomb's plutonium core from Los Alamos to the New Mexico desert site for the first test. He also was at the island air base of Tinian, from which two bombs were launched against Japan. He later witnessed the aftermath of the explosion at Hiroshima in a visit immediately following the war. He went on to become a vocal critic of the nuclear arms race.

In 1946, Morrison joined the physics faculty at Cornell University. He made contributions in quantum electrodynamics, nuclear theory, radiology, and isotope geology. From the 1950s, he investigated cosmic-ray origins and propagation, gamma-ray astronomy, and other topics in high-energy astrophysics and in cosmology.

In 1959, he was among the first scientists to call on the professional community to begin The Search for Extraterrestrial Intelligence (SETI) using microwave technology.

He joined the Massachusetts Institute of Technology faculty in 1964. From 1965, he was a regular reviewer of books on science for *Scientific American*. He also narrated and helped script films on science directed by Charles and Ray Eames. He appeared widely on radio and on British, Canadian, and American television in a number of science programs and series, most visibly as author-presenter of a six-part national Public Broadcasting System series *The Ring of Truth*, which first aired in 1987. Morrison died in Cambridge, Massachusetts on April 22, 2005.

1. Morrison, Philip (1915-2005), The Internet Encyclopedia of Science, www.daviddarling.info/encyclopedia/M/Morrison.
2. Overbye, Dennis, Philip Morrison, 89, Builder of First Atom Bomb, Dies, The New York Times, April 26, 2005.
3. Thompson, Elizabeth A., Institute Professor Philip Morrison dies at 89, MIT News, http://web.mit.edu/newsoffice/2005/morrison, April 25, 2005.

Robert Dicke (1916-1997)

Established the importance of the measurements of cosmic microwave background

Robert Henry Dicke, an American experimental physicist, was born in St. Louis, Missouri on May 6, 1916. He completed his bachelor's degree at Princeton University and his doctorate in nuclear physics at the University of Rochester in 1939. He joined the staff of the Radiation Laboratory at Massachusetts Institute of Technology (MIT) in 1941. He went to Princeton as an assistant professor in 1946, and chaired Princeton's Physics Department from 1967 to 1970. In 1975, he was named the first Albert Einstein University Professor of Science.

He made crucial contributions to the development of radio telescopes and the discovery of thermal background radiation, proposed the concept of super-radiance, predicted the discovery of the Big Bang echo (background microwave radiation), and did extensive work in the experimental and theoretical basis for gravity physics. He investigated topics in atomic physics, quantum optics, and radar technology.

Dicke's work on coherent radiation emission was a precursor to development of the laser. With his graduate student, Carl Brans (1935-), Dicke proposed the **Brans-Dicke theory of gravitation**, which states that gravity has two components—a dominant and familiar force that compels objects to fall downward and a much weaker component that causes objects to shrink or expand.

During World War II, he worked in the Radiation Laboratory at MIT on the development of radar.

In 1965, he proposed that radiation detected near one-centimeter wavelength is left over from the Big Bang start of expansion of the universe. He invented the instrument used to detect this radiation, the **Dicke radiometer**, which is now a standard astronomical tool. The Dicke radiometer detects weak signals in noise by modulating or switching the incoming signal before it is processed by conventional receiver circuits. It has been a key for transforming cosmology from a theoretical to a more experimental science.

Dicke died in Princeton, New Jersey on March 4, 1997. At the time, he was the Albert Einstein Professor of Science, emeritus at Princeton University. He held some 50 patents, from clothes dryers to lasers.

1. Princeton Physicist Robert Dicke Dies, Princeton University, www.princeton.edu/pr/news/97/q1/0304dick, March 4, 1997.
2. Robert H. Dicke, NNDB, www.nndb.com/people/222/000168715.
3. Robert Henry Dicke, Biographical Memoirs, National Academy of Sciences, www.nap.edu/html/biomems/rdicke.

Iosif Shklovsky (1916-1985)

Suggested that the radiation from the Crab Nebula is due to synchrotron radiation

Iosif Samuilovich Shklovsky, a Soviet theoretical astrophysicist, was born in Glukhov, Ukraine on July 1, 1916. In 1933, he entered the Physico-Mathematical Faculty of Moscow State University and studied there until 1938 when he entered a postgraduate course at the Astrophysics Department of Sternberg State Astronomical Institute. He remained working in the Institute the rest of his life.

Shklovsky showed in 1946 that radio-wave radiation from the Sun emanates from the ionized layers of its corona, and he developed a mathematical method for discriminating between thermal and non-thermal radio waves in the Milky Way. He determined that the temperature of the solar corona is of the order of one million Kelvin.

He is best known for his suggestion that radiation from the Crab Nebula is due to synchrotron radiation—radiation that occurs when charged particles are accelerated in a curved path or orbit through a magnetic field.

Shklovsky proposed that cosmic rays from supernova explosions within 300 light years of the Sun could have been responsible for some of the mass extinctions of life on Earth.

In 1959, he examined the orbital motion of Mars's inner satellite, Phobos. He concluded that its orbit was decaying, and noted that if this decay was attributed to friction with the Martian atmosphere, then the satellite must have an exceptionally low density. He suggested that Phobos might be hollow and possibly of artificial origin. This interpretation has since been refuted, but the suggestion of extraterrestrial involvement caught the public's imagination.

Shklovsky 1962 book, *Universe, Life, Intelligence*, was translated into English and expanded on by Carl Sagan (1934-1996) as the best-selling book *Intelligent Life in the Universe.* He won the Lenin Prize in 1960, Jansky Prize in 1968, and the Bruce Medal in 1972. He was a Corresponding Member of Soviet Academy of Sciences.

He died in Moscow on March 3, 1985. His memoir, *Five Billion Vodka Bottles to the Moon: Tales of a Soviet Scientist*, was published posthumously in 1991.

Asteroid 2849 Shklovskij was named in his honor.

1. Iosif S. Shklovskii, NNDB, www.nndb.com/people/655/000173136.
2. Iosif Samuilovich Shklovskii, The Bruce Medalists, www.phys-astro.sonoma.edubruceMedalists/ Shklovskii/Shklovskii.
3. Iosif Shklovsky, Wikipedia, http://en.wikipedia.org/wiki/Iosif_Samuilovich_Shklovskii.

Frederick Reines (1918-1998)

First to detect the neutrino

Frederick Reines, the son of Jewish emigrants from Russia, was born in Paterson, New Jersey on March 16, 1918. His father owned a small country store. Reines was one of the two Nobel Laureates in physics to achieve the level of Eagle Scout. Robert Richardson (1937-) was the other one. Reines attended Stevens Institute of Technology in Hoboken, New Jersey, where he earned his M.E. and M.S. degrees before receiving his Ph.D. from New York University in 1944. He then joined the Manhattan Project in Los Alamos, New Mexico, where he and a colleague, Clyde Cowan (1919-1974), developed a detection procedure for neutrinos.

In 1956, Reines and a team of researchers, colliding atomic particles in water, were the first to detect neutrinos. Neutrinos had been first proposed theoretically by Wolfgang Pauli (1900-1958) 20 years earlier to explain undetected energy that escaped when a neutron decayed into a proton and an electron.

Reines became the head of the physics department of Case Western Reserve University in 1959. There, he led a group that was the first to detect neutrinos created in the atmosphere by cosmic rays.

He had a booming voice, and had been a singer since childhood. Besides being chairman of the physics department, he sang in the Cleveland Orchestra Chorus.

In 1966, Reines took most of his neutrino research team with him to California to become the founding dean of physical sciences at the University of California, Irvine.

In 1987, Supernova 1987A was observed exploding. When a supermassive star collapses and then explodes, it is thought that the resulting jets of neutrinos bombard the escaping masses to create the elements heavier than iron, up through uranium. Without these natural neutrino processes in exploding supermassive stars, these elements would not exist. Reines' work was instrumental in detecting these neutrinos.

Reines and American physicist Martin Perl (1927-) shared the 1995 Nobel Prize in physics for their experimental contributions to lepton physics. A lepton is a particle with spin 1/2 that does not experience the strong interaction. Reines' Prize was for detecting the neutrino.

Reines died in Orange, California on August 26, 1998.

1. Frederick Reines, Autobiography, NobelPrize.org, http://nobelprize.org/nobel_prizes/physics/laureates/1995/reines-autobio.
2. Frederick Reines, Microsoft Encarta Encyclopedia Onlilne, http://encarta.msn.com/encyclopedia_761583320/Frederick_Reines, 2007.
3. Frederick Reines, Wikipedia, http://en.wikipedia.org/wiki/Frederick_Reines.

Gérard de Vaucouleurs (1918-1995)

His survey of bright, relatively nearby galaxies led to the discovery of the local supercluster

Gerard de Vaucouleurs, a French astronomer, was born in Paris on April 25, 1918. His father's name is unknown because early in life he took the maiden name of his mother. He received his undergraduate degree at the University of Paris in 1939. He then began his long association with Julien Péridier (1882-1967), an electrical engineer and amateur astronomer who had built a private observatory at Le Houga in southwest France.

During World War II, de Vaucouleurs served with the artillery corps in the French army from November 1939 to May 1941. He then returned to Le Houga and turned his attention to the measurement of close double stars, variable stars, and the brightness variations of the asteroid Eros.

He returned to Paris in 1943 to work on his doctoral degree at the Laboratoire des Recherches Physiques of the Sorbonne and at the Institut d'Astrophysique Boulevard Arago. Receiving his degree in 1950, de Vaucouleurs continued his association with the Péridier Observatory, applying the research done for his thesis to light scattering in the Earth's atmosphere. During his career, he also worked at a number of important observatories in Europe, the United States, and Australia.

De Vaucouleurs's survey of bright, relatively nearby galaxies led to the discovery that through the interplay of gravity, galaxies themselves are arrayed into superclusters. Among them, a local supercluster includes the Milky Way, its neighbor Andromeda, and a pair of galaxies known as the Magellanic Clouds. His work led him to conclude that the universe was expanding much faster than had been thought, and was therefore much younger, a mere 10 billion years rather than 20 billion or so.

In his *Physics of the Planet Mars* (1954), he argued that while low forms of vegetation remained a possibility, the overwhelming majority of terrestrial organisms would not be able to survive there.

Together with his longtime collaborator, his first wife Antoinette who died in 1988, he produced 400 research and technical papers, 20 books and 100 articles for laymen.

De Vaucouleurs died of a heart attack in Austin, Texas on October 7, 1995. He is almost universally acknowledged to have opened up an entire field of study—that of the galaxies beyond the Milky Way.

1. Burbage, Margaret E., Gérard de Vaucouleurs, Biographical Memoirs, National Academy of Sciences, www.nap.edu/html/biomems/gvaucouleurs.
2. de Vaucouleurs, Gérard Henri (1918-1995), The Internet Encyclopedia of Science, www.daviddarling.info/encyclopedia/D/deVaucouleurs.
3. Thomas, Jr., Robert, Gerard de Vaucouleurs, 77, Galactic Astronomer, Is Dead, New York Times, October 11, 1995.

Martin Ryle (1918-1984)

Greatly improved radio astronomy

Martin Ryle, a British astronomer, was born in Brighton, Sussex, England on September 27, 1918. His father was the first Chair of Social Medicine at Oxford University. Martin was educated at Bradfield College and Oxford, where he graduated in 1939. He worked on radar and radio systems for the Royal Air Force during World War II at the Telecommunications Research Establishment.

After the War, he did a fellowship at Cavendish Laboratory, Cambridge University. He later became Professor of Radio Astronomy at Cambridge, serving from 1957 as director of the University's Mullard Radio Astronomy Observatory. He worked on stellar catalogs, helping to produce the *Third Cambridge Catalogue* in 1959. That same year he was appointed to the new Chair of Radio Astronomy.

Radio astronomy came into being in 1932 when Karl Jansky (1905-1950) built a simple antenna. In 1937, Grote Reber (1911-2002) made a great improvement when he built the first dish antenna, which was 29 feet in diameter. The increase in size was highly relevant to radio astronomy—the larger the telescope, the greater its resolution. Ryle devised a method called "aperture synthesis" (now known as interferometry) to measure weak radio sources. He used two radio telescopes some distance apart to observe a single object. The two beams of light were combined, but were deliberately forced to follow paths of slightly different lengths. The waves were slightly out of phase when combined, which resulted in a pattern of interference fringes. Analysis of the fringes was done by a new phase-switching mechanism. Computers are used for this today. The greater the distance between the radio dishes, the better the resulting detail. Ryle compiled a catalog of over fifty previously undetected radio sources.

The discovery of the first radio galaxy, Cygnus A, located in the constellation Cygnus was confirmed visually in 1955. Ryle was knighted for his work in 1966.

Antony Hewish (1924-) and Jocelyn Burnell (1943-) used Ryle's interferometry technique when they discovered pulsars in 1967. For this discovery, Ryle and Hewish shared the Nobel Prize for physics in 1974.

Ryle served as Astronomer Royal from 1972 to 1982. He died in Cambridge, Cambridgeshire, England on October 14, 1984. The **Ryle Telescope** at Mullard Radio Astronomy Observatory was named for him.

1. Martin Ryle, NNDB, www.nndb.com/people/889/000099592.
2. Sir Martin Ryle, BookRags, www.bookrags.com/research/sir-martin-ryle-scit-0712345.
3. World of Scientific Discovery on Martin Ryle, BookRags, www.bookrags.com/biography/martin-ryle-wsd.

Hendrik van de Hulst (1918-2000)

Predicted the existence of the 21-centimeter hyperfine line of neutral interstellar hydrogen

Hendrik (Henk) Christoffel van de Hulst, a Dutch astronomer and mathematician, was born in Utrecht, Netherlands on November 19, 1918. His father was a well-known writer of children's books. The decision to become an astronomer was made in van de Hulst's second year at the University of Utrecht.

His studies were interrupted in 1939 when he was drafted into military service following the general mobilization. This was shortly before the invasion of the Netherlands and the beginning of the World War II.

He became acquainted with Jan Oort (1900-1992), which led to his interest in light scattering in an astronomical context. This interest was reflected in his doctoral dissertation, *Optics of spherical particles*. After receiving his doctoral degree in 1946, he was awarded a postdoctoral fellowship at the Yerkes Observatory of the University of Chicago.

Van de Hulst had predicted in 1944, while a student, that the amount of neutral atomic hydrogen in interstellar space would be great enough to produce a measurable signal at the radio wavelength of 21 centimeters. The 21-centimeter line of atomic hydrogen was later detected in 1951 by physicists Edward Purcell (1912-) and Harold Ewen (1922-1997).

After the 21-centimeter line was discovered, van de Hulst participated with Jan Oort and Alex Muller in the effort to use radio astronomy to map the neutral hydrogen in the Milky Way. Their results first revealed the galaxy's spiral structure.

Van de Hulst was appointed to the faculty at the University of Leiden in 1948. He remained at Leiden throughout his career, becoming Professor Emeritus in 1984.

He made extensive studies of interstellar grains and their interaction with electromagnetic radiation. Interstellar grains are microscopic solid dust particles in interstellar space which absorb starlight, making distant stars appear dimmer than they truly are. He also investigated the solar corona and the Earth's atmosphere.

He wrote two monographs on light scattering. The first of these, *Light Scattering by Small Particles*, was published in 1957. He died in Leiden on August 1, 2000. **Asteroid 2413 van de Hulst** is named after him.

1. Hendrik C. van de Hulst, Wikipedia, http://en.wikipedia.org/wiki/Hendrik_C._van_de_Hulst.
2. Hulst, van de, Hendrik Christoffel (1918-2000, The Free Dictionary, http://encyclopedia.farlex.com/Hulst,+van+de,+Hendrik+Christoffel.
3. In memoriam: Hendrik Christoffel van de Hulst, Department of Astronomy Leiden Observatory, www.strw.leidenuniv.nl/vdhulst_e.

Margaret Burbidge (1919-)

Collaborated with others on the nucleosynthesis of chemical elements in stars

Eleanor Margaret Burbidge (born Margaret Peachy), a British astrophysicist, was born in Davenport, England on August 12, 1919. Her father was a lecturer in chemistry at the Manchester School of Technology. She studied physics and astronomy at the University of London. After graduation in 1939, she joined the University of London Observatory, where she obtained her Ph.D. in 1943.

In 1948, she married Geoffrey Burbidge, a theoretical astrophysicist. In 1951, she went to the United States, first to the Yerkes Observatory of the University of Chicago (1951-1953) and then to the California Institute of Technology (1955-1957). She returned to Yerkes in 1957 and served as associate professor of astronomy from 1959 to 1962. She then transferred to the University of California, San Diego, where she was professor of astronomy from 1964 until 1990 and emeritus professor after 1990. She also served as director of the Center for Astrophysics and Space Sciences (1979-1988). In 1972, she became the first woman to be appointed Director of the Royal Greenwich Observatory. She held this position only briefly.

Burbidge and Geoffrey collaborated with Fred Hoyle (1915-2001) and William Fowler (1911-1995), and in 1957 published a key paper on the nucleosynthesis of the chemical elements in stars. This gave new insight into the evolution of the stars, and showed how the chemical elements could be built up from the hydrogen and helium with which the stars first formed. They also produced one of the first comprehensive works on quasars in their *Quasi-Stellar Objects* (1967). The first accurate estimates of the masses of galaxies were based on Margaret Burbidge's careful observation of their rotation.

She became one of the most influential personalities in the fight to end discrimination of women in astronomy. Consequently, in 1972 she turned down the Annie J. Cannon Award of the American Astronomical Society because it was awarded to women only. In 1976, she became president of the American Astronomy society. In 1983, she was elected president of the American Association for the Advancement of Science. She was also part of the committee that planned and outfitted the Hubble Space Telescope. **Asteroid 5490 Burbidge** was named for her and Geoffrey.

1. Eleanor Margaret Burbidge, Answers.com, www.answers.com/topic/margaret-burbidge.
2. Eleanor Margaret Peachey Burbidge, The Bruce Medalists, www.phys-astro.sonoma.edu/BruceMedalists/BurbidgeM/BurbidgeM.
3. Margaret Burbidge, Wikipedia, http://en.wikipedia.org/wiki/Margaret_Burbidge.

Hermann Bondi (1919-2005)

Helped develop the steady-state theory of the universe

Hermann Bondi, a Jewish-Austrian mathematician and astronomer, was born in Vienna, Austria on November 1, 1919. His father was a medical heart specialist, who at the time of Hermann's birth was working in a hospital for Russian prisoners of war in a remote part of Austria.

Alarmed by the rise of the Nazis in neighboring Germany, Hermann moved to England, where he enrolled in Trinity College, Cambridge in 1937. He completed a mathematics degree in 1940. He was interned as an "enemy alien" by the British government that same year, and spent over a year at camps on the Isle of Man and in Canada, where he first met Thomas Gold (1920-2004). On his release in 1941, Bondi went back to the England, where he and Gold worked on radar research for the Admiralty under the supervision of Fred Hoyle (1915- 2001).

When the War was over, Bondi, Gold, and Hoyle returned to Cambridge, where they developed their steady-state theory of the universe in 1948. They did not see a need for the Big Bang, and proposed instead that the universe has no beginning or end. To account for the continual expansion of the universe, the theory required that matter be continuously created so that the average density of the universe remained constant. After the 1965 discovery of the cosmic microwave background, Bondi admitted that the steady-state theory was probably incorrect.

In 1954, Bondi accepted a professorship at King's College London, where he carried out pioneering theoretical work on how a black hole, or star, can accrete matter from surrounding gas.

He was one of the first to appreciate the nature of gravitational radiation, introducing **Bondi radiation coordinates**, the **Bondi k-calculus**, and the notion of **Bondi mass**.

Bondi became increasingly attracted to public service, holding a number of positions. These included Director General of the European Space Research, Chief Scientist at the UK Ministry of Defense, Chief Scientist at the Department of Energy, and head of the Natural Environment Research Council. In 1983, he was appointed master of Churchill College, Cambridge, where he remained until 1990.

His books include *Cosmology* (1952) and his autobiography *Science, Churchill and Me* (1990). He died on September 10, 2005.

1. Durrani, Matin, Sir Hermann Bondi: 1919 – 2005, PhysicsWorld.com, http://physicsworld.com/cws/article/news/23134.
2. Hermann Bondi, Wikipedia, http://en.wikipedia.org/wiki/Hermann_Bondi.
3. O'Connor, J.J. and E. F. Robertson, Hermann Bondi, MacTutor, www-groups.dcs.st-and.ac.uk/~history/Biographies/Bondi.

George Herbig (1920-)

Co-discovered nonstellar condensations in high density clouds near regions of recent star formation

George Howard Herbig, an American Astronomer, was born in Wheeling, West Virginia on January 2, 1920. He received his undergraduate degree at the University of California, Los Angeles and his Ph.D. from the University of California, Berkeley in 1948. His dissertation was entitled *A Study of Variable Stars in Nebulosity*.

In 1948, he joined the staff of Lick Observatory, which became part of the University of California, Santa Cruz in the late 1960s. He rose to the rank of professor in 1966. At Lick, he designed the Coudé spectrograph for the Shane three-meter telescope. In 1988, he moved to the University of Hawaii

Herbig's main area of research was the nebular variables, of which the prototype is T Tauri. It is believed that the members of this group are in an early stage of stellar evolution. Most of them are red and fluctuate in light intensity. In 1960, Herbig drew attention to the fact that many of them have a predominance of lithium lines, similar to the abundance of this element on Earth and in meteorites, and concluded that this might represent the original level of lithium in the Milky Way. In the Sun and other stars, lithium may have largely been lost through nuclear transformation.

He is best known for his discovery, with Guillermo Haro (1913-1988), of the **Herbig-Haro objects**, which are bright patches of nebulosity excited by bipolar outflow from a star being born.

Herbig has also made prominent contributions to the field of diffuse interstellar band research, especially through a series of nine articles published between 1963 and 1995 entitled *The Diffuse Interstellar Bands*.

He has shown that lithium abundance is correlated with age in young stars, and has investigated rotation rates of stars of different spectral class. He has also investigated the spectra of atoms and molecules that originate in interstellar space. During his career, Herbig has been awarded the Helen B. Warner Prize for Astronomy of the American Astronomical Society (1955), Médaille, Université de Liège (1969), Bruce Medal of the Astronomical Society of the Pacific (1980), and Petrie Prize and Lectureship of the Canadian Astronomical Society (1995). Also named after him are **Asteroid 11754 Herbig** and **Herbig Ae/Be stars**.

1. George Herbig, Wikipedia, http://en.wikipedia.org/wiki/George_Herbig.
2. George Howard Herbig, The Bruce Medalists, www.phys-astro.sonoma.edu/BruceMedalists/Herbig/Herbig.
3. Herbig-Haro Object, Wikipedia, http://en.wikipedia.org/wiki/Herbig-Haro_object.

Thomas Gold (1920-2004)

Co-proposed the steady-state theory of the universe

Thomas Gold, an Austrian astrophysicist, was born into a Jewish family in Vienna, Austria on May 22, 1920. He was educated at Zuoz College in Switzerland and Trinity College, Cambridge. At the beginning of World War II, he was held in a British internment camp as a suspected enemy alien for a year. When released, he worked with Hermann Bondi (1919-2005) and Fred Hoyle (1915-2001) on the development of radar for the British Admiralty.

Gold is best known for developing, along with Bondi and Hoyle, the steady-state theory of the universe. It assumes that we live in a universe that has no beginning or end, and in which matter is constantly being created. The theory initially seemed plausible, but the discovery of the cosmic microwave background by Arno Penzias (1933-) and Robert Wilson (1936-) in 1965 dealt it its first big blow.

He later worked at the Royal Greenwich Observatory in Herstmonceux, Sussex, England and at Harvard University in Cambridge, Massachusetts.

Gold, in 1955, suggested that the Moon's surface is covered with a fine rock powder. It was not until 1969 that he was proven correct when the Apollo 11 crew brought the first samples of lunar soil back to Earth. Analyses revealed that it is indeed powdery, with each grain covered in a thin metal coating caused by the penetration of the solar wind.

In 1967, he argued that pulsars are neutron stars that emit radio waves as they spin. His view was initially considered to be implausible. However, the discovery of a pulsar in the Crab Nebula later led to the theory being accepted.

Gold's most recent idea, which he discussed in his 1998 book *The Deep Hot Biosphere*, is that oil and coal are not remnants of ancient surface life that became buried and subjected to very high temperatures and pressures as most believe. He instead argues that these deposits are produced from primordial hydrocarbons dating back to when the Earth was formed.

He also designed the stereo camera that was carried on the lunar surface by the U.S. astronauts. He died of heart failure in Ithaca, New York on June 22, 2004.

1. Duranni, Matin, Thomas Gold: 1920 – 2004, PhysicsWorld.com, http://physicsworld.com/cws/article/news/19733, June 23, 2004.
2. Gold, Thomas (1920-2004), The Internet Encyclopedia of Science, www.daviddarling.info/encyclopedia/G/Gold.
3. Thomas Gold, Wikipedia, http://en.wikipedia.org/wiki/Thomas_Gold.

Chushiro Hayashi (1920-)

Discovered the almost temperature-independent early evolution of pre-main-sequence stars

Chushiro Hayashi, a Japanese astrophysicist, was born in Kyoto, Japan on July 25, 1920. He received a B.S. from the University of Tokyo in 1942 and his Ph.D. from Kyoto University in 1954. He subsequently taught astrophysics at Osaka Prefecture University and Kyoto University from 1945 to 1984.

He is best known for his discovery of the almost temperature-independent early evolution of pre-main-sequence stars along what has become known as the **Hayashi track.** He made the discovery while studying under Hideki Yukawa (1907-1985). The Hayashi track is a nearly vertical path of stellar evolution on the Hertzsprung-Russell diagram down which an infant star progresses on its way to the main sequence. While on the Hayashi track, a star is largely or completely in convective equilibrium. As it progresses, its luminosity, initially very high, decreases rapidly with contraction, but its surface temperature remains almost the same.

Hayashi limit is a constraint upon the maximum radius of a star for a given mass. When a star is fully within hydrostatic equilibrium—a condition where the inward force of gravity is matched by the outward pressure of the plasma—then the star cannot exceed the radius defined by the Hayashi limit. This has important implications for the evolution of a star, both during the formulative contraction period and later when the star has consumed most of its hydrogen supply through nuclear fusion.

In 1950, Hayashi was the first to offer a variant of the hot Big Bang model as put forward in 1946 by George Gamow (1904-1968) and colleagues. He also showed that electron-positron pair production must be taken into account in considering the early neutron-proton ratio in the aftermath of the Big Bang. He is also known for his advanced models showing star formation, stellar evolution, and pre-main sequence evolution.

He published *Origin of the Solar System* in 1985 with Kiyoshi Nakazawa and Shoken Miyama. He was awarded the Eddington Medal in 1970, Kyoto Prize in 1995, and Bruce Medal in 2004. **Asteroid 12141 Chushayashi** was named for him.

1. Chushiro Hayashi, NNDB, www.nndb.com/people/158/000169648.
2. Hayashi Limit, Wikipedia, http://en.wikipedia.org/wiki/Hayashi_limit.
3. Hayashi Track, The Internet Encyclopedia of Science, www.daviddarling.info/encyclopedia/H/Hayashi_track.
4. Hayashi, Chushiro (1920-), The Internet Encyclopedia of Science, www.daviddarling.info/encyclopedia/H/Hayashi.

Ralph Alpher (1921-2007)

Suggested that the abundances of chemical elements could be explained by a thermonuclear processes immediately after the Big Bang

Ralph Asher Alpher, a Russian-American cosmologist, was born in Vitebsk, Russia on February 3, 1921. He studied at George Washington University at night and worked during the day. After graduating, he became a Ph.D. student of George Gamow (1904-1968). Together, they began calculating the relative abundance of elements that would be produced in a hot Big Bang. The pair assumed that the early universe was very hot and full of neutrons. Nuclei then formed by capturing neutrons one at a time, with the occasional nucleus decaying to produce a heavier nucleus plus an electron and a neutrino. However, this early version of Big Bang nucleosynthesis did not explain the origin of all the chemical elements. We now know that elements heavier than lithium are produced in the interior of stars. Alpher and Gamow reported their calculations in a paper published in *Physics Review* in 1948.

Several months later, Alpher and Robert Herman (1914-1997) from Johns Hopkins University published a separate paper predicting that the radiation left over from the Big Bang would have a temperature of 5K. Arno Penzias (1933-) and Robert Wilson (1936-) of Bell Labs later shared the 1978 Nobel Prize for physics for discovering this cosmic microwave background, which they found to have a temperature of 2.7K.

Alpher's contribution went largely unrecognized, partly because he left cosmology and joined General Electric's research center in Schenectady, New York in 1955. At General Electric, he worked on energy conversions, gas dynamics, and color television. He wrote almost 100 papers for General Electric. He taught at Union College, also in Schenectady, from 1986 to 2004, and was director of the Dudley Observatory. He and Herman wrote a book in 2001 about their early work entitled *Genesis of the Big Bang*.

In 2005, President Bush announced that Alpher had been awarded the National Medal of Science, which is administered by the National Science Foundation and is the highest honor for science. Alpher had been in failing health since falling and breaking his hip. His son, Victor, attended the ceremony at the White House to receive his father's medal. Alpher died in Austin, Texas on August 12, 2007.

1. Big Bang pioneer Ralph Alpher dies following a long illness, Union News, www.union.edu/N/DS/s.php?s=7332, August 13, 2007.
2. Cain, Jeanette, Alpher, Ralph Asher: 1921-, Light-Science.com, www.light-science.com/alpher.
3. Durrani, Mati, Ralph Alpher: 1921 – 2007, PhysicsWorld.com, http://physicsworld.com/cws/article/ news/30915, August 23, 2007.

Freeman Dyson (1923-)

Proposed several novel ideas in the search for extraterrestrial intelligence

Freeman John Dyson, a British-born American theoretical physicist, was born in Crowthorne, England on December 15, 1923. He worked as an analyst for Royal Air Force Bomber Command during World War II. After the War, he obtained a B.A. in mathematics from Cambridge University (1945) and was a Fellow of Trinity College, Cambridge from 1946 to 1949. In 1947, he moved to the United States on a fellowship at Cornell University, and joined the faculty there as a physics professor in 1951. He moved to the Institute for Advanced Study at Princeton University in 1953.

During the years just after World War II, new experimental evidence raised questions about how quantum theory might be extended to cover the interactions of matter and light. But the mathematical techniques used to extend quantum theory had internal difficulties. In about 1949, two seemingly unrelated, theoretical solutions to these mathematical problems emerged. About 1950, Dyson showed that both theories were reducible to a single formalism. The resulting theory and the mathematical techniques associated with it became central to modern theoretical physics during the second half of the 20th century.

From 1957 to 1961, he worked on the Orion Project, which proposed the possibility of space-flight using nuclear pulse propulsion.

In 1960, Dyson wrote a paper for the journal *Science* entitled *Search for Artificial Stellar Sources of Infrared Radiation*. In it, he theorized that a technologically advanced society might completely enclose their star with what has come to be known as a **Dyson Sphere**. The purpose would be to maximize the capture of the star's available energy by intercepting electromagnetic radiation with wavelengths from visible light downwards and radiating waste heat outwards as infrared radiation. Therefore, one method of searching for extraterrestrial civilizations would be to look for large objects radiating in the infrared range of the electromagnetic spectrum. In 2003, he became president of the Space Studies Institute, whose mission is to open the energy and material resources of space for human benefit within our lifetime.

His books that discuss nuclear strategy and arms control include his autobiography, *Disturbing the Universe* (1979) and *Weapons and Hope* (1984). Dyson received an honorary Sc.D. from Bates College in 1990.

1. Dyson, Freeman John (1923-), The Internet Encyclopedia of Science, www.daviddarling.info/encyclopedia/D/DysonF.
2. Freeman Dyson, Microsoft Encarta Online Encyclopedia, http://encarta.msn.com/encyclopedia_761579461/freeman_dyson, 2008.
3. Freeman Dyson, Wikipedia, http://en.wikipedia.org/wiki/Freeman_Dyson.

Antony Hewish (1924-)

Identified pulsars as a new class of stars

Anthony Hewish, a British radio astronomer, was born in Fowey, Cornwall, England on May 11, 1924. His father was a banker. Antony's undergraduate work at Gonville and Caius College, Cambridge was interrupted by war service at the Royal Aircraft Establishment and at the Telecommunications Research Establishment, where he worked with Martin Ryle (1918-1984).

Returning to Cambridge in 1946, Hewish completed his degree in 1948, and immediately joined Ryle's research team at the Cavendish Laboratory, where he obtained a Ph.D. in 1952. He joined the faculty at Cambridge in 1961, and became professor of radio astronomy in 1971. He became professor emeritus in 1989.

At first, Hewish used radio telescopes to study the atmosphere of the Sun. He determined the electron density of the Sun's corona and studied the hot plasma that comprises this part of the Sun's atmosphere.

In 1965, he designed a new kind of radio telescope, and assigned Jocelyn Bell (1943-), one of his graduate students, the task of bringing the telescope on line. As her work progressed, she began receiving unwanted radio noise that Hewish thought stemmed from local ham operators or other electrical interference. After eliminating that possibility, he found that the source was located outside the Solar System, and that the signals arrived at regular intervals of 1.337 seconds.

Hewish thought they might come from pulsating white dwarf stars. Others suggested the signals were due to a rapidly rotating neutron star. The current view is that a large star near the end of its life explodes, leaving a small, extremely dense star behind. This remnant has a strong magnetic field that captures radiation and channels it into space along narrow beams that may sweep past the Earth if the neutron star's axis rotation is inclined at the proper angle. The pulses correspond to the speed of the star's rotation and can range from 0.033 seconds to 4.0 seconds.

Work on pulsars has led to research on other aspects of stellar evolution, including white dwarfs, collapsars, frozen stars, and black holes.

For his work in the field, Hewish shared with Ryle the 1974 Nobel Prize for physics.

Hewish is currently on emeritus status at the Cavendish Laboratory's Astrophysics Department.

1. Anthony Hewish, Answers.com, www.answers.com/topic/antony-hewish.
2. Anthony Hewish, Wikipedia, http://en.wikipedia.org/wiki/Antony_Hewish.
3. World of Scientific Discovery on Antony Hewish, BookRags, www.bookrags.com/biography/antony-hewish-wsd.

Audouin Dollfus (1924-)

Discovered one of Saturn's Moons, Janus

Audouin Charles Dollfus, a French astronomer, was born on in Paris, France on November 12, 1924. Beginning in 1946, he worked as an astronomer at the Paris Observatory. He studied at the University of Paris, obtaining a doctorate in physical sciences in 1955. Most of his work was carried out based on observations from the Pic du Midi Observatory operated by the University of Toulous in the French Pyrenees. His preferred research method was the use of polarized light to ascertain the properties of Solar System objects.

Before the Viking spacecraft landed on Mars, the composition of the Martian surface was the subject of many debates. Dollfus attempted to determine the composition of the Martian soil by comparing it with the appearance of several hundred terrestrial minerals in polarized light. He found that only pulverized limonite (Fe_2O_3) corresponded with the appearance of Mars, and correctly concluded that the Martian surface is composed of iron oxide.

Dollfus announced that he had detected a very small atmosphere on Mercury, again using polarization measurements made at the Pic du Midi Observatory. His discovery contradicted the previous theoretical predictions based on the kinetic theory of gases. He estimated that the atmospheric pressure at the surface of Mercury was only about one millimeter of mercury. Currently, it is known that the atmosphere of Mercury is indeed very thin.

He studied the possible presence of an atmosphere around the Moon, again using the polarization of light in an attempt to detect it. He found that there was no detectable polarization, thereby confirming the theoretical prediction that the Moon lacks an atmosphere.

In 1966, Dollfus discovered Janus, a small inner moon of Saturn. He made this discovery by observing Saturn at a time when the rings very close to Janus were nearly edge-on to the Earth, and thus practically invisible.

With his father, the aeronautical pioneer Charles Dollfus (1893-1981), Dollfus holds several world records in ballooning, including the first stratospheric flight in France. He was the first to carry out astronomical observations from a stratospheric balloon, in particular the study of Mars.

Asteroid 2451 Dollfus was named in his honor.

1. Audouin Dollfus, Wikipedia, http://en.wikipedia.org/wiki/Audouin_Dollfus.
2. Dollfus, Audouin Charles (1924-), The Internet Encyclopedia of Science, www.daviddarling.info/encyclopedia/D/Dollfus.
3. Hamilton, Calvin J., Janus, Solar Views, www.solarviews.com/eng/janus.

Edwin Salpeter (1924-)

Suggested stars could burn helium into carbon with the Triple-alpha process

Edwin Ernest Salpeter, an Austrian-Australian-American theoretical astrophysicist, was born in Austria on December 3, 1924. He emigrated from Austria to Australia while in his teens. He received his B.Sc. and M.Sc. at the University of Sydney, and earned his Ph.D. at the University of Birmingham in England in 1948. He then joined the faculty at Cornell University, and later became the James Gilbert White Distinguished Professor of the Physical Sciences, Emeritus.

In 1951, he suggested that stars could burn helium into carbon with the triple-alpha process, in which three helium nuclei are transformed into carbon. He later derived the initial mass function for the formation rates of stars of different mass in the Galaxy. That same year, he wrote, with Hans Bethe (1906-2005), two articles which introduced the equation bearing their names, the **Bethe-Salpeter equation.** This equation describes the interactions between a pair of fundamental particles under a quantum field theory.

In 1964, Salpeter and Yakov Zel'dovich (1914-1987) independently suggested that accretion discs around massive black holes are responsible for the huge amounts of energy radiated by quasars. This is currently the most accepted explanation for the physical origin of active galactic nuclei and the associated extragalactic relativistic jets.

Two of Salpeter's interests in astrophysics were high velocity gas clouds and pairs of galaxies and even larger structures—galaxy clusters and superclusters.

In two quite different areas, he has been involved in the study of synapses in neurobiology and in epidemiology and Meta-analysis in Medicine. Although these fields are very different from Astrophysics, similar techniques of mathematics and statistics are used.

His awards include the Royal Astronomical Society Gold medal (1973), American Astronomical Society Henry Norris Russell Lectureship (1974), Astronomische Gesellschaft Karl Schwarzschild Medal (1985), Bruce Medal (1987), and American Physical Society Hans A. Bethe Prize (1999). Also named after Salpeter are minor planet **11757 Salpeter** and the Cornell University **Salpeter Lectureship**.

1. Edwin E. Salpeter, Cornell University Department of Astronomy, www.astro.cornell.edu/people/facstaff-detail.php?pers_id=110.
2. Edwin Ernest Salpeter, The Bruce Medalists, www.phys-astro.sonoma.edu/BruceMedalists/Salpeter/Salpeter.
3. Edwin Ernest Salpeter, Wikipedia, http://en.wikipedia.org/wiki/Edwin_Ernest_Salpeter.

Alastair Cameron (1925–2005)

Proposed that the Moon formed after a giant object collided with the Earth

Alastair Graham Walter Cameron, a Canadian-American astronomer was born in Winnipeg, Manitoba on June 21, 1925. As a teenager, he was the class bookie for student betting on horse races. He once made a bet with a classmate, for $25 or so, that man would land on the Moon by 1970. Cameron studied physics and mathematics at the University of Manitoba, and earned his Ph.D. in nuclear physics, studying photonuclear reactions, at the University of Saskatchewan in 1952.

After graduating, he took a position at Iowa State College in Ames, Iowa, working at the Ames Laboratory of the U.S. Atomic Energy Commission.

In 1954, he moved back to Canada to work for the Canadian Atomic Energy Project at Chalk River, Ontario. There, he first developed his numerical models for equilibrium burning and the s-process inside stars.

In 1959, he spent a sabbatical year at the California Institute of Technology, where he worked with Fritz Zwicky (1898-1974) on the then-controversial idea of neutron stars. This led him to consider the physics of what is now known as r-process nucleosynthesis in the late stages of a star's life.

Cameron left Chalk River in 1961 for the newly formed Goddard Institute for Space Physics in New York City.

He accepted an appointment at Harvard in 1972, and assumed the role of associate director for planetary sciences (and later of theoretical astrophysics) at the Harvard-Smithsonian Center for Astrophysics. By 1976, he was chairman of the astronomy department at Harvard.

Cameron's research led him to dispute the claim of Otto Struve (1897-1963) that stars which spin slowly, like the Sun, do so because they have transferred angular momentum to a planetary system. Instead, he proposed that virtually all single stars, irrespective of their rotational speed, have planets, and that loss of angular momentum occurs when young stars eject matter into space in the form of vigorous stellar winds.

Cameron also proposed the now-accepted theory that the Moon formed after a giant object collided with the Earth. He died of heart failure in Tucson, Arizona, on October 3, 2005.

1. Al Cameron, 1925-2005: Planetary science 'giant' dies at 80, Campus News, Arizona Daily Wildcat, Wednesday, October 26, 2005.
2. Cameron, Alastair G. W. (Graham Walter) (1925-2005), The Internet Encyclopedia of Science, www.daviddarling.info/encyclopedia/C/Cameron.
3. Consolmagno, Guy, Bruce Fegley, and David King, The University of ArizonaMemorial, Alastair Graham Walter Cameron (1925–2005), *Meteoritics & Planetary Science* 41, No. 1, 151–153, http://meteoritics.org/Online%20Supplements/Memorial%20Cameron.pdf, 2006.

Geoffrey Burbidge (1925-)

Collaborated with others on the nucleosynthesis of chemical elements in stars

Geoffrey Ronald Burbidge, a British-American astrophysicist, was born in Chipping Norton, Oxfordshire, England on September 24, 1925. He studied physics at Bristol and University College, London before going to the United States, first to Harvard and then the University of Chicago.

From 1953 to 1955, he worked at the Cavendish Laboratories, Cambridge, but then returned to the United States, where he was professor of physics at the University of California, San Diego (1963 to 1978) and director of the Kitt Peak National Observatory in Arizona (1978 to 1984).

Burbidge began his research career studying particle physics, but after his marriage in 1948 to astronomer Margaret Peachey (1919-) he turned to astrophysics and began a productive research partnership.

In 1957, with the British astronomer Fred Hoyle (1915-2001) and the American nuclear physicist William Fowler (1911-1995), the Burbidges wrote their famous paper on stellar nucleosynthesis. This gave new insight into the evolution of the stars, and showed how the chemical elements could be built up within stars from the hydrogen and helium with which the stars first formed. This idea is now firmly established.

In 1965, Burbidge proposed with Hoyle that quasars were perhaps comparatively small objects ejected at relativistic speeds from highly active radio galaxies, such as Centaurus A. The effect of this would be to place the main body of quasars only three to 30 million light years from our Galaxy and not the three billion light years or more demanded by the generally accepted view.

In 1970, using evidence gained from observations, Burbidge calculated that the stars emitting light in elliptical galaxies could not account for more than 25 percent of the mass. He argued that black holes are the most likely source of the missing mass.

He is known mostly for his alternative cosmology theory, which contradicts the Big Bang theory. According to him, the universe is oscillatory, and as such expands and contracts periodically over infinite time. He received the Warner Prize, with his wife (1959); Bruce Medal (1999), and Gold Medal of the Royal Astronomical Society (2005). **Asteroid 11753 Geoffburbidge** is named for him.

1. Burbidge, (Eleanor) Margaret and Geoffrey, Microsoft Encarta Online Encyclopedia, http://au.encarta.msn.com/encyclopedia_781533493/Burbidge_(Eleanor)_Margaret_and_Geoffrey, 2008.
2. Geoffrey Burbidge, Answers.com, www.answers.com/topic/geoffrey-burbidge.
3. Geoffrey Ronald Burbidge, The Bruce Medalists, www.phys-astro.sonoma.edu/BruceMedalists/BurbidgeG/BurbidgeG.

Allan Sandage (1926-)

Produced the first spectrographic image of quasars

Allan Rex Sandage, an American astronomer, was born in Iowa City, Iowa on June 18, 1926. His father was a business professor at Miami University (Ohio) and his mother was the daughter of the president of a Church of Jesus Christ of Latter Day Saints school. After studying physics and philosophy at Miami University, Sandage served in the U.S. Navy as an electronics specialist during World War II. After the war, he earned a bachelor's degree in physics from the University of Illinois in 1948 and a Ph.D. from the California Institute of Technology in 1953.

While still a student, he worked at the Palomar Observatory with Edwin Hubble (1889-1953) and Walter Baade (1893-1960) using the world's largest telescope at that time. In 1952, he joined Carnegie Observatories in Pasadena, California, where he built on the work Hubble began in the 1920s and 1930s, becoming involved in investigating the origins of the universe. During his first year there, he equated the luminosity of the globular clusters M92 and M3 to the luminosity of the Sun. He found that stars in those globular clusters were as much as 12 billion years old.

To determine the age of a star, Sandage plotted the brightness of stars against their colors or temperatures. How bright a star is depends on its age, mass, and chemical makeup. He looked at the relationships between stars that belong to younger clusters and stars that belong to older clusters to find clues to stellar evolution. Working with Gustav Tammann (1861-1938) of the University of Basel, Switzerland and Abhijit Saha of Kitt Peak National Observatory, Sandage found that the universe is expanding at a speed that indicates that it is about 14 billion years old.

In 1964, Sandage and a young radio astronomer, Thomas Matthews, discovered sources of concentrated radio energy in distant space. They called them quasars, short for quasi stellar radio sources. The center of a quasar is thought to be a black hole that sucks in gases and other materials that form the discus shape associated with them. Quasars are about 1,000 times brighter than the Milky Way Galaxy. They are thought to be the most distant objects in the universe.

Sandage's book, *Lonely Hearts of the Cosmos*, was published in 1991. **Asteroid 9963 Sandage** is named after him.

1. Alan Rex Sandage, Answers.com, www.answers.com/topic/allan-sandage.
2. Alan Rex Sandage, The Bruce Medalists, www.phys-astro.sonoma.edu/BruceMedalists/Sandage/Sandage.
3. Encyclopedia of World Biography on Allan Rex Sandage, BookRags, www.bookrags.com/biography/allan-rex-sandage.

Masatoshi Koshiba (1926-)

Co-detected cosmic neutrinos

Masatoshi Koshiba, a Japanese physicist, was born in Toyohashi, Aichi Prefecture, Japan on September 19, 1926. He graduated from the University of Tokyo School of Science in 1951, and received a Ph.D. in physics at the University of Rochester in New York in 1955. From 1955 to 1958, he conducted research as a Research Associate in the Department of Physics at the University of Chicago. From 1958 to 1963, he was an associate professor at the Institute of Nuclear Study at the University of Tokyo. Also, from 1959 to 1962, while on leave from the University of Tokyo, he was a senior research associate with the honorary rank of Associate Professor and Acting Director of the Laboratory of High Energy Physics and Cosmic Radiation at the University of Chicago.

From 1963 to 1970, Koshiba was Associate Professor in the Department of Science at the University of Tokyo. He remained there for the next 17 years, until 1987. He then became a professor of Tokai University until 1997.

Since the 1920s, it had been suspected that the Sun generates heat and light because of nuclear fusion reactions that transform hydrogen into helium and release energy. Later, theoretical calculations indicated that countless neutrinos from the Sun must constantly flood the Earth; however, they were thought by many to be undetectable.

Drawing on the work of Raymond Davis Jr. (1914-2006), Koshiba constructed an underground neutrino detector in a zinc mine in Japan in the 1980s. It consisted of an enormous water tank surrounded by electronic detectors to sense flashes of light produced when neutrinos interacted with atomic nuclei in water molecules. He was able to confirm Davis's results. In 1987, he also detected neutrinos from a supernova explosion outside the Milky Way.

After building a larger, more sensitive detector, which became operational in 1996, Koshiba found strong evidence that neutrinos come in three types, and change from one type to another in flight.

In 2000, he shared the Wolf Prize in Physics with Davis. He also shared the Nobel Prize in physics with Davis and Riccardo Giacconi (1931-) in 2002. His and Davis's part was for pioneering contributions to astrophysics, in particular for the detection of cosmic neutrinos. Giacconi's part was for work that led to the discovery of cosmic X-ray sources.

1. Masatoshi Koshiba, Japan Society for Promotion of Science, www.jspsusa.org/FORUM2003/bio.koshiba.
2. Masatoshi Koshiba, NNDB, www.nndb.com/people/030/000027946.
3. Masatoshi Koshiba, Wikipedia, http://en.wikipedia.org/wiki/Masatoshi_Koshiba.

Halton Arp (1927-)

Challenged the theory that the large redshifts of quasars and other active galaxies are an indication of great distance

Halton Christian (Chip) Arp, an American astronomer, was born in New York City on March 21, 1927. He was awarded a bachelor's degree by Harvard University in 1949 and a Ph.D. by the California Institute of Technology in 1953. Afterward, he became a Fellow of the Carnegie Institution of Washington, performing research at the Mount Wilson and Palomar Observatories. He became a Research Assistant at Indiana University in 1955, and 12 years later became a staff member at Palomar Observatory, where he worked for 29 years. In 1983 he joined the staff of the Max Planck Institute for Astrophysics in Germany.

Arp hypothesized that quasars (quasi-stellar objects) are local objects which have been ejected from the core of active galactic nuclei. The theory was originally proposed in the 1960s as an alternative to the explanation of Maarten Schmidt (1929-) for quasars, which stated that they were very distant galaxies that appeared to be highly redshifted because of the expansion of the universe. Arp made the argument that, in some photographs, quasars seems to be in the foreground of galaxies that, according to the Hubble's law, are significantly closer to earth than most think. Hubble's law states that redshift in light coming from a distant galaxy is proportional to the galaxy's distance.

Arp also claimed that quasars are not evenly spread over the sky, but tend to be more commonly found in positions of small angular separation from certain galaxies. The implication of the hypothesis of local quasars is that most of the observed redshift must have a non-cosmological, or intrinsic, origin. He suggested that quasar emission may instead be ejecta from active galactic nuclei. Nearby galaxies with both strong radio emission and peculiar morphologies, particularly M87 and Centaurus A, appear to support Arp's hypothesis. In his books, Arp provides his reasons for believing that the Big Bang theory is incorrect, citing his research into quasars. Instead, Arp supports the redshift quantization theory as an explanation of the redshifts of galaxies.

Arp compiled a catalog of unusual galaxies, titled *Atlas of Peculiar Galaxies*, which was first published in 1966. This atlas was intended to provide images that would give astronomers data from which they could study the evolution of galaxies.

1. Halton Arp, Wikipedia, http://en.wikipedia.org/wiki/Halton_Arp.
2. Halton C. Arp, ApologeticsPress.org, www.apologeticspress.org/articles/2473.
3. Seeing Red: Intrinsic redshifts, stable universe, QuackGrass Roots, www.quackgrass.com/roots/arp.

Vainu Bappu (1927-1982)

Co-discovered a relationship between the luminosity of certain kinds of stars and some of their spectral characteristics

Manali Kallat Vainu Bappu, an Indian astronomer, was born in Madras, India on August 10, 1927. His father was an astronomer in the Nizamiah Observatory, Hyderabad. Bappu not only excelled in studies, but took active parts in debates, sports, and other extra-curricular activities. However, astronomy became his primary interest. He attended Harvard University on a scholarship after receiving his Master's degree in physics from Madras University.

Shortly after arriving at Harvard, Bappu discovered a comet, which was named **Bappu-Bok-Newkirk comet** after him and his colleagues, Bart Bok (1906-1983) and Gordon Newkirk. He completed his Ph.D. in 1952 and joined the Mount Palomar Observatory.

During his short stay as Carnegie Fellow at the Hale Observatories, he jointly discovered with Colin Wilson an important phenomenon in stellar chromospheres, which came to be known as the **Wilson-Bappu Effect**. This is a linear relation between the absolute magnitudes of late-type stars and the width of the K_2 emission core in the resonance line of ionized calcium at a wavelength of 3933 nanometer.

Bappu returned to India in 1953 and helped set up the Uttar Pradesh State Observatory in Nainital.

In 1960, he left Nainital to take over as the director of the Kodaikanal Observatory in south India. He modernized the facilities. Today the observatory is an active center of astronomical research.

Bappu realized that the Kodaikanal Observatory was inadequate for making stellar observations, and began searching for a good site for a stellar observatory. As a result of his efforts, a totally indigenous 2.3 meter telescope was designed, fabricated, and installed in Kavalur, Tamil Nadu. Both the telescope and the observatory were named after Bappu when it was commissioned in 1986.

He was awarded the Donhoe Comet Medal by the Astronomical Society of the Pacific in 1949, and he was elected president of the International Astronomical Union in 1979. He was also elected Honorary Foreign Fellow of the Belgium Academy of Sciences, and was an Honorary Member of the American Astronomical Society.

1. Bappu, (Manali Kallat) Vainu (1927-1982), The Internet Encyclopedia of Science, www.daviddarling.info/encyclopedia/B/Bappu.
2. M. K. Vainu Bappu (1927-1982), ScienticIndia.net, www.scientificindia.net/scientists/mkvbappu.
3. M.K. Vainu Bappu, India's Who is Who?, www.mapsofindia.com/who-is-who/science-technology/m-k-vainu-bappu.

Robert Wilson (1927-2002)

Father of the International Ultraviolet Explorer Satellite

Robert Wilson, A British astronomer, was born at South Shields, County Durham, England on April 16, 1927. He was the son of a miner. He studied physics at King's College, Durham, and obtained his Ph.D. in Edinburgh, where he worked on stellar spectra at the Royal Observatory.

In 1959, Wilson joined the Plasma Spectroscopy Group at Harwell, England, where he was responsible for measuring the temperature in the Zeta fusion experiment. He confirmed that it had not been hot enough to produce thermonuclear fusion. As head of the same group at Culham, he led a program of rocket observations of ultraviolet spectra of the Sun and other stars. By placing telescopes on rockets, it was possible to avoid the absorption of the ultraviolet light by the Earth's atmosphere and gain a great deal of information about the hot plasmas, especially in the Sun's chromosphere and corona.

Wilson then became involved in the first astronomy satellite, the TD-1 mission of the European Space Research Organization (ESRO), and led the British collaboration with Belgium in the S2/68 experiment, which in 1972 conducted the first all sky survey in the ultraviolet range.

Wilson is best known for his role as father of the International Ultraviolet Explorer Satellite. This work began in 1964 as a proposal for a large astronomical satellite, which proved too expensive for the ESRO, and in 1967 the project was abandoned. Wilson achieved a radical redesign, which had greater capability and was cheaper. It was again submitted to ESRO in 1968, but despite a favorable assessment report, it was again rejected. Convinced of the soundness of the concept, Wilson offered the design to the National Aeronautics and Space Administration (NASA) in the United States. This ultimately led to an international project among NASA, the European Space Agency, and the United Kingdom. The launch took place from Cape Canaveral in 1978.

In 1972, Wilson relinquished his post as Director of the Science Research Council's Astrophysics Research Unit in Culham to become Perren Professor of Astronomy at University College, London. He was knighted in 1989 for his services to astronomy. He was one of the pioneers who laid the ground work for the Hubble Space Telescope. He died on September 2, 2002.

1. Robert Ian, Obituary: Sir Robert Wilson, Independent, The (London), http://findarticles.com/p/articles/mi_qn4158/is_20020924/ai_n12641213, September 24, 2002.
2. Robert Wilson (Astronomer), Wikipedia, http://en.wikipedia.org/wiki.
3. Sir Robert Wilson, Telegraph.co.uk, www.telegraph.co.uk/news/obituaries/1408657/Sir-Robert-Wilson, September 2002.

Eugene Shoemaker (1928-1997)

His research led to an appreciation of the role of asteroid and comet impacts as a primal and fundamental process in the evolution of planets

Eugene (Gene) Merle Shoemaker, an American astrogeologist, was born in Los Angeles, California on April 28, 1928. He graduated from the California Institute of Technology at the age of 19, and earned a master's degree a year later, at which point he joined the United States Geological Survey (USGS). His first work for the USGS involved searching for uranium deposits in Colorado and Utah. While doing this, he became interested in the Moon, the possibility of traveling there, and of establishing the relative roles of asteroidal impacts and volcanic eruptions in forming the lunar craters. He then embarked on work for a Ph.D. at Princeton University.

He married Carolyn Spellman (1929-) in 1951. A visit to Arizona's Meteor Crater the following year began to direct Shoemaker toward the view that both it and the lunar craters were due to asteroidal impacts. His research at the crater led to an appreciation of the role of asteroid and comet impacts as a primal and fundamental process in the evolution of planets.

Shoemaker contributed greatly to space science exploration, particularly of the Moon. He was part of a leading comet-hunting team, which also included his wife Carolyn and David Levy (1948-). The team discovered comet **Shoemaker-Levy 9,** and charted the object's breakup. Pieces of the comet slammed into Jupiter in July 1994—an unprecedented event in the history of astronomical observations. That same year, Shoemaker led the U.S. Defense Department's Clementine mission, which first detected the possibility of pockets of water ice at the Moon's south pole. He was instrumental in establishing the discipline of planetary geology. He founded the U.S. Geological Survey's Branch of Astrogeology, which mapped the Moon and prepared astronauts for lunar exploration.

While carrying out research on impact craters on July 18, 1997, Shoemaker's car collided head on with another vehicle on an unpaved road in the Tanami Desert in Australia. He was killed and Carolyn was injured. A small vial of his ashes were scattered on the lunar surface by the Lunar Prospector spacecraft, which was purposely crashed on the Moon on July 31, 1999 after completing its mission.

1. Marsden, Brian, Eugene M. Shoemaker (1928-1997), Comet Shoemaker-Levy Collision with Jupiter, www2.jpl.nasa.gov/sl9/news81.
2. Shoemaker, Eugene, BookRags, www.bookrags.com/research/shoemaker-eugene-spsc-02.
3. Shoemaker, Eugene, BookRags, www.bookrags.com/research/shoemaker-eugene-spsc-02.

Carolyn Shoemaker (1929-)

Discovered the most comets of any individual

Carolyn Jean Shoemaker, an American astronomer, was born in Gallup, New Mexico on June 24, 1929. Her maiden name was Spellmann. Her father owned a clothing store business and her mother was a schoolteacher. Her family later moved to Chico, California, where she attended Chico State College, receiving a B.S. and M.S. in history and political science. She became a junior high school teacher, which she found disappointing. In 1950, she met Gene Shoemaker (1928-1997) at her brother's wedding. They were married on August 18, 1951.

Gene encouraged her to fly and become a pilot in 1960, because he knew she wanted to, as he did but could not due to health problems.

In 1980, Shoemaker took up astronomy at the age of 51, after her three children had left home. She was looking for something fulfilling to do, so she asked Gene for advice, and he suggested that she might be interested in the telescopic search for near-Earth asteroids. She began using film taken at the wide-field telescope at Palomar, combined with a stereoscope, to find objects which moved against the background of fixed stars.

Shoemaker holds the record for most comets discovered by an individual. As of 2002, she had discovered 32 comets and over 800 asteroids.

Carolyn and Gene Shoemaker are most famous for their joint discovery, with David Levy (1948-), of **Comet Shoemaker-Levy 9** in 1993. The comet had already been ripped apart by Jupiter's gravitational force. The comet orbited Jupiter in 21 icy fragments until they crashed into the planet in July 1994 as the world observed from telescopes and the Voyager 2 spacecraft views.

She received an honorary doctorate from the Northern Arizona University at Flagstaff, Arizona and the NASA Exceptional Scientific Achievement Medal in 1996. She and Gene were awarded the James Craig Watson Medal in 1998.

While carrying out research on impact craters on July 18, 1997, the Shoemaker's car collided head on with another vehicle on an unpaved road in the Tanami Desert in Australia. Gene was killed and Carolyn sustained many serious injuries, but she eventually recovered and continued observation work.

1. Carolyn and Gene Shoemaker, Astronomy, http://pvastro0714.blogspot.com/2008/05/carolyn-and-gene-shoemaker.
2. Chapman, Mary G., Carolyn Shoemaker, USGS, http://astrogeology.usgs.gov/About/People/CarolynShoemaker, May 17, 2002.
3. Laing, Jennifer, Comet Hunter, www.universetoday.com/html/articles/2001-1211a, December 11, 2001.

Maarten Schmidt (1929-)

Identified the first quasar

Maarten Schmidt, a Dutch-American astronomer, was born in Groningen, Netherlands on December 28, 1929. He attended the University of Leiden, where he studied under Jan Oort (1900-1992), earning his Ph.D. in 1956. In 1959, he moved to the California Institute of Technology (Caltech), where he initially studied the mass distribution and dynamics of galaxies. In the early 1960s, however, Rudolph Minkowski (1895-1976) retired, and Schmidt assumed leadership of the project he had directed—the analysis of spectra of radio-emitting objects.

In 1963, Schmidt was the first astronomer to describe a quasar. While its star-like appearance suggested it was relatively nearby, its spectrum proved to have a very high redshift, indicating that it lay far beyond the Milky Way, and thus possessed an extraordinarily high luminosity. Schmidt termed it a "quasi-stellar object" or quasar.

He identified different wavelengths of quasar radiation, helping to establish quasars as among the oldest and most distant objects yet known in the universe. He theorized that if quasars are visible despite their great distance they must be almost unimaginably powerful. He found that there were far more quasars in an earlier epoch. This finding helped lead to the decline of the steady-state theory, which had once competed with the Big Bang theory as a model for how the universe came into being.

Theoretical astrophysicist Donald Lynden-Bell (1935-) later proposed that massive black holes exist at the center of galaxies, and that these black holes provide the energy for quasars.

Schmidt, who retired as a professor in 1996, served in a variety of roles at Caltech—Executive Officer for Astronomy (1972 to 1975); Chairman of the Division of Physics, Mathematics, and Astronomy (1976 to 1978); and Director of the Hale Observatories (1978 to 1980).

Winner of the 1992 Bruce Medal, he continued to investigate quasars with a high red shift, his aim being to find the red shift cutoff above which no quasars exist. Later, Schmidt joined teams finding X-ray and gamma ray sources from orbiting observatories. He then helped obtain their optical spectra at the Keck Observatory.

In 2008, Schmidt was named a co-winner of the first Kavli Prize in astrophysics, a $1-million award. **Asteroid 10430 Martschmidt** is named for him.

1. Maarten Schmidt, BookRags, www.bookrags.com/research/maarten-schmidt-scit-0712345.
2. Maarten Schmidt, The Bruce Medalists, www.phys-astro.sonoma.edu/BruceMedalists/Schmidt/Schmidt.
3. Maarten Schmidt, Wikipedia, http://en.wikipedia.org/wiki/Maarten_Schmidt.

Roger Penrose (1931-)

Proved that, according to relativity, singularities must form when massive stars collapse gravitationally

Roger Penrose, a British mathematical physicist, was born in Colchester, Essex, England on August 8, 1931. His father was a medical geneticist and his mother was a physician. Roger was educated at University College, London University and St. John's College, Cambridge University. In 1973, he became Rouse-Ball Professor of Mathematics at the University of Oxford. He was knighted in 1994 for services to science.

Penrose introduced many of the techniques characteristic of present-day relativity theory. In a number of theorems, developed alone or later with Stephen Hawking (1942-), Penrose proved that, according to relativity, singularities must form when massive stars collapse gravitationally, and that the Big Bang must have been a singularity. A singularity is a place where the density of matter and the curvature of space become infinite. He proposed the "cosmic censorship hypothesis," which states that such singularities when they form are not "naked" in the sense of being visible to outside observers, but instead are hidden within the horizon of a black hole.

Penrose has criticized the notion that human thinking is basically the same as the action of a very complicated computer. A computer can only carry out algorithms, which are systematic calculational procedures. It is possible to prove for certain classes of mathematical problems that there cannot be an algorithm for finding the solution. Therefore, Penrose argues that in forming mathematical judgments the brain must be acting in a way that does not follow any algorithm. A proper explanation of consciousness, according to him, needs some kind of non-algorithmic physics, which has yet to be found.

In 1996, Penrose and Hawking published *The Nature of Space and Time*. This book is a record of a debate between the two at the Isaac Newton Institute of Mathematical Sciences at the University of Cambridge in 1994.

Several universities have awarded Penrose an honorary degree, including New Brunswick (1992), Surrey (1993), Bath (1994), London (1995), Glasgow (1996), Essex (1996), St. Andrews (1997), and Santiniketon (1998).

1. O'Connor J.J. and E. F. Robertson Roger Penrose, MacTutor, www-groups.dcs.st-and.ac.uk/~history/Biographies/Penrose.
2. Penrose, Sir Roger, Microsoft Encarta Online Encyclopedia, http://au.encarta.msn.com/encyclopedia_781534837/penrose_sir_roger, 2008.
3. World of Mathematics on Roger Pemrose, BookRags, www.bookrags.com/biography/roger-penrose-wom.

Riccardo Giacconi (1931-)

A pioneer of X-ray astronomy

Riccardo Giacconi, an Italian-born American astrophysicist, was born in Genoa, Italy on October 6, 1931. His mother taught mathematics and physics at the high school level and his father owned a small business. Riccardo earned his Ph.D. in cosmic ray physics at the University of Milan, and then spent brief postdoctoral periods at Indiana and Princeton Universities. In 1959, he joined American Science and Engineering, a Massachusetts research firm, where he began work on X-ray astronomy.

Giacconi's team developed grazing incidence X-ray telescopes and launched them on rockets. In 1962, they discovered Sco X-1, the first known X-ray source outside the Solar System. They then built the UHURU orbiting X-ray observatory, and made the first surveys of the X-ray sky. They discovered 339 X-ray "stars," most of which turned out to be matter falling into black holes and neutron stars. Among these was Cygnus X-1, the first object to be widely accepted as a black hole. They also discovered X-ray emission by hot gas in clusters of galaxies.

Joining the Harvard-Smithsonian Center for Astrophysics in 1973, Giacconi led the construction and successful operation of the powerful X-ray observatory, HEAO-2, also known as "Einstein."

In 1982, he joined the faculty at Johns Hopkins University. From 1993 to 1999, he was also Director General of the European Southern Observatory, which is composed and supported by 14 countries from Europe. From 1999 to 2004, he served as president of Associated Universities, Inc., which is the operator of the National Radio Astronomy Observatory. In this position he was involved in the development of the Atacama Large Millimeter Array being built at high altitude in Chile by a team of European, American, and Japanese institutions.

Giacconi is currently principal investigator for the Chandra Deep Field-South project with NASA's Chandra X-ray Observatory, whose elliptical orbit takes the satellite to an altitude of approximately 86,500 miles, more than a third of the distance to the Moon.

He was awarded the Nobel Prize in Physics in 2002 for the discovery of cosmic X-ray sources. In 2008, he was awarded the National Inventors Hall of Fame's Lifetime Achievement Award. **Asteroid 3371 Giacconi** was named for him.

1. Riccardo Giacconi, Autobiography, NobelPrize.org, http://nobelprize.org/nobel_prizes/physics/laureates/2002/giacconi-autobio.
2. Riccardo Giacconi, The Bruce Medalists, www.phys-astro.sonoma.edu/BruceMedalists/Giacconi/Giacconi.
3. Riccardo Giacconi, Wikipedia, http://en.wikipedia.org/wiki/Riccardo_Giacconi.

Arno Penzias (1933-)

Co-discovered cosmic microwave background radiation

Arno Allan Penzias, a Jewish German-born American radio engineer, was born in Munich, Germany on April 26, 1933. His parents fled Nazi Germany in 1940 and immigrated to the United States.

Studying at the City College of New York, Penzias earned his bachelor's degree in physics in 1954. He served in the U.S. Army Signal Corps for two years, and then returned to New York. He continued his studies at Columbia University, where he was awarded his master's degree in 1958 and his doctorate in 1962, both in physics.

In 1961, Penzias became associated with the Radio Research Laboratories at Bell Laboratories in New Jersey. From 1961 to 1972 he was a staff member of the Radio Research Department, and from 1972 to 1976 he was head of that department. In 1976, he became the director of the Radio Research Laboratory, and he served as director of the Communications Sciences Research Division from 1979 to 1981. From 1981 to 1995, he was Vice President of Research for Bell Labs. He served as Vice President and Chief Scientist from 1995 to 1998. He retired in 1998 to work as an advisor and spokesperson for Lucent Technologies.

Penzias and American engineer Robert Wilson (1936-) were the first to detect the cosmic microwave background radiation. This radiation seems to be evenly distributed throughout the universe, leading scientists to believe that it is the cooling remains of energy released at the Big Bang. The background radiation represents some of the strongest evidence in favor of the Big Bang theory.

Penzias and Wilson shared half of the 1978 Nobel Prize in physics for their discovery. The other half of the prize went to Soviet physicist Peter Kapitza (1894-1984) for his work in low-temperature physics.

In addition to his posts in the telecommunications industry, Penzias also held a series of academic positions. The first of these was as lecturer in the Department of Astrophysical Science at Princeton University from 1967 to 1982. After that, he held several year-long lecturer positions at other institutions, including the National Radio Astronomical Observatory in West Virginia and Stanford University in California.

Penzias was an avid skier, swimmer, and runner, with an interest in kinetic sculpture and writing limericks.

1. Arno Penzias, Autobiography, NobelPrize.org, http://nobelprize.org/nobel_prizes/physics/laureates/1978/penzias-autobio.
2. Arno Penzias, Microsoft Encarta Online Encyclopedia, http://encarta.msn.com/encyclopedia_761579468/arno_penzias, 2008.
3. World of Physics on Arno Penzias, BookRags, www.bookrags.com/biography/arno-penzias-wop.

Carl Sagan (1934-1996)

Helped show that seasonal changes on Mars are due to windblown dust

Carl Edward Sagan, an American astronomer, was born in Brooklyn, New York on November 9, 1934. His father was a Jewish garment worker. Carl studied astronomy at the University of Chicago, receiving his undergraduate degree in 1954 and his doctorate in 1960. After holding a number of posts, he became director of Cornell University's Laboratory for Planetary Studies in 1970. In addition to his academic appointments, he served as a consultant to the National Aeronautics and Space Administration, and was closely associated with the unmanned space missions to Venus, Mars, Jupiter, and Saturn.

His first major research effort was an investigation of the surface and atmosphere of Venus. In the late 1950s, the surface of Venus was thought to be relatively cool and might support life. The observed radio emissions were thought to come from the activity of charged particles in an atmospheric layer. In 1961, Sagan showed that the emissions could be explained by simply assuming that the Venusian surface was very hot, and therefore hostile to life. He accounted for the high temperatures by postulating the existence of a greenhouse effect. This was confirmed by an exploratory space vehicle sent to Venus by the Soviet Union in 1967.

Telescopic observation of Mars revealed distinctive bright and dark areas on its surface. This led some to speculate that large regions of Mars were covered with vegetation, subject to seasonal changes. Reviewing available data, Sagan concluded that the bright regions were lowlands filled with sand and dust blown by the wind and that the dark areas were elevated ridges or highlands.

Sagan's scientific interest in planetary surfaces and atmospheres led him to investigate the origins of life on Earth, and to champion the study of the biology of extraterrestrial life.

In 1973, he published *The Cosmic Connection*, an introduction to space exploration and the search for extraterrestrial life. His book on the evolution of human intelligence, *The Dragons of Eden*, won him the Pulitzer Prize in 1978. Another of his books, *Cosmos* (1980), was written in conjunction with the television series of the same name, which he hosted. Sagan died in Seattle on December 20, 1996 of a rare bone marrow disease, myelodysplasia. In July of the following year, the Mars Pathfinder Lander was renamed the **Dr. Carl Sagan Memorial Station**.

1. Carl Sagan Dies at 62, CNN, www.cnn.com/US/9612/20/sagan.
2. Carl Sagan, A Life in the Cosmos, Crystalinks.com, www.crystalinks.com/sagan.
3. Encyclopedia of World Biography on Carl E. Sagan, BookRags, www.bookrags.com/biography/carl-e-sagan.

John Bahcall (1934-)

Predicted the flux of solar neutrinos

John Norris Bahcall, an American astrophysicist, was born in Shreveport, Louisiana on December 30, 1934. He became state tennis champion and a national debate team champion. He started his university career at Louisiana State University as a philosophy student on a tennis scholarship, and considered becoming a rabbi. After a year, he moved to the University of California, Berkeley, where he graduated with an A.B. in physics in 1956. He obtained his M.S. in 1957 from the University of Chicago and his Ph.D. in physics from Harvard University in 1961. After posts at Indiana University and the California Institute of Technology, he moved to The Institute for Advanced Study at Princeton in 1968, where he spent the rest of his career. He married Princeton University professor of astrophysics, Neta Bahcall.

Bahcall spent much of his career participating in the pursuit of an answer to the solar neutrino problem with physical chemist Raymond Davis, Jr. (1914-2006). They collaborated on the Homestake Experiment, which was the creation of an underground detector for neutrinos in a gold mine in South Dakota. The detector consisted of a very large tank filled with cleaning fluid. The flux of neutrinos found by the detector was 1/3 the amount predicted by Bahcall, a discrepancy that took over 30 years to resolve. The 2002 Nobel Prize in physics was awarded to Davis and Masatoshi Koshiba (1926-) for their pioneering work in observing the neutrinos predicted from Bahcall's solar model.

Bahcall also contributed to the development and implementation of the Hubble Telescope, along with Lyman Spitzer, Jr. (1914-1997), from the 1970s to after the telescope was launched in 1990. NASA recognized his achievements by awarding him the 1992 Distinguished Public Service Medal for his observations and leadership with Hubble.

He worked in many other areas, including the standard model of a galaxy with a massive black hole surrounded by stars, known as the **Bahcall-Wolf model.** The **Bahcall-Soneira model** was for many years the standard model for the structure of the Milky Way. He also contributed to accurate models of stellar interiors.

Bahcall died at the age of 70 from a rare blood disorder on August 17, 2005. He published over 600 scientific papers and five books in the field of astrophysics.

1. John N. Bahcall, Wikipedia, http://en.wikipedia.org/wiki/John_N._Bahcall.
2. John Bahcall Dies, PhysicsWorld.com, http://physicsworld.com/cws/article/news/22917.
3. Hendrix, Susan, Astrophysics Pioneer John Bahcall Dies, Goddard Space Flight Center, www.nasa.gov/centers/goddard/news/topstory/2005/bahcall_passing.

Robert W. Wilson (1936-)

Co-detected the cosmic background radiation

Robert Woodrow Wilson was born in Houston, Texas on January 10, 1936. His father worked for an oil well service company. In high school Robert played trombone in the marching band. He attended Rice University, where he received a B.A. in physics in 1957. He then received his Ph.D. in 1962 from the California Institute of Technology (Caltech). Wilson's thesis and post-doctoral research involved making radio surveys of the Milky Way Galaxy.

When he heard of the existence of specialized radio equipment at Bell Laboratories, he left Caltech and accepted a job at Bell's research facility at Crawford Hill, New Jersey.

At Bell, Wilson and Arno Penzias (1933-) studied the possible causes of static interference that impaired the quality of radio communications. The source of the radiation was unknown. Finally, they made observations of the radio flux from the sky. The intensity of the radio noise was what would be expected from a source with a very low temperature, almost near absolute zero. Furthermore, it was not coming from a discrete source, but was emanating uniformly from every direction in the sky.

Meanwhile, Robert Dicke (1916-1997) and his colleagues at Princeton University, unaware of the project at Bell Labs, were building a radio receiver of their own, designed to look for the radiation effects of the Big Bang. If the remnant of this energy flash had survived after several billion years, it would be detected as a very weak signal in a radio telescope, and would be present in nearly equal intensities in every direction, forming a cosmic background radiation. When Dicke heard the details of Wilson and Penzias's' findings, he knew that they had discovered what he was looking for.

In 1965, Wilson and Penzias published their results, and a companion paper written by Dicke and colleagues explained the cosmological implications of the finding—support for the Big Bang theory. The discovery of the expansion of the universe added additional support.

In 1976, Wilson was named head of the Radio-Physics department of Bell Telephone. In 1978, Wilson and Penzias shared the Nobel Prize in physics with Pyotr Kapitsa (1894-1984), whose work was in low-temperature physics.

1. Robert Woodrow Wilson, Autobiography, NobelPrize.org, http://nobelprize.org/nobel_prizes/physics/laureates/1978/wilson-autobio.
2. Robert Woodrow Wilson, IEEE Virtual Museum, www.ieee-virtual-museum.org/collection/people.php?id=1234738&lid=1.
3. World of Scientific Discovery on Robert Woodrow Wilson, BookRags, www.bookrags.com/biography/robert-woodrow-wilson-wsd.

James Christy (1938-)

Discovered that Pluto has a moon, which he named Charon

James Walter Christy, an American astronomer, was born in Wisconsin in 1938. As a child, he contracted a kidney infection, and doctors said he would be much better off living in a warm climate, so his family moved to Arizona when he was 16. In 1962, even before he had graduated from the University of Arizona, Christy took a job at the U.S. Naval Observatory in Flagstaff where a new telescope had been put into operation that could measure the positions of stars with great accuracy.

The Naval Observatory sent him to back Tucson in 1966 to complete his master's degree. Shortly after resuming his studies, his only brother was killed in a car crash.

In 1972, Christy was transferred to the headquarters of the Naval Observatory in Washington, D.C. Afterward, when he was examining an enlargement of a photographic plate of Pluto, he noticed that Pluto had a very slight bulge on one side. This plate and others had been marked "poor" because the elongated image was thought to be a defect resulting from improper alignment.

Christy alertly noticed that only Pluto was elongated; the background stars were not. It occurred to him that this bulge might be a companion of Pluto, but everyone knew that Pluto didn't have a moon; astronomers had searched for one with the world's largest telescope and found nothing.

After examining images from observatory archives dating back to 1965, on June 22, 1978 Christy concluded that the bulge was indeed a moon. It was quite close in and had an orbital period of a little more than six days, exactly the same as Pluto's rotation rate. He named the moon Charon.

The photographic evidence was considered convincing, but not conclusive. However, based on Charon's calculated orbit and a series of mutual eclipses of Pluto, the discovery of Charon was confirmed.

In more modern telescopes, such as the Hubble Space Telescope or ground-based telescopes using adaptive optics, separate images of Pluto and Charon can very easily be seen.

Christy left the Naval Observatory in 1982 and moved his family to Tucson, where he worked for 17 years in the physics department of Hughes Missile Systems.

1. 25th Anniversary of the Discovery of Pluto's moon CHARON, U.S. Naval Observatory, www.usno.navy.mil/pao/press/charon, June 20, 2003.
2. A Bump in the Night, Sky and Telescope, www.allesoversterrenkunde.nl/cgi-bin/scripts/db.cgi?db=default&uid=default&ID=806&ww=1&view_records=1, June 2008.
3. James W. Christy, Wikipedia, http://en.wikipedia.org/wiki/James_W._Christy.

Walter Alvarez (1940)

Co-located the Cretaceous-Tertiary boundary

Walter Alvarez, an American geologist, was born in Berkeley, California in 1940. His father was Nobel Prize winning physicist, Luis Alvarez. Walter received a B.A. in geology in 1962 from Carleton College in Minnesota and his Ph.D. in geology in 1967 from Princeton University. In 1977 he became a professor in the Earth and Planetary Science Department at the University of California, Berkeley.

He worked for American Overseas Petroleum Limited in Holland and in Libya at the time of Gadaffi's revolution. After developing an interest in archaeological geology, he left the oil company and spent some time in Italy, studying the Roman volcanoes and their influence on patterns of settlement in early Roman times.

Alvarez then moved to Lamont-Doherty Geological Observatory of Columbia University and began studying the Mediterranean tectonics. His work on tectonic paleomagnetism in Italy led to a study of the geomagnetic reversals recorded in Italian deep-sea limestone. He and his colleagues were able to date the reversals for an interval of more than 100 million years of Earth history.

Alvarez and his father are most widely known for their discovery that a clay layer occurring at the Cretaceous-Tertiary (K-T) boundary is highly enriched in the element iridium. Since iridium enrichment is common in asteroids, but very uncommon on the Earth, they postulated that the layer had been created by the impact of a large asteroid, and that this impact event was the likely cause of the Cretaceous-Tertiary extinction event, which occurred 65 million years ago, eliminating 85 percent of all species and all the dinosaurs.

The iridium enrichment has now been observed in many other sites around the world. The very large Chicxulub crater buried underneath the Yucatán Peninsula in Mexico is regarded as evidence of such a large impact. Alvarez's book, *T. Rex and the Crater of Doom*, details the discovery of the K-T extinction event.

Alvarez is the recipient of numerous awards and honors, including the 2006 Nevada Medal and the Penrose Medal, which is the Geological Society of America's highest award. In 2005, he received an honorary doctorate in Geological Sciences from the University of Siena, Italy.

1. Brusatte, Steve, A Portrait of Walter Alvarez, DinoData, www.dinodata.org/index.php?option=com_content&task=view&id=691&Itemid=25.
2. Walter Alvarez, University of California, Berkeley, http://eps.berkeley.edu/cgi-bin/faculty.cgi?name=alvarez.
3. Walter Alvarez, Wikipedia, http://en.wikipedia.org/wiki/Walter_Alvarez.

Kip Thorne (1940-)

A leading researcher in the area of gravitational waves

Kip Stephen Thorne, an American theoretical physicist, was born in Logan, Utah on June 1, 1940. He is the son of Utah State University professors Wynne and Alison Thorne. He graduated from the California Institute of Technology (Caltech) with a B.S. in 1962, and was awarded an M.A. in 1963 and a Ph.D. in 1965 from Princeton University. He wrote his Ph.D. dissertation, *Geometrodynamics of Cylindrical Systems*, under the supervision of John Wheeler (1911-2008).

Thorne's research has principally focused on relativistic astrophysics and gravitation physics, with emphasis on relativistic stars, black holes, and gravitational waves. He is best known for his controversial theory that wormholes can conceivably be used for time travel. However, his scientific contributions, which center on the general nature of space, time, and gravity, span the full range of topics in general relativity.

He is considered one of the world's authorities on gravitational waves. In part, his work has dealt with the prediction of gravity-wave strengths and their temporal signatures as observed on Earth.

Thorne has made significant contributions to black hole cosmology. He proposed his Hoop Conjecture that cast aside the thought of a naked singularity. The Hoop Conjecture states that if an object gets compressed in a highly nonspherical manner, then the object will form a black hole around itself when and only when its circumference in all directions becomes less than the critical circumference.

LIGO (Laser Interferometer Gravitational-Wave Observatory), cofounded in 1992 by Thorne and Ronald Drever of Caltech and Rainer Weiss of the Massachusetts Institute of Technology, is a joint project sponsored by the National Science Foundation. Its mission is to directly observe gravitational waves of cosmic origin. These waves were first predicted by Einstein's Theory of General Relativity in 1916.

In 1995, Thorne published the best-selling book, *Black Holes & Time Warps, Einstein's Outrageous Legacy*, which outlines the efforts of many of his colleagues, past and present. His presentations on subjects such as black holes, gravitational radiation, relativity, time travel, and wormholes have been included in PBS shows in the U.S. and in the United Kingdom. He is currently the Feynman Professor of Theoretical Physics at Caltech.

1. Bennett, Clark, Kip Thorne (1940-Present), Founding Fathers of Relativity, www.usd.edu/phys/courses/phys300/gallery/clark/thorne.
2. Kip S. Thorne, Biographical Sketch, California Institute of Technology, www.its.caltech.edu/~kip/scripts/biosketch.
3. Kip Thorne, Wikipedia, http://en.wikipedia.org/wiki/Kip_Thorne.

James Young (1941-)

Prolific asteroid discoverer

James Whitney Young, an American astronomer, was born in Portland, Oregon on January 24, 1941. He was the lead technical guide at the NASA exhibit of the Seattle World's Fair during 1962. It was there he was encouraged to apply for an assistant observer and darkroom technician position at the recently developed Table Mountain Observatory (TMO) of the Jet Propulsion Laboratory (JPL) near Wrightwood, California. He got the job.

In the 1970s and 1980s, Young collaborated with Alan Harris and Ellis Miner in the study of the rotational rates of asteroids. At one time, he contributed almost one half of all known asteroidal rotational rate data.

He was the telescope observer responsible for the successful aiming of lasers to the surface of the Moon in 1968, as well as to Earth-orbiting satellites and the Galileo spacecraft when it was six million kilometers from Earth during the 1990s.

A very prolific asteroid observer of both physical properties and astrometric positions, he discovered some 390 main belt asteroids in a six-year period, as well as two Near Earth Objects, two Trojan asteroids, three Mars crossers, and one extra-galactic supernova.

Along with Charles Capen, Jr., Young carried out photographic synoptic patrols, using specific electromagnetic wavelengths (ultraviolet through infrared), of Venus, Mars, Jupiter, and Saturn. Color astrophotography was investigated for planetary imaging using recently developed high speed color film emulsions.

Spectroscopic studies of the planet Venus were carried out by JPL astronomers, with Young assisting with hypersensitization of Eastman Kodak infrared spectroscopic glass plates. He developed a new technique of cold storage for these extremely sensitive plates. His experimentation of clean and properly washed plates, stored at minus 70 °C for over two years, were without increased noise or loss of sensitivity. Previous experimenters could manage only about a two-month reliability.

Young taught an astronomy extension course for the University of California at Riverside, California in 1969 and 1970, specifically for high school teachers and educators.

Asteroid 2874 Jim Young was named in 1985 in honor of his contributions to the physical study of asteroids. He is currently the resident astronomer at TMO.

1. James Whitney Young, Wikipedia, http://en.wikipedia.org/wiki/James_Whitney_Young.
2. Young, James W. (1941-), The Internet Encyclopedia of Science, www.daviddarling.info/encyclopedia/Y/Young_James.

Joseph Taylor, Jr. (1941-)

Co-discovered the first binary pulsar

Joseph Hooton Taylor, Jr., an American astrophysicist, was born in Philadelphia, Pennsylvania on March 29, 1941. He graduated from Haverford College in 1963 with a B.A. in physics. He received a Ph.D. from Harvard University in 1968, and spent the next year as a research fellow and lecturer in astronomy at Harvard. In 1969 he joined the faculty at the University of Massachusetts in Amherst. In the fall of 1980, he left Massachusetts to become professor of physics at Princeton University.

In 1970, while at the University of Massachusetts, Taylor and one of his graduate students, Russell Hulse (1950-), in search of a dissertation project decided to search the skies for the weak radio signals emitted by pulsars. They used the 300-meter diameter Arecibo telescope in Puerto Rico, the world's largest single-element radio telescope,

Pulsars were first discovered in 1967 by Jocelyn Burnell (1943-) and Antony Hewish (1924-). They are neutron stars whose diameters are extremely small. Their masses, on the other hand, are as great, or greater, than that of the Sun. As a result of their extremely strong gravitational pull, radio waves are released from pulsars only at the poles. The beams reach Earth in pulses as the star spins. In analyzing the results of a pulsar detected in 1974, Taylor and Hulse noticed an unexpected variation in the pulsar's period. The bursts were not perfectly regular like those of known pulsars, and the irregularity revealed that there were actually two pulsars orbiting each other—the first binary pulsar..

The discovery of these stars gave scientists an opportunity to study the effects of gravity outside the gravitational field of our Solar System. Over a period of almost 20 years, Taylor and Hulse made detailed observations of the behavior of these stars in orbit. They discovered that the path the pulsars follow is changing, their orbit is contracting, and the two stars are rotating at greater speeds as they grow closer to each other.

Taylor and Hulse's examination of the timing of the pulses provided the first evidence for the existence of the magnetic aspect of gravity. In 1916, Einstein predicted that two masses in orbit around each other would emit what he called gravitational waves, and thus lose energy. For their discovery, Taylor and Hulse were awarded the 1993 Nobel Prize in physics.

1. Joseph H. Taylor, Jr., Microsoft Encarta Online Encyclopedia, http://encarta.msn.com/encyclopedia_761583370/Taylor_Joseph_H_Jr_.html#461529904, 2008.
2. Joseph H. Taylor, Jr., NobelPrize.org, http://nobelprize.org/nobel_prizes/physics/laureates/1993/taylor-autobio.
3. World of Scientific Discovery on Joseph H. Taylor, Jr., www.bookrags.com/biography/joseph-h-taylor-jr-wsd.

Stephen Hawking (1942-)

Showed that the only properties particles of matter keep once they enter a black hole are mass, angular momentum, and electric charge

Stephen William Hawking, a British theoretical astrophysicist and cosmologist, was born in Oxford, England on January 8, 1942. His father was a research biologist. Almost immediately after Stephen entered Trinity Hall, Cambridge to study theoretical astronomy and cosmology he started developing symptoms of amyotrophic lateral sclerosis, a type of progressive motor neuron disease. After receiving his Ph.D., he joined the faculty at Cambridge, where he has remained.

During the late 1960s, Hawking proved that if the general theory of relativity is correct, then a singularity must also have occurred at the Big Bang. In 1970, his research turned to the examination of the properties of black holes. He realized that the surface area of the event horizon around a black hole could only increase or remain constant with time. This meant, for example, that if two black holes merge, the surface area of the new black hole would be larger than the sum of the surface areas of the two original black holes.

From 1970 to 1974, he and his associates provided mathematical proof for the hypothesis formulated by American physicist John Wheeler (1911-2008) that the only properties that particles of matter keep once they enter a black hole are mass, angular momentum, and electric charge. Since 1974, he has studied the behavior of matter in the immediate vicinity of a black hole from a theoretical basis in quantum mechanics.

Throughout the 1990s, Hawking sought to produce a single theory that could connect several theories used by scientists to explain the universe. This theory would combine quantum mechanics and relativity to form a quantum theory of gravity.

On his 65th birthday, he announced his plans for a zero-gravity flight. The event took place on April 26, 2007. He experienced weightlessness eight times. He became the first quadriplegic to float free in a weightless state. It was the first time in 40 years that he moved freely beyond the confines of his wheelchair.

Hawking has written a series of popular science books, including *A Brief History of Time* (1988), *The Universe in a Nutshell* (2001), and a collection of essays *Black Holes and Baby Universes* (1993.

1. Stephen Hawking, Microsoft Encarta Online Encyclopedia, http://encarta.msn.com/encyclopedia_761556019/stephen_hawking, 2008.
2. Stephen Hawking, Wikipedia, http://en.wikipedia.org/wiki/Stephen_Hawking.
3. World of Scientific Discovery on Stephen William Hawking, BookRags, www.bookrags.com/biography/stephen-william-hawking-wsd.

Charles Bolton (1943-)

The first to present irrefutable evidence of the existence of a black hole

Charles Thomas Bolton, an American astronomer, was born at Camp Forrest, a World War II military base in Tennessee, in 1943. He received his Bachelor's degree in 1966 from the University of Illinois, followed by a 1968 Master's degree and a 1970 doctoral degree, both from the University of Michigan.

From 1970 to 1972, he was a postdoctoral fellow at the David Dunlap Observatory of the University of Toronto Astronomy Department. At the same time, he taught at the Observatory and at Scarborough College from 1971 to 1972. From 1972 to 1973, he taught at Erindale College. He then became a professor in the Astronomy Department at the University of Toronto in 1973.

In 1970, Bolton was the first to develop a computer model for stellar atmospheres that was able to generate large regions of the modeled spectrum with enough precision to allow comparison with the spectra from real stars.

In 1971, while studying binary systems at the Dunlap Observatory, he observed star HDE 226868 wobble as if it was orbiting around an invisible, but massive companion which emitted powerful X-rays. He estimated that the amount of mass needed for the observed gravitational pull was too much for a neutron star. Further observations confirmed the results in 1973. The object was black hole, Cygnus X-1, lying in the center of the Milky Way galaxy. Louise Webster and Paul Murdin at the Royal Greenwich Observatory independently made the same discovery.

In 1978, Bolton demonstrated that the nitrogen anomalies observed in the spectra of OBN stars are due to the transfer of material to a nearby neighboring star. OB stars are very hot, blue stars with thick atmospheres. OBN stars are a subclass of OB stars characterized by nitrogen anomalies in their spectra.

In 2008, the University of Toronto decided to sell the David Dunlap Observatory. Bolton and the Richmond Hill Naturalists contended that the Observatory and 190 acres of parkland and arboretum should all be designated as a Provincial Heritage site, and not allowed to be used for suburban housing or apartment development. Despite this, the University announced an imminent sale of the land to local developers. Frustrated, Bolton was said to have stood crying while helpers carried out boxes with 37 years of his work.

1. C. T. Bolton, University of Toronto, www.astro.utoronto.ca/staff.html#Bn.
2. Charles Thomas Bolton, Astrolab, http://astro-canada.ca/_en/a2214.
3. Charles Thomas Bolton, Wikipedia, http://en.wikipedia.org/wiki/Charles_Thomas_Bolton.

Aleksander Wolszczan (1943-)

Co-discovered the first confirmed planets beyond the Solar System

Aleksander Wolszczan, a Polish astronomer, was born in Szczecinek, Poland on April 29, 1946. He received his master's degree in astronomy in 1969 and his doctoral degree in physics in 1975, both from North Copernicus University in Toruń, Poland.

He was on the faculty of North Copernicus University from 1969 to 1979 and a research associate at Copernicus Astronomical Center from 1979 to 1983. He was a visiting scientist at the University of California, Berkeley in 1988, a visiting professor at Princeton University in 1992, and a research associate at Cornell University from 1983 to 1992. He joined the astronomy faculty at Pennsylvania State University (PSU) in 1992, and since 1997 he simultaneously has been on the faculty of North Copernicus University.

Wolszczan and Dale Frail (1961-) carried out astronomical observations from the Arecibo Observatory in Puerto Rico, which led to the discovery of the pulsar PSR B1257+12 in 1990. The data analysis showed that the pulsar is orbited by two planets with masses at least 3.4 and 2.8 times that of Earth. Their orbits were determined to be 0.36 and 0.47 astronomical units, respectively. An astronomical unit is a length of approximately 150 million kilometers, and is based on the distance from the Earth to the Sun.This planetary system was the first extra-Solar System discovered in the universe whose existence was verified. They published their findings in 1992 and 1994.

They accomplished the discovery while the telescope was undergoing repair for a series of cracks and was locked in a fixed position. So basically they were the only ones who wanted to use the broken telescope.

In 1996, Wolszczan was awarded the Beatrice M. Tinsley Prize by the American Astronomical Society. In 2002, Poland honored him by having his likeness featured on a special set of 16 postage stamps. Also featured with Wolszczan were Nicolaus Copernicus and the Arecibo radio telescope

In 2003, Wolszczan and Maciej Konacki (1972-) determined the orbital inclinations of the two planets, showing that the correct masses are approximately 3.9 and 4.3 Earth masses. In 2005, he and Konacki discovered the smallest extra-solar planet to date. Wolszczan is currently the Evan Pugh Professor of Astronomy and Astrophysics at PSU.

1. Aleksander Wolszczan, Wikipedia, http://en.wikipedia.org/wiki/Aleksander_Wolszczan.
2. Dr. Aleksander Wolszczan, Lifeboat Foundation, http://lifeboat.com/ex/bios.alex.wolszczan.
3. Sampsell, Steve, Alexander Wolszczan, Faces of Penn State, Science Journal, www.science.psu.edu/journal/sum2000/WolszczanSum2000, Summer 2000 -- Vol. 17, No. 1.

Jocelyn Burnell (1943-)

Co-discovered pulsars

Susan Jocelyn Burnell, a British astronomer, was born in Belfast, Northern Ireland near Armagh Observatory on July 15, 1943. Her maiden name was Bell. She graduated from Glasgow University in Scotland with a B.Sc. in physics in 1965 and received a Ph.D. in radio astronomy from the University of Cambridge in England in 1968. That same year, she married Martin Burnell.

From 1974 to 1982, she worked in X-ray astronomy at Mullard Space Science Laboratory at University College in London. In 1982 she became a senior research fellow at the Royal Observatory in Edinburgh, Scotland, where she was responsible for British research with the James Clerk Maxwell Telescope in Hawaii. She also did astrophysical research in the optical and infrared parts of the electromagnetic spectrum. She is currently a professor at the Open University in Milton Keynes, England.

In 1967, while a graduate student at Cambridge, she shared with her advisor, Antony Hewish (1924-), and other colleagues at the radio astronomy observatory the finding of a source of regular, intense pulses of radio waves that emitted a burst every 1.337 seconds. Within a few months, the astronomers had discovered a number of other sources in distant space, and concluded that from their dissimilar locations and other characteristics they must be naturally occurring. They soon realized that the pulse patterns came from a special type of star that they subsequently termed a pulsar.

Astronomers have since discovered more than 400 pulsars, but only two, the Crab Pulsar and the Vela Pulsar, emit visibly detectable pulses.

Pulsar emissions consist of periodic sequences of brief pulses of phenomenal regularity. Each pulsar has a different period, ranging from a few milliseconds to as long as several seconds. In this way, a pulsar behaves like a precise clock. Pulsars are also powerful electric generators, capable of accelerating charged particles to energies of a trillion volts.

Hewish was awarded the Nobel Prize, along with Martin Ryle (1918-1984), without the inclusion of Burnell as a co-recipient, which was somewhat controversial. She has been awarded a number of other prizes, and in 2007 she was awarded an honorary doctorate by Harvard University.

1. Jocelyn Bell Burnell, Wikipedia, http://en.wikipedia.org/wiki/Jocelyn_Bell_Burnell.
2. Susan Jocelyn Bell Burnell, Microsoft Encarta Online Encyclopedia, http://encarta.msn.com/encyclopedia_761583106/Bell_Burnell_(Susan)_Jocelyn, 2008.
3. World of Scientific Discovery on Jocelyn Susan Bell Burnell, BookRags, www.bookrags.com/biography/jocelyn-susan-bell-burnell-wsd.

Jill Tarter (1944-)

Co-compiled a catalog of nearby habitable systems

Jill Cornell Tarter, an American astronomer, was born in 1944. She is the current director of the Center for SETI (Search for Extra-Terrestrial Intelligence) Research, holding the Bernard M. Oliver Chair for SETI. The Center is located in Mountain View, California. Its mission is to explore, understand, and explain the origin, nature, and prevalence of life in the universe. Tarter also serves on the management board for the Allen Telescope Array, a joint project between the SETI Institute and the University of California, Berkeley Radio Astronomy Laboratory.

Tarter received her bachelor of engineering physics degree from Cornell University and her Ph.D. in astronomy from the University of California, Berkeley. As a graduate student, she worked on the radio-search project SERENDIP (Search for Extraterrestrial Radio Emissions from Nearby Developed Intelligent Populations).

She was project scientist for the National Aeronautics and Space Administration's High Resolution Microwave Survey (HRMS) in 1992 and 1993, and subsequently director of Project Phoenix (HRMS reconfigured) under the auspices of the SETI Institute. She was co-creator of the *HabCat* in 2002, a principal component of Project Phoenix. *HabCat* is a catalog of star systems which conceivably have habitable planets

Tarter has published dozens of technical papers, and lectures extensively both on the search for extraterrestrial intelligence and the need for proper science education. She was elected a Fellow of the American Association for the Advancement of Science in 2002 and a Fellow of the California Academy of Sciences in 2003. She was awarded the Telluride Tech Festival Award of Technology in 2001.

Tarter's astronomical work is illustrated in Carl Sagan's novel *Contact*. In the film version of *Contact*, the protagonist, Ellie Arroway, is played by Jodie Foster. Tarter conversed with the actress for months before and during filming. The character of Samantha Crowe in Frank Schätzing's novel *The Swarm* is strongly based on Tarter.

Tarter is a frequent speaker for science teacher meetings and at museums and science centers, bringing her commitment to science and education to both teachers and the public. In 2004, *Time Magazine* named her one of the Time 100 most influential people in the world.

1. Dr. Jill Tarter, SETI Institute, www.seti.org/about-us/people/staff/tarter-jill.php, December 18, 2006.
2. Jill Tarter, Wikipedia, http://en.wikipedia.org/wiki/Jill_Tarter.
3. Martin, Allison M., Searching the Sky: Jill Tarter, *AWIS Magazine, Volume 34, Number 2 Spring 2005*, www.seti.org/pdfs/awis%20-%20tarter.pdf.

George Smoot (1945-)

Co-discovered the anisotropy of the cosmic microwave background radiation

George Fitzgerald Smoot, III, an American astrophysicist, was born in Yukon, Florida on February 20, 1945. His father was a hydrologist and his mother was a science teacher. Smoot earned his doctorate in physics at the Massachusetts Institute of Technology in 1970. He then went to work at the Lawrence Berkeley Laboratory. He became a professor of physics at Berkeley in 1994.

Smoot's research focused on aspects of the Big Bang, which theorizes that the entire universe began as an explosion from an extremely compact, hot, dense state of pure energy. As the universe expanded, the energy cooled and some of it began to condense into matter, which took the form of free particles, such as protons, neutrons, and electrons. The free electrons interacted with the remaining energy, absorbing photons at all wavelengths of light. As the universe cooled, protons began to combine with electrons to form hydrogen atoms. The event filled the universe with light, which became the cosmic background radiation. Over time, the released light shifted from visible wavelengths to microwaves as the universe further expanded and cooled.

In 1974, the National Aeronautics and Space Administration (NASA) began work on a satellite to study the cosmic background radiation. The project, called the Cosmic Background Explorer (COBE), was headed by John Mather (1946-) at NASA's Goddard Space Flight Center in Maryland. Smoot developed an instrument for the COBE satellite that could detect temperature variations down to a few hundred-thousandths of a degree Kelvin. Finding differences in the temperature of the background radiation (anisotropy) would indicate that matter in the early universe was slightly clumped. The clumping of matter would have provided the seeds for the first galaxies.

In April 1992, Smoot announced that his instrument had indeed found lumps in the background radiation, a finding that supported the Big Bang. He recounted his experience with COBE and other research related to cosmology in the book *Wrinkles in Time* (1993), co-written with Keay Davidson. Smoot shared the of 2006 Nobel Prize in physics with John Mather for their work.

1. George Smoot, Microsoft Encarta Online Encyclopedia, www.encarta.net/s encyclopedia_701835566/George_Smoot, 2008.
2. Prof. Dr. George Fitzgerald Smoot, The Nobel Laureate Meetings at Lindau, www.lindau-nobel.de/LaureateDetails.AxCMS?UserID=6951.
3. World of Physics on George F. Smoot, BookRags, www.bookrags.com/biography/george-f-smoot-wop.

John Mather (1946-)

Co-discovered the anisotropy of the cosmic microwave background radiation

John Cromwell Mather, an American astrophysicist, was born in Roanoke, Virginia on August 7, 1946. He grew up in New Jersey. His father was a faculty member at what is now Virginia Tech. John completed a bachelor's degree in physics in 1968 at Swarthmore College in Pennsylvania. He earned his Ph.D. in physics at the University of California, Berkeley in 1974. That same year, he began post-doctorate work at NASA's Goddard Space Flight Center in Greenbelt, Maryland.

Mather's early research focused on aspects of the Big Bang, which theorizes that the entire universe began as an explosion from an extremely compact, hot, dense state of pure energy. As the universe expanded, the energy cooled and some of it began to condense into matter, which took the form of free particles, such as protons, neutrons, and electrons. The free electrons interacted with the remaining energy, absorbing photons at all wavelengths of light. When the universe cooled, protons began to combine with electrons to form hydrogen atoms. The event filled the universe with light, which became the cosmic background radiation. Over time, the released light shifted from visible wavelengths to microwaves as the universe further expanded and cooled.

Mather first proposed a space probe to measure the cosmic microwave background radiation in 1974. He was hired by NASA to pursue the project, which became the Cosmic Background Explorer (COBE). The three main instruments on the COBE satellite were designed to study electromagnetic radiation in infrared and microwave wavelengths.

In particular, Mather was responsible for the instrument that measured blackbody radiation. It was launched onboard COBE in 1989, and Mather announced the first results in 1990. The 2.7°K above absolute zero that instrument detected matched precise predictions for the Big Bang as the source of the cosmic background radiation. He later told the history of the COBE project in the book The *Very First Light* (1996), written with John Boslough. In 2007, Mather was listed among *Time magazine*'s 100 Most Influential People in The World.

Mather shared the 2006 Nobel Prize in physics with George Smoot (1945-) for research that provided strong support for the Big Bang theory.

1. John C. Mather, Autobiography, NobelPrize.org, http://nobelprize.org/nobel_prizes/physics/laureates/2006/mather-autobio.
2. John C. Mather, Wikipedia, http://en.wikipedia.org/wiki/John_C._Mather.
3. John Mather, Microsoft Encarta Online Encyclopedia, http://encarta.msn.com/encyclopedia_701835565/John_Mather, 2008.

Alan Guth (1947-)

Proposed the inflationary universe model

Alan Harvey Guth, an American theoretical physicist and cosmologist, was born in New Brunswick, New Jersey on February 27, 1947. He was educated at Massachusetts Institute of Technology (MIT), where he obtained his Ph.D. in physics in 1969. His dissertation involved the exploration of an early model of how quarks combine to form the elementary particles that scientists observe. During the next nine years, Guth held postdoctoral appointments at Princeton, Columbia, Cornell, and Stanford. He returned to MIT in 1980.

Initially, Guth worked as a theorist in elementary particle physics, but, stimulated by the work of Steven Weinberg (1933-) he began to consider problems of cosmology, including a number of difficulties with the standard interpretation of the Big Bang account of the origin of the universe. He felt that the Big Bang was in need of revision.

In 1980, he proposed the "inflationary universe" model in which the universe expands for a fleeting instant at its beginning at a much higher rate than expected for the Big Bang. This period, which is called the inflationary epoch, is a consequence of the nuclear force breaking away from the weak and electromagnetic forces that it was unified with at higher temperatures in what is called a phase transition.

This phase transition is thought to have happened about 10^{-35} seconds after the creation of the universe. It filled the universe with a kind of energy called the vacuum energy, and as a consequence of this vacuum energy density, gravitation effectively became repulsive for a period of about 10^{-32} seconds. During this period, the universe expanded at an astonishing rate, increasing its size scale by about a factor of 1050. Then, when the phase transition was complete, the universe settled down into the Big Bang evolution. This theory predicts that the entire volume of the universe out to a distance of about 18 billion light years expanded from a volume that was only a few centimeters across when inflation began.

Guth has been awarded the Dirac Prize of the International Center for Theoretical Physics in Trieste and the Cosmology Prize of the Peter Gruber Foundation. He is the author of *The Inflationary Universe: The Quest for a New Theory of Cosmic Origins* (1998). He is now the Victor F. Weisskopf Professor of Physics at MIT.

1. Alan H. Guth, Faculty and Staff, MIT, http://web.mit.edu/physics/facultyandstaff/faculty/alan_guth.
2. Alan Harvey Guth, Answers.com, www.answers.com/topic/alan-guth.
3. The Inflationary Universe, Astronomy 162 Stars, Galaxies, and Cosmology, http://csep10.phys.utk.edu/ astr162/lect/cosmology/inflation.

David Levy (1948-)

Jointly discovered comet Shoemaker-Levy 9

David H. Levy, a Jewish Canadian amateur astronomer, was born in Montreal, Quebec, Canada on May 22, 1948. He developed an interest of astronomy at an early age; however, he studied English literature, receiving a bachelor's degree from Acadia (Nova Scotia) University and a master's degree from Queen's University, Kingston, Ontario.

Levy, Carolyn Shoemaker (1929-), and her husband Eugene Shoemaker (1928-) jointly discovered the fragmented comet, **Shoemaker-Levy 9**, in orbit around Jupiter in March 1993. For six days, between July 16 and 22, 1994, they looked through telescopes to watch the estimated 21 major fragments of the comet pummel Jupiter. From the atmosphere of Jupiter arose tall, bright plumes that left broad, dark stains beneath them, providing a spectacular show for astronomers around the world. Levy had met the Shoemakers in 1988 as a result of a comet he had discovered and they were tracking.

A science writer by trade, Levy has written over 30 books, mostly on astronomical subjects, such as *The Quest for Comets* and his tribute to Shoemaker, *Shoemaker by Levy*. Also, he has provided periodic articles for such magazines as *Sky and Telescope* and *SkyNews*. He is the science editor for *Parade Magazine*.

He was awarded the C.A. Chant Medal of the Royal Astronomical Society of Canada in 1980. In 1993, he won the Amateur Achievement Award of the Astronomical Society of the Pacific. In 2007, he was awarded the Smithsonian Astrophysical Observatory's Edgar Wilson Award for the discovery of comets. He won an Emmy in 1998 as part of the writing team for the Discovery Channel documentary, *Three Minutes to Impact*. He is President of the National Sharing the Sky Foundation, an organization intended to inspire new generations to develop an inquiring interest in the sciences. He now lives in Vail, Arizona. He and his wife, Wendee, host a weekly radio talk show on astronomy on the internet.

Levy has discovered 22 comets, nine of them using his own backyard telescopes –a quest he began at age 15. **Asteroid 3673 Levy** was named in his honor. He is currently involved with the Jarnac Comet Survey, which is based at the Jarnac Observatory in Vail, Arizona but which has telescopes planned for locations around the world.

1. David H. Levy's Home Page, www.jarnac.org/index.
2. Friedich, Mary Jane, David H. Levy, Britannica, www.britannica.com/#tab=active~home%2Citems ~home&title=Britannica%20Online%20Encyclopedia.
3. Knapp, Tom, David Levy: Falling Star, Rambles, www.rambles.net/levy_jupiter, November 1995.

Marc Aaronson (1950-1987)

Found that the universe is much younger and smaller than once believed

Mark Arnold Aaronson was born in Los Angeles, California on August 24, 1950. He received a B.Sc. degree from the California Institute of Technology in 1972 and a Ph.D. from Harvard University in 1977. His dissertation was on the near-infrared aperture photometry of galaxies. He joined Steward Observatory at the University of Arizona as a Postdoctoral Research Associate in 1977, and became an Associate Professor of Astronomy in 1983. He is best known for his research on the age and size of the universe.

Aaronson's work concentrated on three fields—the determination of the Hubble constant using the Tully-Fisher relation, the study of carbon rich stars, and the velocity distribution of those stars in dwarf spheroidal galaxies. The Hubble constant is proportionality constant in Hubble's law, which is the statement that the redshift in light coming from distant galaxies is proportional to their distance. The Tully-Fisher relation is an empirical relationship between the intrinsic luminosity (proportional to the stellar mass) of a spiral galaxy and its velocity width (the amplitude of its rotation curve). The rotation curve of a galaxy can be represented by a graph that plots the orbital velocity of the stars in the galaxy on the abscissa against the distance from the center of the galaxy on the ordinate.

Aaronson and Jeremy Mould (1949-) won the George Van Biesbroeck Prize in 1981 and the Newton Lacy Pierce Prize in Astronomy from the American Astronomical Society in 1984 for their discovery dating the universe at only 12 billion years, rather than 20 billion as once thought.

On April 30, 1987, Aaronson was killed in a tragic accident in the dome of the 4-meter Mayall Telescope on Kitt Peak in Arizona. Due to a malfunction of the emergency stop, he became trapped in the catwalk door of the 150-ton rotating telescope dome when the outer stepladder closed it. When the door leading outside is opened, the telescope's dome automatically stops turning, but the dome coasts five to 10 feet before coming to a full stop. He was leaving the 18-story building that houses the telescope to check the weather when he was killed. He was 37 years old.

The **Marc Aaronson Memorial Lectureship**, promoting and recognizing excellence in astronomical research, is held every 18 months by the University of Arizona and Steward Observatory as a tribute to his memory. **Asteroid 3277 Aaronson** is named in his honor.

1. Aaronson, Marc (1950-1987), The Internet Encyclopedia of Science, www.daviddarling.info/encyclopedia/ A/Aaronson.
2. Marc A. Aaronson, Astronomer, Killed by Revolving Dome, New York Times, May 2, 1987.
3. Marc Aaronson, Wikipedia, http://en.wikipedia.org/wiki/Marc_Aaronson.

Russell Hulse (1950-)

Co-discovered the first binary pulsar

Russell Alan Hulse, an American physicist and astronomer, was born in New York City on November 28, 1950. He received his bachelor's degree from Cooper Union College in 1970 and his Ph.D. in physics in 1975 from the University of Massachusetts, where he was a graduate student under Joseph Taylor (1941-).

In the early 1970s, Hulse and Taylor undertook an extensive search for new pulsars using the reflector of the Arecibo Observatory in Puerto Rico. By using computers to carefully process the signals received from space, they were able to pinpoint the locations of various new pulsars.

Pulsars are highly-magnetized, rapidly-rotating neutron stars that periodically emit a beam of electromagnetic radiation in the form of radio waves. Pulsars are extremely dense, and are surrounded by tremendously powerful magnetic fields. The beams of radiation sweep out through space as the pulsar spins. Astronomers detect these moving beams as short, regular pulses of radiation

In 1974, Hulse observed a pulsar whose characteristics were unlike any other known pulsar. He and Taylor correctly concluded that they had found a binary pulsar, a pair of pulsars orbiting each other. The strong gravitational field of the binary system and the acceleration of the pulsars as they orbited each other provided the researchers with a new opportunity to test and verify aspects of Einstein's general theory of relativity. Einstein predicted in his theory that accelerating bodies would radiate energy in the form of gravitational waves. This seemed to be confirmed by observations that showed that this binary pulsar was indeed losing energy. Since then, several dozen binary pulsars have been found and studied, further supporting Hulse and Taylor's findings.

After postdoctoral work at the National Radio Astronomy Observatory in Virginia, Hulse changed his research focus from astronomy to plasma physics, and in 1977 joined the staff of the Princeton Plasma Physics Laboratory.

Hulse and Taylor shared the 1993 Nobel Prize in physics for their discovery of the first binary pulsar. In 2004, Hulse became affiliated with The University of Texas at Dallas as a visiting professor of physics and of science and math education.

1. 1993 Nobel Laureate Dr. Russell Hulse To Join U. T. Dallas as Visiting Professor, News Release, The University of Texas at Dallas, www.utdallas.edu/news/archive/2003/hulse_joins.
2. Russell A. Hulse, Autobiography, NobelPrize.org, http://nobelprize.org/nobel_prizes/physics/laureates/1993/hulse-autobio.
3. Russell A. Hulse, Microsoft Encarta Online Encyclopedia, http://encarta.msn.com/encyclopedia_761583370/Taylor_Joseph_H_Jr_, 2008.

Geoffrey Marcy (1954-)

Shares credit for the discovery of well over half the presently known extrasolar planets

Geoffrey W. Marcy, an American astrophysicist, was born on September 29, 1954. He grew up in the San Fernando Valley in the suburbs of Los Angeles. His father was an aerospace engineer and his mother was an anthropologist. When Geoffrey was 14 years old, his parents bought him a used 4-1/4-inch Newtonian telescope, which he climbed out the window of his bedroom onto the patio roof to use.

Marcy received a B.A. in physics and astronomy at UCLA in 1976. He then went to the University of California, Santa Cruz for a Ph.D. in astrophysics, which he received in 1982.

From 1982 to 1984, he was a fellow at the Carnegie Institution of Washington. From 1984 to 1999, he was on the faculty of San Francisco State University.

Marcy is famous for discovering more extrasolar planets than anyone else. He and his colleagues discovered 70 of the first 100 known extrasolar planets. He also confirmed the discovery of the first extrasolar planet, 51 Pegasi b, by Michel Mayor (1942-) and Didier Queloz (1966-).

Other achievements include discovering the first multiple planet system revolving around a star, Upsilon Andromedae, similar to our own; the first transiting planet around another star, HD209458b; the first extrasolar planet orbiting beyond five astronomical units, 55 Cancri d; and the first Neptune-sized extrasolar planets, Gliese 436b and 55 Cancri e. A transit is the astronomical event that occurs when one celestial body appears to move across the face of another celestial body.

Marcy and Michel Mayor (1942-) shared the $1-million Shaw Prize in astronomy in 2005 for their work. Marcy also received the NASA Medal for Exceptional Scientific Achievement. He was named *Discovery Magazine*'s Space Scientist of the Year in 2003, and in 2006 he received an honorary doctorate in science from the University of Delaware.

Currently, Marcy is an adjunct professor at San Francisco State University and a Professor of Astronomy at the University of California, Berkeley. He is also Director of Berkeley's Center for Integrative Planetary Science, a research unit designed to study the formation, geophysics, chemistry, and evolution of planets.

1. Geoff Marcy, Berkeley Astronomy Department, University of California, http://astro.berkeley.edu/people/ faculty/marcy.
2. Geoffrey Marcy, Astronomer, UCLA.edu Spotlight, http://spotlight.ucla.edu/alumni/geof-marcy_astro.
3. Geoffrey Marcy, Wikipedia, http://en.wikipedia.org/wiki/Geoffrey_Marcy.

Michael Mayor (1959-)

Co-discovered the first extrasolar planet orbiting a Sun-like star

Michel G. E. Mayor, a Swiss astronomer, was born on January 12, 1942. After studying physics at the University of Lausanne, Mayor obtained his doctorate in astronomy at the Geneva Observatory in 1971. His research interests include galactic structure and evolution, globular cluster dynamics, stellar duplicity, stellar rotation, and extrasolar planets.

From 1989 to 1992 he was involved in scientific research at the European Southern Observatory in Chili, from 1988 until 1991 he studied galactic structure with the International Astronomical Union, and from 1990 until 1993 he was with the Swiss Society for Astrophysics and Astronomy. From 1998 to 2004, he was Director of the Geneva Observatory.

He is the principal investigator on the High Accuracy Radial velocity Planetary Search (HARPS) spectrograph project, which since 2003 has conducted radial velocity searches for extrasolar planets at the European Southern Observatory's 3.60 meter telescope at La Silla, Chile.

In 1995, Mayor and Didier Queloz (1966-), a graduate student, discovered 51 Pegasi b, the first extrasolar planet orbiting a Sun-like star. His team has since discovered more than 100 additional planets and planetary systems. Among his team's recent discoveries is a planetary system with three large planets, the so-called super-Earths. The planets have minimum masses between 10 and 18 times the mass of the Earth. The innermost planet is most probably rocky, while the outermost is the first known Neptune-mass planet to reside in the habitable zone.

Mayor is interested in detecting signs of extraterrestrial life. He predicts that top researchers are less than two decades away from being able to detect real signs of such life—if it exists. He said scientist will be able to directly look for signatures of life on other planets, similar to the high presence of oxygen in our atmosphere, within 15 to 20 years.

He has had a number of awards, including the Swiss Marcel Benoist Prize in recognition of his work and its significance for human life (1998), the Balzan Prize (2000), the Albert Einstein Medal (2004), and the Shaw Prize in Astronomy (2005). He is currently professor in the Department of Astronomy at the University of Geneva.

1. Klapper, Bradley S., Swiss scientist: Search for life next, USA Today, www.usatoday.com/tech/science/2007-04-25-2459586131_x, April 25, 2007.
2. Michel Mayor, International Geographic Congress Oslo 2008, www.33igc.org/coco/LayoutPage.aspx
3. Three Neptune-Size Planets Found In Nearby Star System, Science Daily, www.sciencedaily.com/ releases/2006/05/060518222901, May 18, 2006.

Index

CPSIA information can be obtained at www.ICGtesting.com
Printed in the USA
BVOW06s0802160816

459118BV00021B/172/P